Creating Katrina, Rebuilding Resilience

Creating Katrina, Rebuilding Resilience
Lessons From New Orleans on Vulnerability and Resiliency

Edited by

Michael J. Zakour
West Virginia University, Morgantown, WV, United States

Nancy B. Mock
Tulane University, New Orleans, LA, United States

Paul Kadetz
Drew University, Madison, NJ, United States

Butterworth-Heinemann is an imprint of Elsevier
The Boulevard, Langford Lane, Kidlington, Oxford OX5 1GB, United Kingdom
50 Hampshire Street, 5th Floor, Cambridge, MA 02139, United States

Copyright © 2018 Elsevier Inc. All rights reserved.

No part of this publication may be reproduced or transmitted in any form or by any means, electronic or mechanical, including photocopying, recording, or any information storage and retrieval system, without permission in writing from the publisher. Details on how to seek permission, further information about the Publisher's permissions policies and our arrangements with organizations such as the Copyright Clearance Center and the Copyright Licensing Agency, can be found at our website: www.elsevier.com/permissions.

This book and the individual contributions contained in it are protected under copyright by the Publisher (other than as may be noted herein).

Notices
Knowledge and best practice in this field are constantly changing. As new research and experience broaden our understanding, changes in research methods, professional practices, or medical treatment may become necessary.

Practitioners and researchers must always rely on their own experience and knowledge in evaluating and using any information, methods, compounds, or experiments described herein. In using such information or methods they should be mindful of their own safety and the safety of others, including parties for whom they have a professional responsibility.

To the fullest extent of the law, neither the Publisher nor the authors, contributors, or editors, assume any liability for any injury and/or damage to persons or property as a matter of products liability, negligence or otherwise, or from any use or operation of any methods, products, instructions, or ideas contained in the material herein.

British Library Cataloguing-in-Publication Data
A catalogue record for this book is available from the British Library

Library of Congress Cataloging-in-Publication Data
A catalog record for this book is available from the Library of Congress

ISBN: 978-0-12-809557-7

For Information on all Butterworth-Heinemann publications
visit our website at https://www.elsevier.com/books-and-journals

Publisher: Candice Janco
Acquisition Editor: Candice Janco
Editorial Project Manager: Hilary Carr
Production Project Manager: Maria Bernard
Cover Designer: Matthew Limbert

Typeset by MPS Limited, Chennai, India

*In loving memory of Lori Ann Zakour, who passed away
in Hurricane Katrina.*

Contents

List of Contributors xix

PART I INTRODUCTION AND THEORETICAL FRAMEWORK

CHAPTER 1 Editors' introduction: The voices of the barefoot scholars 3
Michael J. Zakour, Nancy B. Mock and Paul Kadetz

Classifying Disasters 4
New Orleans and the Livin' Ain't Easy 6
    Vulnerable Lives 6
    Linking Vulnerability and Resilience 7
    Working With Cultural Assets 9
Classifying Resilience 10
    Social Infrastructure and Capacity 10
Frameworks Utilized in This Volume 11
    Disasters Require Multiple Perspectives and Paradigms 12
    The Not-So-Hidden Agendas of Post-Katrina Research 13
    Hybridizing Knowledge for Disaster Research 13
    Barefoot Scholars 14
    A Complex Systems Approach: The Umbrella of Our Theoretical Framework 15
    Vulnerability and Resilience as a Systems Problem 16
    Structural Violence and Intersectionality 17
Book Sections and Chapters 18
References 21

CHAPTER 2 Settlement shifts in the wake of catastrophe 25
Richard Campanella

Introduction 25
Prologue: New Orleans' Historical Settlement Patterns 26
Post-Katrina Residential Settlement Patterns 28
    Resettlement in Vertical Space 28
    Resettlement by FEMA Flood Zones 32
    Resettlement in Horizontal Space 36
Conclusions 40
Epilogue 42

Acknowledgments 43
References 43

CHAPTER 3 Vulnerability-plus theory: The integration of community disaster vulnerability and resiliency theories 45
Michael J. Zakour and Charles M. Swager

Vulnerability-plus Theory 46
The Development Perspective 47
    Chain of Causality: Root Causes, Structural Constraints, and Unsafe Conditions 48
    Access to Resources Model 50
The Resilience Perspective 51
    Attributes of Resilience Resources 52
    Types of Resources 53
Comparing and Contrasting Vulnerability and Resiliency 56
    Vulnerability and Resilience Theories Are Complementary 57
    Continuity Between Vulnerability and Resilience Theories .. 59
Vulnerability-plus Theory as Integration of Theories 61
    Assumptions of Vulnerability-plus Theory 62
    Empirical Support 64
Causal Chains in V + Theory 68
    Root Causes 68
    Structural Constraints 69
    Unsafe Conditions 69
    Hazard Types 70
    Disaster Characteristics 71
    Resources 71
Summary and Conclusions 72
References 73

CHAPTER 4 A systems approach to vulnerability and resilience in post-Katrina New Orleans 79
Nancy B. Mock, Melissa Schigoda and Paul Kadetz

Systems Approach to Vulnerability and Resilience 79
    Key Features of Complex Systems in Post-Katrina Recovery 81
Recovery as a Complex Adaptive Social System Problem 85

Signals and Information in Post-Katrina Recovery 86
    Information Blackout 87
    Monitoring Recovery: *The New Orleans Index* 87
    Resiliency, Recovery Planning and Collective
    Action 92
    Aid, Culture, and Resilience 93
    Resilience and Footprint 93
    In Need of a "New Deal" 93
Conclusion: The Road to Complex Recovery 94
References 95
Further Readings 96

CHAPTER 5   "Built-in" structural violence and vulnerability: A common threat to resilient disaster recovery 99

Shirley Laska, Susan Howell and Alessandra Jerolleman

Introduction 100
Important Terms and Concepts 100
    Agency 100
    Resiliency 102
    Agency Disrespect as Structural Violence 103
    Trauma 104
    Delay 105
Examples of Structural Violence 106
    "Louisiana Road Home" and New York City's
    "Build It Back" 106
    Shrinking the Footprint or Not?—How *Not* to Have
    That Discussion 108
The Extreme Structural Violence on the Economically
and Politically Powerless 111
    Evacuation Nightmare 111
    Health Care Scarcity 112
    Postdisaster Housing 113
How to Stop the Violence 114
    Owner-Occupied Housing Recovery 114
    Recovery Planning 117
    Successful Evacuation That Supports Return 118
    Expediently Provided, Postdisaster Health Care 121
    Survivor Aid and Return of Agency to Economically
    and Politically Disenfranchised 121

Concluding Remarks 123
References... 125

PART II DISASTER VULNERABILITY 131

CHAPTER 6 Setting the Stage for the Katrina Catastrophe: Environmental Degradation, Engineering Miscalculation, Ignoring Science, and Human Mismanagement 133

Ivor L. van Heerden

Introduction.. 134
Origin and Function of Louisiana's Coastal Wetlands............. 134
    The "Natural" Cycle of Wetland Development and Maintenance..134
    Subsidence and Relative Sea-Level Rise..............................137
    Surge and Storm Wind Reduction 138
Management of the Coastal Wetlands and the Resultant Dilemma... 139
    Control of the Lower Mississippi River 139
    Oil and Gas Extraction and the Disruption of Natural Hydrology; Enhanced Subsidence..141
    The Mississippi River Gulf Outlet Navigation Channel 141
    The MRGO "Funnel" 144
New Studies—Wetland Storm Reduction Value 145
    Wetland Role in Surge Reduction...145
    Computer Modeling to Reconstruct Surge and Waves Along the MRGO 145
    Effects of MRGO Reach 2 on Waves....................................146
Hurricane Betsy and the 1965 Flood Control Act 148
Forensic Investigations—Levee Failures 149
    Polder Levee Failures..149
    Accounting for Subsidence and Sea-Level Rise..................150
    Recommendations for Sustainability 151
Conclusions: Habitat Sustainability Is Needed to Support Human Resilience 154
References... 155
Further Reading 158

CHAPTER 7 Three centuries in the making: Hurricane Katrina from an historical perspective **159**
Michael J. Zakour and Kayla Grogg
Significance of an Historical Perspective 160
Conceptual and Theoretical Framework 162
    Vulnerability and Resilience 162
    Political Ecology of New Orleans 163
Historical Events Creating Katrina 165
    The Progression to Vulnerability 165
    Political and Economic Marginalization of People of Color 165
    Oil, Canals, and Environmental Degradation 169
    Migration and Population Displacement Since 1960 171
    Results of the Neglect of New Orleans 172
Vulnerability and Resilience During Katrina 173
    Katrina's Natural, Technological, and Organizational Failure Hazards 173
    The Nature of the Hurricane Katrina Disaster 175
    The Pattern of Disaster Damage and Loss 176
    Katrina as Catastrophe 177
Resilience Resources 178
    Environmental Justice: Recovery by Race, Income, Gender, and Age 178
    Economic Resources 178
    Information and Communication 180
    Social Capital 181
    Collective Action 182
Summary and Conclusions 183
    The Political Ecology of Katrina and Southeast Louisiana . 183
    Implications for Vulnerability-plus Theory 184
    Substantive Implications of Katrina 188
    Policy Implications 189
References 191

CHAPTER 8 The resilience in the shadows of catastrophe: Addressing the existence and implications of vulnerability in New Orleans and Southeastern Louisiana **193**
Regardt J. Ferreira and Charles R. Figley
Introduction 193

Vulnerability Defined 194
Resilience Defined 196
The New Orleans and Southeastern Louisiana
"*Catastrophe*" 198
The New Orleans Vulnerability and Resilience Paradigm 200
Causal Process: *The New Orleans and Southeastern
Louisiana "Catastrophe"* 201
    Root Causes 203
    Dynamic Pressures...204
    Unsafe Conditions 205
Predictors of Social Vulnerability in Louisiana:
A Multilevel Analysis 207
Addressing Vulnerability in New Orleans and
Southeastern Louisiana.. 208
Conclusion 209
References... 210
Further Readings... 213

CHAPTER 9 Problematizing vulnerability: Unpacking gender, intersectionality, and the normative disaster paradigm.. 215

Paul Kadetz and Nancy B. Mock

Introduction... 215
Problematizing the Essentializing of Female Vulnerability
in Disaster Research 218
    Fitting One Size of Vulnerability to All 218
Vulnerabilities in the Context of New Orleans 220
    Framework for Understanding Urban Vulnerability
    in New Orleans 220
The Intersectionality of Gendered Vulnerability in New
Orleans 222
    Financial Vulnerability 223
    Housing Vulnerability 223
    Health Care, Education, and Transportation..........................224
    Political Economy and Neoliberal Vulnerabilities 225
    Other Intersectionalities of Gendered Vulnerability
    in New Orleans 226
Conclusion 228
References... 229
Further Reading 230

PART III DISASTER RESILIENCE

CHAPTER 10 Culture and resilience: How music has fostered resilience in post-Katrina New Orleans 233
James R.G. Morris and Paul Kadetz

Introduction 233
    Background: Supporting Resilience From the Outside 234
Understanding Resilience and Designing the Research 236
    The Context for This Research 236
    Questions to be Answered 238
Methodology 239
    Study Design 239
    Sample 239
    Data Collection 240
    Data Analysis 240
    Ethics 241
Locating Resilience in Musical Performance 241
    Risk Factors, Stress, and Mental Health 241
    Protective Factors and Assets 247
    Social Support 247
    Connections to Community and Mentoring 248
    The Impact of Music Performance on Performers
    and Audience Members 249
    Hobfoll's Conservation of Resources Theory 251
Conclusion 251
References 253

CHAPTER 11 Resilience among vulnerable populations: The neglected role of culture 257
Mark VanLandingham

Introduction 257
The Resilience and Recovery Frameworks 258
    Resilience 258
    Recovery 259
    Resilience and Recovery 259
Application of Current Frameworks to the
Vietnamese-American Community in Post-Katrina
New Orleans 260
    Recovery 260
    Resilience 260

The Missing Piece: Culture 261
Application of an Expanded Framework to the
Vietnamese-American Community in Post-Katrina
New Orleans 262
    Culture Confounders 262
    Cultural Influences on Post-Katrina Recovery 263
Conclusions 264
References 264

CHAPTER 12 Faith-based organizations in Katrina: The United Methodist Church 267

Sarah Kreutziger, Ellen Blue and Michael J. Zakour

Theoretical and Conceptual Foundations 268
The United Methodist Church in Katrina 269
    Katrina-Related Outreach and Ministries 271
    Cursillo 271
    The United Methodist Church and Transformative
    Resilience 271
    The Decision Process of Mergers and Closures 273
    Outreach Ministry at First Grace 274
    Outreach at Hagar's House 274
    Outreach Through Luke's House 275
    Mt. Zion UMC Outreach 276
    United Methodist Women 276
Effects on Individual Katrina Survivors 277
    A Recovery Imbued With Spirituality 277
    Individuals and Resilient Recovery 278
Conclusion 280
References 280

CHAPTER 13 Collective efficacy, social capital and resilience: An inquiry into the relationship between social infrastructure and resilience after Hurricane Katrina 283

Paul Kadetz

Introduction 284
    The Context of This Analysis 285
Identifying and Fostering Resilience: The Essential Lens
of an Assets-Based Approach 286
    Do Needs-Based Approaches to Change Create Need? 286

The Relationship Between Neoliberalism and the
Creation of Need 287
"Acts of Faith" or Tyranny: From Neoliberalism to
Normative Community Development 287
Assets-Based Approaches: The Difference Between
Listening and Telling..290
A Tale of Two Cities: Two Studies of Resilience in
Post-Hurricane Katrina New Orleans.. 291
They're Called "Evacuees": Semantics or
Representations of Structural Violence?...............................291
Disaster Capitalism and Forced Disaster Migration:
The Merger of Neoliberalism and Structural
Violence 292
Positive Deviance and the Language of Resilience..............295
The Relationship of Structural Violence to Resilience 297
The Impact of Inequality: Neoliberalization, Individual
Competition, and the Erosion of Resilience 298
Conclusion: Understanding Resilience in the Complex
System of a Community.. 300
References... 301

CHAPTER 14 Dynamics of early recovery in two historically low-income New Orleans' neighborhoods: Tremé and Central City 305

*Nancy B. Mock, Paul Kadetz, Adam Papendieck
and Jeffrey Coates*

Background.. 305
Focusing Recovery Efforts: Theoretical Underpinnings 306
Assessing Information and Data at Multiple Levels
to Aid Recovery..307
RALLY and Neighborhood Action Research........................... 308
The Neighborhood Context 310
Neighborhood Dynamics and Change in Tremé
and Central City..312
Neighborhood Residents' Perceptions and Social
Infrastructure... 317
Significant Findings of RALLY.. 321
Conclusion 323
References... 325
Further Reading 328

PART IV CONCLUSION AND LESSONS LEARNED 329

CHAPTER 15 The Katrina catastrophe and science: Does experiencing a catastrophe at "*ground zero*" have impacts on the professional performance/identity of social scientist survivors? 331
Shirley Laska

Ground Zero Manifested 331
The *Ground Zero* Impact: Reasoning for the Study 332
Study Methods 333
Findings 335
    Trajectory of Research 335
    Assessment of Own Research 336
    Research Challenges and Benefits of Being Ground Zero Survivor Researchers 337
    Witnessing 337
    If/How Ground Zero Experience Changed Respondents Professionally 338
Implications for Researchers and Discipline 340
References 342

CHAPTER 16 How barefoot scholars were deployed: The good, the bad, the ugly 345
Nancy B. Mock

Introduction 345
The Lead-up to Katrina 346
Katrina and Early Recovery Efforts 347
Developing the Barefoot Scholar Initiative 348
The Return to New Orleans and the Evolution of RALLY 352
Lessons Learned 353
Post Script: The Crowd and the Cloud 354

CHAPTER 17 Lessons learned from New Orleans on vulnerability, resilience, and their integration 357
Michael J. Zakour

Economic Inequality Causal Chain 359
    Migration of the Poor 359
    Housing and Health Crises 360
    Lack of Economic Resources 360
Social Stratification Causal Chain 361

       Lack of Formal Education 361
       Lack of Human Capital 362
       Lack of Economic Growth 362
   Structures of Domination Causal Chain 362
       Trusted Media 363
       Media Controlled by Elites 363
       Lack of Political Partnerships 364
   Racial Ideology Causal Chain 365
       Marginalization 365
       Segregation 366
       Separate and Unequal 366
   Geographic Distance Causal Chain 367
       Smaller Geographic Service Ranges 367
       Low Volunteer Capacity 368
   Environmental Ideology Causal Chain 368
       Population Growth 369
       Rapid Urbanization 369
   Community Empowerment and Social Development 370
       Community Empowerment Causal Chain 370
       Evacuation Experience 372
       Client-Centered Services 372
       Place Attachment 373
       Human and Social Capital 374
       Flexible Disaster Plans 374
   Social Development Causal Chain 375
       Public Funding 375
       Population Programs 376
       Health Programs 377
       Building Codes 377
       Economic Growth 378
   Assumptions of V + Theory 378
       Unsafe Conditions Assumption 379
       Root Causes Assumption 379
       Disaster Causal Chain Assumption 380
       Assumption of Capabilities, Liabilities, and Susceptibility
       Relationships 381
   Conclusion 381
   References 382

Epilogue: Back to the future? **385**
Index 391

List of Contributors

Ellen Blue
Phillips Theological Seminary, Tulsa, OK, United States

Richard Campanella
Tulane School of Architecture, New Orleans, LA, United States

Jeffrey Coates
National Conference on Citizenship, Washington, DC, United States

Regardt J. Ferreira
Tulane University, New Orleans, LA, United States

Charles R. Figley
Tulane University, New Orleans, LA, United States; University of the Free State, Bloemfontein, South Africa

Kayla Grogg
West Virginia University, Morgantown, WV, United States

Susan Howell
University of New Orleans, New Orleans, LA, United States

Alessandra Jerolleman
Lowlander Center, Gray, LA, United States

Paul Kadetz
Drew University, Madison, NJ, United States

Sarah Kreutziger
Tulane University, New Orleans, LA, United States

Shirley Laska
University of New Orleans, New Orleans, LA, United States; Lowlander Center, Gray, LA, United States

Nancy B. Mock
Tulane University, New Orleans, LA, United States

James R.G. Morris
Stephen F. Austin State University, Nacogdoches, TX, United States

Adam Papendieck
Tulane University, New Orleans, LA, United States

Melissa Schigoda
City of New Orleans, New Orleans, LA, United States

Charles M. Swager
Essential Foundations PLLC, Morgantown, WV, United States

Ivor L. van Heerden
Agulhas Ventures, Inc., Reedville, VA, United States

Mark VanLandingham
Tulane University, New Orleans, LA, United States

Michael J. Zakour
West Virginia University, Morgantown, WV, United States

PART I

Introduction and Theoretical Framework

CHAPTER 1

Editors' introduction: The voices of the barefoot scholars

Michael J. Zakour[1], Nancy B. Mock[2] and Paul Kadetz[3]

[1]*West Virginia University, Morgantown, WV, United States* [2]*Tulane University, New Orleans, LA, United States* [3]*Drew University, Madison, NJ, United States*

CHAPTER OUTLINE

- Classifying Disasters 4
- New Orleans and the Livin' Ain't Easy 6
  - Vulnerable Lives 6
  - Linking Vulnerability and Resilience 7
  - Working With Cultural Assets 9
- Classifying Resilience 10
  - Social Infrastructure and Capacity 10
- Frameworks Utilized in This Volume 11
  - Disasters Require Multiple Perspectives and Paradigms 12
  - The Not-So-Hidden Agendas of Post-Katrina Research 13
  - Hybridizing Knowledge for Disaster Research 13
  - Barefoot Scholars 14
  - A Complex Systems Approach: The Umbrella of Our Theoretical Framework ...15
  - Vulnerability and Resilience as a Systems Problem 16
  - Structural Violence and Intersectionality 17
- Book Sections and Chapters 18
- References 21

This book examines one of the most devastating, deadly, and costly disasters in the history of the United States. Hurricane Katrina resulted in the near-total destruction of a major US metropolitan area (Knabb, Rhome, & Brown, 2005), with over 1500 deaths in New Orleans alone in the immediate impact period (Osofsky, Osofsky, Kronenberg, Brennan, & Hansel, 2009) and nearly 4000 more deaths in its wake (Stephens et al., 2007). Eighty percent of the city's area and built structures were flooded (Cigler, 2007). The disaster in the wake of Hurricane Katrina, in August 2005, is examined in this volume through an historical perspective with the goal of providing more clarity of how vulnerability and resilience interacted to affect the postdisaster recovery trajectory, as well as the sustainability of New Orleans and coastal Louisiana. Studying the (ongoing) recovery from Hurricane Katrina via a long-term analysis provides important new insights, theories, and knowledge pertinent to postdisaster recovery and resilience studies. Ongoing disaster vulnerability and resilience are often best understood by studying recovery and reconstruction. Sustainability is only evident over longer periods of time. Post-Katrina recovery has slowly progressed, yet after more than 12 years it is still not completed. Some have called the years after Katrina, "a failed recovery" (Adams, 2013). Failed recovery and reconstruction is actually a "second disaster," which can lead to as much misery and uncertainty as the original disaster.

The decade following Katrina offers many lessons concerning the growing risks associated with sea-level rise, climate extremes, wetland loss, coastal development, river delta subsidence, and long-term levee management. These growing risks represent a movement toward greater vulnerability. Globally progressing risks challenge group resilience and sustainable human development that are needed to mitigate new hazards. The examination of the precursors and sequelae of Hurricane Katrina through a complex systems lens, with a strong emphasis on local knowledge and capacity, offers important lessons for urban and coastal regions globally, particularly given the ongoing outcomes of climate change. In this introductory chapter we develop the theoretical and conceptual framing for this volume. We then describe how these concepts and theories apply to the Katrina event. Finally, we provide a brief overview of this volume's chapters.

CLASSIFYING DISASTERS

Disasters are the socially constructed human reactions to natural, geological, or meteorological events (Echterling & Wylie, 2013). According to Noji (2005), the human-influenced components of disasters can be classified as technological or complex humanitarian emergencies. Following Noji, we distinguish between natural and human-caused hazards. Hurricane Katrina would most accurately be considered a human-caused hazard. Although the Katrina event has often been classified as a natural disaster, the politics of this designation and the shirking from human accountability is problematic. The actual source of the city's flooding lies in the structural failure of the levees, worsened by the subsiding coastline, rising sea levels, collapse of the emergency management system, and associated technological disasters, such as the Murphy Oil spill in St. Bernard Parish. All of these causes of flooding were outcomes of human decisions (for a full discussion, see Chapter 6).

The wetlands surrounding New Orleans serve as a last line of defense from a storm surge, yet 2000 of coastal Louisiana's original 7000 square miles of wetlands have been lost (Freudenburg, Gramling, Laska, & Erikson, 2009). This loss is due, in part, to the environmental damage caused by offshore oil drilling, urban and rural development, and the building of navigation canals and other transportation projects (Bullard & Wright, 2009). Because these fresh-water marshes are at sea level, the increasing rate of sea level rise in the 21st century threatens to destroy additional areas of Louisiana's wetlands.

September 2017 marked the busiest month of Atlantic Ocean hurricanes on record, commencing with Hurricane Harvey that devastated parts of Texas (particularly Houston) and followed in rapid succession by two category five Hurricanes. Irma wreaked havoc in the Caribbean and Florida, and Maria caused further damage throughout the Caribbean, especially to Puerto Rico and the U.S. Virgin Islands. And this only covers recent weather-based disasters that affected the United States and its territories. Current disaster recovery must consider the uncertainty introduced by climate change, particularly for urban settlements on river deltas throughout the world, for it is not clear how rapidly sea levels will rise, or how prevalent or severe weather extremes will become in the future. Furthermore, "Effective risk reduction and adaptation strategies [need to] consider the dynamics of vulnerability and exposure and their linkages with socioeconomic processes, sustainable development, and climate change" (IPCC, 2014, p. 25).

The combined disaster of Hurricane Katrina and the floods that followed was produced by a situation in which both basic priorities that ensured safe infrastructure prior to the storm and effective humanitarian-oriented relief programs after the storm were largely absent (Adams, 2013). The current political milieu in the United States, embracing climate change denial, renders the ongoing lessons from this disaster even more imperative as we brace for more frequent and damaging climate events (see the epilogue of this volume for a fuller discussion). The United States is at an urgent crossroads in climate change policies and interventions. Climate change has already begun to demonstrate markedly negative impacts throughout the world, particularly for the well-being of marginalized populations. Increased risks are especially salient for ethnic and racial minorities in both rural and urban areas in the United States (IPCC, 2014). In New Orleans and Southeast Louisiana, marginalized communities were both disproportionately harmed by Hurricane Katrina and were largely excluded from the recovery process. African-Americans, women, the poor, and other marginalized populations were most impacted by the flooding of New Orleans. Pre-hurricane vulnerabilities limited the participation of many in recovery, rebuilding, and reconstruction efforts. The recovery for many marginalized groups in New Orleans has been delayed and perhaps permanently disrupted (Bullard & Wright, 2009).

The often complex relationships between vulnerability and resilience, as well as environmental liabilities and capabilities, are important lessons to share from this disaster. Although a majority of studies on Hurricane Katrina focused on the physical and geographic vulnerability of New Orleans and South Louisiana, knowledge of social and cultural aspects of vulnerability, susceptibility, and resilience is prioritized in this volume.

NEW ORLEANS AND THE LIVIN' AIN'T EASY
VULNERABLE LIVES

The New Orleans area is a nexus for disasters. Economic, social, cultural, and other forms of capital, as well as institutional contexts, are critical capacities for managing disaster risk. Resources, capital, and capacity have long been inequitably distributed in New Orleans and Southeast Louisiana. This uneven and inequitable distribution was essential for the creation of vulnerability (Oliver-Smith, 2004).

The immediate impact of Hurricane Katrina and the subsequent uneven recovery have resulted in a devastating cycle of impoverishment and vulnerability for many households and families. A new cycle of impoverishment has been created by a combination of

- environmental injustice, including the inequitable accessibility to pre-Katrina disaster preparedness and post-Katrina recovery resources;
- damage to social infrastructure, including health and human services provisions; and
- a markedly neoliberal approach to recovery aid.

This cycle has trapped low-income and former middle-class households and individuals in greater disaster vulnerability and thwarted resilient recovery. The political economic restructuring of postdisaster places, a phenomenon Klein (2007) describes as a form of disaster capitalism, has proved particularly destructive. In disaster capitalism, a disaster facilitates a "blanking of the beach," or a tabula rasa upon which a markedly neoliberal restructuring can occur across multiple sectors for the sole enrichment of the power elite. This restructuring has increased vulnerability and compromised the human agency upon which resilience depends (Adams, 2013).

Post-Katrina New Orleans reflects the political economic trends and resulting disaster vulnerabilities encroaching on much of the United States and the world. Larger structural trends in the political economy of the nation have created an increasing and pervasive vulnerability and fragility among Americans. Vulnerability has increased among many populations because of growing environmental injustices associated with increased disaster exposure and reduced access to response and recovery resources (Adams, 2013). According to the U.S. Environmental Protection Agency (EPA, 2017, para. 1), "Environmental justice is the fair treatment and meaningful involvement of all people regardless of race, color, national origin, or income, with respect to the development, implementation, and enforcement of environmental laws, regulations, and policies." As discussed in Chapter 7 of this volume, environmental justice includes

- protection from environmental degradation;
- prevention of adverse effects from environmental deterioration before harm occurs;

- mechanisms for assigning culpability to the agents of environmental degradation, rather than blaming the residents affected; and
- equitably redressing the impacts of environmental damage through remedial action and needed resources (Cutter, 1995).

The response, recovery, and reconstruction phases for the Hurricane Katrina event provide an important analysis of environmental justice in a major metropolitan area. Disaster vulnerability was inequitably distributed across the residents of New Orleans and concentrated in low-income, female, and African-American populations, as well as in rural areas of cultural and ethnic minorities (Adams, 2013; Bullard & Wright, 2009).

Vulnerabilities can be identified across all sectors of society in New Orleans, particularly in the health sector. The health sector includes health service provision that may be impacted by extreme events, including structural failures in hospitals and health centers and inability to access health services because of storms and floods (IPCC, Chap. 2, 2012). The provision of health and human services was badly damaged by Hurricane Katrina, and these service sectors will not fully recover or will remain inoperable. This was the case with the closure of Charity Hospital and the false rhetoric of irreparable damage that prevented its reopening. Furthermore, the tremendous damage to health and human services compounded the negative impacts that neoliberal approaches to recovery placed on resilience.

LINKING VULNERABILITY AND RESILIENCE

In general, vulnerability and resilience theory and research have largely been portrayed as different and distinct streams of knowledge. In examining the past decade's lessons from Katrina, this volume attempts to provide an integrated understanding of community disaster vulnerability and resilience capacities. Vulnerability and resilience theories identify a great deal of overlap in the classification of types of resources. Vulnerability theory identifies economic, social, human, physical, natural, and political resources that determine the progression to vulnerability (Wisner, Gailard, & Kelman, 2012).

Resilience theory specifies a similar network of resources that must be mobilized if a community is to experience adaptive resilience. Resilience capacity can be broadly defined as the ability of a system (including individuals, households, communities, and societies) to absorb, adapt, and transform in the face of hazards, shocks, and stresses in order to promote sustainable human development. General or inherent resilience capacities can exist predisaster, as identified in evacuation systems, social capital, and robust built infrastructure. But resilience can also emerge after a disaster, which is known as "adaptive resilience" (Tierney, 2014). All of the constituent populations of a community must achieve resilience in order for community recovery to be considered resilient (Norris, Stevens, Pfefferbaum, Wyche, & Pfefferbaum, 2008).

We extend the conceptualization of Norris et al. (2008) by distinguishing geographic resilience from human/community-centered resilience (the focus of this volume). A geographical approach to resilience favors the identification of vulnerable populations according to place and location to assess the level of resilience. Human- and community-centered resilience approaches seek to identify vulnerable populations and to foster the resilience of these populations. The latter is more pertinent to a given sociocultural context. A human- or community-centered approach focuses on identifying and reducing inequities in agency and across social institutions to promote resilience. The chapters in this volume are more concerned with building and rebuilding the capacity of social infrastructure, than with a sole focus on the postdisaster management of physical infrastructure and the built environment.

The level of vulnerability in a social system prior to a disaster ultimately affects the likelihood of resilience after the disaster (Zakour & Gillespie, 2013). Disaster vulnerability and general (or inherent) resilience are inversely related, because many environmental liabilities that increase vulnerability also decrease resilience (Cutter et al., 2006). Many capabilities are aspects of both vulnerability and resilience, although there can be cases in which a capability can only be applied to one and not the other. For example, community narratives celebrate stories of survival and successful coping in disasters. A constructive community narrative represents an example of a capability that is neither part of general resilience, nor does it contribute to the predisaster progression to vulnerability. Vulnerability can also affect resilience in terms of how the vulnerability of a social system determines the level of loss experienced in a disaster. After a disaster, if losses are great, then a resilient trajectory of recovery is less likely. Vulnerability determines the postdisaster baseline of functioning. The lower the baseline, the greater the resiliency needed to be transformative and foster a level of functioning exceeding predisaster levels (Laska, 2012).

Just as Katrina and the subsequent flooding revealed a high level of vulnerability, the nature of the recovery revealed that the level of resilience was not equal among the populations of Southeast Louisiana. Despite some success in terms of building resilience, the case of New Orleans should give us pause in considering how we intervene with vulnerable communities. To insure effective recovery, we also need to more fully consider the political economic context of pre- and postdisaster societies. Although many risk drivers were present in New Orleans before, during, and after Hurricane Katrina, perhaps the foremost consideration among these risk drivers is income and socio-economic inequity. Katrina and New Orleans should help us recognize that a system that is built on neoliberalism, prioritizes profit above all else, and relies on the market for recovery, will not prioritize the care of vulnerable populations (Adams, 2013). Instead, profit will be sought from the vulnerability of others (Klein, 2007). In the end, neoliberalism not only slowed recovery, but markedly augmented and reinforced class disparities that existed prior to the disaster.

However, community capacity and social capital can also overcome predisaster vulnerabilities. For example, before Hurricane Katrina, New Orleans lacked an evacuation system that included and targeted vulnerable populations. A fundamental capacity in New Orleans' communities has been a culture of humanitarian aid, particularly from faith-based organizations and their volunteers, and the emergence of large numbers of community organizations seeking to help in the recovery effort. In the specific case of evacuation, a volunteer organization was created post-Katrina to evacuate those without personal transportation.

In general the social welfare policies and programs already in place before Hurricane Katrina helped enhance resilience among the poor, powerless, and marginalized of New Orleans. Since Katrina, numerous social movements and nonprofit organizations have focused on rebuilding community resilience. Social protections, including insurance and disaster risk management, enhanced the long-term resilience of the poor and other marginalized groups, particularly when the policies have addressed poverty and multidimensional inequalities (IPCC, 2014). The many recovery groups that arose after Katrina, constituted new social movements and spawned the birth of numerous nonprofits and NGOs. They became important conduits for labor, legal support, and financial aid. Many organizations not only helped people recover, but also revisited the past context of structural inequality and social injustice in New Orleans. They focused not only on rebuilding and restoring New Orleans to its pre-Katrina state, but also on rebuilding within a social justice framework (Adams, 2013).

WORKING WITH CULTURAL ASSETS

Part of a social justice framework for disasters includes designing a recovery that is appropriate to a given community and their culture. To understand vulnerability and resilience, as well as their intersections, it is important to know and work within the cultures and cultural diversity of a community and region. The rich cultural diversity of New Orleans and Southeast Louisiana provides an ideal setting for understanding the influence of culture on vulnerability and resilience. Culture, as a dimension of both vulnerability and resilience, includes elements such as way of life, behavior, taste, ethnicity, ethics, values, beliefs, customs, ideas, institutions, art, and intellectual achievements. Elements of culture affect, are produced, and are shared by a particular society (for a full discussion of how social values and distinctions are constructed, see Bourdieu, 1989). Social capital, social cohesiveness, and social support networks are fostered by societies and their cultures and shape the distribution of vulnerability and resilience within a group.

A group's culture is an important determinant of both vulnerability and resilience. Both etic and emic perspectives are employed throughout this volume, as both are needed to fully understand the role of culture in resilience. The concepts of emic and etic are derived from the linguistic terms phonemic and phonetic (Pike, 1967). The terms are used here to distinguish the perspectives and paradigms originating from within a given group (emic), from those which originate outside of that group and is often used to interpret and/or explain the group and their meanings (etic). This differentiation is central to understanding the difference between subjective and objective framings (see Headland, Pike, & Harris, 1990). Etic perspectives are very useful in examining behaviors including social capital, norms, and social structures. Emic perspectives are useful in understanding the rules and behaviors that govern society and foster social organization. These cultural codes include ethics, values, ideas, and information.

A community or region can be classified as either a culture of vulnerability, with large disparities in political and economic power, or as a culture of safety and resilience. Priority 3 of the Hyogo Framework for Action 2006–2015 recommends using knowledge, innovation, and education to build a culture of safety and resilience (IPCC, 2012). For example, as discussed in depth in Chapters 11 and 13, the social capital and community capacity of the Vietnamese community in New Orleans East facilitated their postdisaster recovery and resilience.

Institutions also shape the distribution of vulnerability and resilience. Broadly defined, institutions include patterns of behavior (social structures) and rules and norms that govern society. This conceptualization of institutions allows us to examine social networks, social cohesiveness, and social capital. Social capital and related mobilization through networks of social support can reduce vulnerability and increase resilience. Institutional structures that govern natural resource use and management can determine if development is ultimately sustainable (IPCC, Chap. 2, 2012).

CLASSIFYING RESILIENCE
SOCIAL INFRASTRUCTURE AND CAPACITY

Examining the Katrina disaster and recovery through a social infrastructure lens is essential for establishing an equitable and socially just recovery, particularly given the magnitude and long-term repercussions of the disaster. Social infrastructure is concerned with the development of several types of human capacity. Capacity refers to present ability, skills, or resources, including those associated with human and social capital. The presence and availability of community capacities make up a large part of both vulnerability and general (inherent) resilience. In reducing vulnerability and building inherent resilience, it is important to identify and amplify drivers of capacity. According to the Intergovernmental Panel on Climate Change (IPCC, Chap. 2, 2012), drivers of capacity can include

- an integrated economy,
- information technology,
- human rights,
- access to insurance,
- health and well-being,
- income and socio-economic equality,
- access to public health,

- community organizations,
- decision-making frameworks,
- warning and protection from hazards, and
- good governance.

Capacity can be classified in different ways. Absorptive capacity is the ability to maintain essential social structures and continue to meet human needs, without adverse impacts from disasters. Adaptive capacity enables socio-ecological systems to adjust to hazards. Transformative capacity involves fundamentally altering the attributes of a socio-ecological system to greatly reduce and eliminate risks. Hence, transformative capacity involves fundamental changes in the system and its components. Transformative capacity often requires change at varying levels in socio-ecological systems. For example, the US federal emergency response was tested and failed in the Katrina disaster. Katrina brought about the revision of federal disaster policies and procedures that had previously constrained the capacity of local communities to recover. Much of the work of the Federal Emergency Management Agency (FEMA) actually compromised local capacity through its lack of coordination among national, state, and local response. However, with the post-Katrina implementation of the National Response Plan, disaster response organizations at various levels of government were better coordinated.

When a community is able to mobilize a capacity for recovery, it is called a capability. Capability refers to the potential to mobilize skills and abilities in order to reduce the level of vulnerability and increase resilience. Capabilities also reduce susceptibility. Susceptibility is the degree to which socio-ecological systems can be disrupted and damaged by hazards. Combined with a system's general or inherent resilience, susceptibility determines the level of vulnerability. The most important types of capabilities concern access to resources. Increasing a community's access to resources is critical to reducing vulnerability and to building resilience. Capabilities and capacities lead to a progression toward safety, as well as toward adaptive resilience.

Adaptive resilience is an especially important consideration, given the uncertainty and dynamic impact of climate change (IPCC, 2012, 2014). Adaptive capacities are cornerstones of vulnerability reduction, as well as of sustainability and resilience. Disaster risk management and vulnerability reduction can be addressed through the enhancement of generic adaptive capacity alongside hazard-specific response strategies. Investing in generic adaptive capacity to reduce vulnerability is an investment in the foundations of resilience and sustainability (IPCC, Chap. 8, 2012).

FRAMEWORKS UTILIZED IN THIS VOLUME

There are several frameworks employed in this volume, including postfoundationalism, complex systems thinking, a barefoot scholars approach, and a social justice approach.

DISASTERS REQUIRE MULTIPLE PERSPECTIVES AND PARADIGMS

Since disasters are socially constructed processes resulting in multidimensional social shifts that will vary widely from one sociocultural setting to another, it would seem irrefutable that disasters must integrate multiple perspectives for appropriate and effective interventions. Yet disasters are normatively approached through a single, predictable foundation or framework. As exhibited by elites in almost any postdisaster context, elites in post-Katrina New Orleans determined whose knowledge would count and thereby provided a means for these elites to achieve their own goals and interests (Adams, 2013; Mercer, 2012). Hence, although "equally legitimate parallel knowledge systems exist" that people may "move easily between, one kind of knowledge gains ascendancy and legitimacy" (Jordan, 1997, p. 56). Anthropologist Bridget Jordan identifies this phenomenon as the domination of authoritative knowledge, with its devaluation or dismissal "of all other kinds of knowing...as backward, ignorant, naive" (Jordan, 1997). Authoritative knowledge is similar to Gramci's (1994) concept of "hegemony", defined as the dominance and control "of a social group over the entire national society, exercised through the so-called private organizations, such as the Church, the unions, the schools" (1994, p. 67). "Elites control the 'ideological sectors' of society and thereby engineer consent for their rule" (Scott, 1985, p. 39). Authoritative knowledge is also consistent with Foucault's concept of discourse, defined as "a system of statements within which the world can be known, by which dominant groups in society constitute the field of truth by imposing specific knowledges, disciplines, and values upon dominated groups [and thereby] works to constitute reality" (Ashcroft, Griffiths, & Tiffin, 1998, p. 42).

Normative etic perspectives for knowledge building are closely associated with positivism. Positivist approaches are useful for building generalizations about such phenomena as disaster vulnerability and resilience capacities. In the positivist view, the scientific method relies on clarity of language and validation of claims by rational means of logic and empirical inquiry (Schweizer, 1998). The positivist approach is nomothetic (employing abstract, universal, or general statements). Positivist approaches seek generalities (e.g., laws) about societies and cultures. All positivist epistemological approaches rely on some type of direct observation that builds an empirical foundation.

Etic perspectives employ variables that are conceptually and operationally defined by the investigator. In disaster vulnerability research, investigators using an etic perspective often solely derive variables and measurement from theory or theoretical models. For example, the use of the Pressure and Release (PAR) model (Wisner, Blaikie, Cannon, & Davis, 2004) or the resilience models of Norris et al. (2008) for research design and measurement are part of the etic perspective in disaster research. In the etic perspective, direct observation signifies accurate reporting of observed scenes and activities. Experimental designs, in which a supposition is tested by a laboratory type model and a specific intervention is compared with a counterfactual through randomized assignment, are the holy grail of positivist analytical tools. Although experimental design has strong internal validity, it understandably has many critics for its application to socio-ecological systems, whose components cannot be separated and reduced to core parts. Disaster socio-ecological research favors natural experiments and multiple methods research, including participatory action research, with its collective inquiry grounded in experience and social history.

THE NOT-SO-HIDDEN AGENDAS OF POST-KATRINA RESEARCH

Although local and expert knowledge are both needed in disaster research and recovery, authoritative expert knowledge will invariably dominate, at least in part, to camouflage the exploitations of outside academics and protect elite accountability. Collectively, this volume's contributors have experienced Hurricane Katrina through the progression of disaster vulnerability to resilience building, encompassing disaster response, recovery, and reconstruction. Yet disaster researchers living in New Orleans at the time of Katrina, with their intimate knowledge of the regional culture and intuitive understanding of New Orleans' vulnerability, often had difficulty conducting their own disaster research because outside experts were prioritized for research funding. However, local scholars and local knowledge persevered. Although some of these researchers experienced losses from Katrina, barriers to conducting their research programs largely originated from elite internal and external interests attempting to extend their influence in the region.

Corporate interests active in the Katrina recovery have sought to create "official" narratives that protect the ability of corporations to profit via the recovery (Button, 2010). In the decade since Hurricane Katrina, knowledge has been contested, as witnessed in other disasters. Powerful governmental and business interests often prevailed in this contestation. Government sought to reinforce knowledge that reduced their liability. This is readily apparent, for example, in the official language used to label internally displaced persons as "evacuees." FEMA issued statements to Congress, the media, and the public that they were unaware of any significant health hazards present in FEMA-supplied trailers (Button, 2010), even though in 2008, the Centers for Disease Control (CDC) identified "higher-than-typical indoor levels of formaldehyde." In short, business interests and elites sought to emphasize knowledge that improved their chances of earning profits, either despite the disaster or through business participation in recovery (see Klein, 2007).

HYBRIDIZING KNOWLEDGE FOR DISASTER RESEARCH

Hybrid knowledge combines the positivist and interpretive knowledge of stakeholders both inside and outside of a community. Hybrid knowledge can be generated by scientists, policymakers, nonprofits, and community activists, among others; at local, state, and national levels. Actions taken to understand and reduce disaster risk would effectively access and integrate relevant and applicable local and outside knowledge (Mercer, 2012). Hybrid knowledge helps disaster practitioners and scholars understand the local "social dialect," and thereby avoid significant issues of cultural incompatibility, as is common with the universal "one-size-fits-all" approach to disaster intervention. Hybrid knowledge avoids the limitation of official narratives promoted through the electronic media after a disaster (Button, 2010). Local scholars working within the complex social web of New Orleans have been the primary source of hybrid knowledge and are of primary importance in generating the understanding and evidence base for localizing disasters.

The contributors to this book combine both positivist and interpretive epistemologies to illuminate the complex, multiscalar, and multidimensional nature of disasters. Their hybrid knowledge is essential for understanding the complex political-economic systems and the dynamics of intervention that affect both disaster vulnerability and resilience capacity in the aftermath of the Katrina event. Without local knowledge, it is impossible to truly understand the effects, motivations, goals, and subjective meanings of disaster survivors, and whether disaster victims experience acceptable recovery interventions and improved conditions or long-term despair (Adams, 2013; Hoffman, 1999).

In addition to integrating explicit with tacit knowledge, this volume is also a hybridization of markedly different traditions of social and human science. This book provides an opportunity to listen to the voices of scholars whose knowledge and "insider" experiences with this disaster have not yet been fully recorded or heard.

BAREFOOT SCHOLARS

Given their intimate understanding and lived knowledge of the social conditions, culture, and political and economic context of New Orleans and Southeast Louisiana, the editors and chapter authors can be considered "barefoot scholars." Barefoot scholars possess a comprehensive and postfoundational knowledge of a field of inquiry, discovered through a combination of different epistemological frameworks. As members of community systems in New Orleans and Southeast Louisiana, the authors collectively have acquired in-depth knowledge of these systems and their dynamics. To fully understand the meaning of Hurricane Katrina for the people most affected, it is of critical importance for these "insider" accounts to be heard.

We borrow the term barefoot scholars from the so-called barefoot doctors of China; the community-picked health workers in the rural villages of China throughout the late Maoist period of the 1970s (Kadetz, 2016). Through their widespread health care to rural communities, these health workers were instrumental in providing access to health care and benefiting the health of communities across China. Their success stemmed in large part from the fact that they were from the communities in which they worked, and therefore possessed an intimate knowledge of the social and cultural complexities of their communities. Through being a part of the systems they treated, and through a daily examination of their communities, barefoot doctors were able to understand the complex interplay of factors that affected the health of their specific community.

Similarly, local scholars were able to understand and experience firsthand the role of being disaster-affected and vulnerable, as well as experience the importance of agency and resilience capacities to their well-being. Their experience in the complex power dynamics of disaster response and recovery is also unique. Most of the contributors to this volume were disaster scholars before Hurricane Katrina. Yet they identified post-Katrina as an enormous period of growth in understanding the power dynamics surrounding vulnerability, disaster response, recovery, and the nature of resilience capacities among residents, identity groups, and neighborhoods. They lived through the complex emergence of recovery and the longer term sustainability of the region.

Barefoot scholars bring great added value to multiple methods of research because they may operate and produce a quantitative analytical, and also an interpretive, or hermeneutic, epistemology. This epistemology is ideographic, describing the particular social and historical context of the Katrina event and the subsequent recovery period. Barefoot scholars understand emic perspectives as epistemological tools, as well as how to harvest emic perspectives in knowledge generation. One notable example of an interpretative and emic approach is found in the ethnographic study of psychological distress, found to accompany structural inequities unmasked by disasters during the recovery period (Adams, 2013; Oliver-Smith & Hoffman, 2002).

A COMPLEX SYSTEMS APPROACH: THE UMBRELLA OF OUR THEORETICAL FRAMEWORK

Barefoot scholars approach disasters as complex, open, multidirectional systems. No social event, least of all the dynamic changes accompanying disasters, can be addressed as a simple system. Simple, closed, unidirectional systems analyses will never completely capture the complex social, cultural, and environmental interactions that have shaped the outcomes of this (or any) disaster. Disasters are exceptional events that are precipitated and modulated by a multidimensional, multiscale, and multilevel interplay of vulnerability and resilience capacities of socio-ecological systems, which simultaneously examine the social, cultural, built, and natural environments. Multilevel analysis offers simultaneous inquiry across different levels of influence. All of these elements require a complex systems approach for full understanding of how the factors affect one another.

By employing a systems thinking lens, this book embraces the complexity of Katrina's impact on the social system of New Orleans and the multidimensionality of the Katrina disaster affecting individual, household, neighborhood, metropolitan area, state, and national levels over time. The complex systems approach addresses the complexity of perspectives between local and "expert" knowledge and of different stakeholders in disaster recovery, including government, nonprofits, corporations, faith-based volunteers, and at-risk communities. Finally, a complex systems approach is required for the transdisciplinarity that is a feature of this volume in its integration of disaster knowledge from a variety of disciplines and professions.

Most disaster researchers to date have focused on the immediate disaster response period, leaving the rebuilding and reconstruction periods of disasters among the least studied. Systems thinking facilitates a dynamic approach before, during, and after the disaster. Temporal scale is significant because it is consistent with a disaster loss reduction framework that requires situating risk reduction approaches within a broader time frame.

VULNERABILITY AND RESILIENCE AS A SYSTEMS PROBLEM

A working definition of vulnerability from a systems perspective is the propensity of a system to be disrupted and experience severe harm when a natural or human-caused shock occurs. Vulnerability originates in the complex interaction of factors in the social, built, and natural environments. Thus, use of the label "vulnerable" is not meant as pejorative or synonymous with weaknesses or deficiencies in a population. This unfortunate and inaccurate label, resulting from the Katrina experience, effectively acts to reduce the agency of affected people.

Whole societies can be resilient, as can different sectors and systems (e.g., local retailers and faith-based systems). In order for a community, region, or society to understand vulnerability reduction and resilience building, many different components of a community and its resources need to be examined and addressed. Disasters affect community systems and subsystems (such as households, local organizations, and neighborhoods) at many levels. Resources including economic development, social capital, information, communication, and collective action must be networked to support the timely and complete recovery of each of these subsystems (Norris et al., 2008; Zakour & Gillespie, 2013).

A systems perspective permits the analysis of vulnerability and resilience as complex multilevel interactions of liabilities and capabilities. The systems perspective enables the understanding of the simultaneous and lagged (cross-scale and cross-level) interactions that lead to resilience pathways toward sustainable development. Systems thinking also helps simplify disaster complexity by addressing key drivers of vulnerability and resilience, as opposed to addressing all factors that could affect them, by differentiating "signals" from "noise" in systems.

Complex systems analysis is needed to understand how vulnerability and resilience can impact one another. Cycles and feedback loops of vulnerability, susceptibility, and the lack of resilience have been revealed after Hurricane Katrina. Feedback can be negative or positive. Negative feedback aims to maintain system states, such as community sustainability, while positive feedback aims to change system states, rendering them more or less sustainable. Drivers of resilience, such as social protection systems and income and resource equity, are critical for achieving sustainable development (i.e., an improved systems state). In the absence of these critical drivers, many individuals, neighborhoods, and institutions can become resilient through positive feedback loops. In other words, the chaos around the Katrina event created space for those affected to innovate, empowering them to further innovation and resilience capability. These loops illustrate the dynamic and cyclical relationships between vulnerability and resilience. During the Katrina disaster, poor households quickly expended limited resources in coping actions, and undermined household sustainability in the long run. In a vicious cycle of decline, households spent savings and even sold productive assets. Households began to splinter, individuals were forced to migrate, and household members sometimes entered into culturally inappropriate, dangerous, or illegal livelihoods. The poverty and vulnerability trap made resilience and recovery to predisaster levels of well-being increasingly difficult (Adams, 2013; IPCC, Chap. 9, 2012).

Information and knowledge concerning disasters comes from a variety of sources including the media, government, nonprofits, social support networks, and the altruistic community (Zakour & Gillespie, 2013). The complexity of disasters in the context of socio-ecological systems points to the need for an equally diverse knowledge base (Wisner et al., 2012). With increasing sophistication of analytical tools and "big" databases, systems analysis is coming of age. Access to reliable information and knowledge, as well as disaster narratives, is important for understanding both disaster vulnerability and resilience capacity (Norris et al., 2008). For example, using ethnographic and other qualitative methods, Adams (2013) successfully delineated the neoliberal influences on recovery, making an important contribution to the links between disaster recovery and political economy.

The multidimensionality of disasters, and the numerous disciplines and professions studying disasters, signal the importance of mixed-methods research (see Aldrich, 2012). Theoretical models integrating vulnerability and resilience, such as Vulnerability-plus (V+) theory (see Chapter 3), include both quantitative and qualitative variables (Zakour & Gillespie, 2013). For example, research on economic inequality, a root social determinant of the progression to vulnerability, incorporates the measurement of incomes and the skewness of the income distribution. In this causal chain, causality can be better understood through ordinary least squares and other types of regression. At the end of the causal chain that begins with the root cause of economic inequality, community action serves as a type of collective action capability. Collective action is very similar to collective agency (Norris et al., 2008), and measurement of collective action includes both qualitative and quantitative methods (Wisner et al., 2004).

STRUCTURAL VIOLENCE AND INTERSECTIONALITY

Finally, to fully understand the social quality of resilience and the landscape of the social infrastructure, it is imperative to contextualize the particular social circumstances contributing to group vulnerability and resilience from the perspective of social justice. The perpetuation of inequity and exclusion toward any given group of society via social structures and institutions has been categorized by the anthropologist Paul Farmer (2004, after Galtung) as structural violence. However, structural violence does not fully address the complexity of bias that may be directed either toward multiple groups in societies, or toward the multiple groups to which individuals belong and identify, such as, for example, single, poor, African-American mothers in New Orleans. The social justice concept of intersectionality facilitates a real world, complex-open-system analysis of structural violence that is directed toward multiple groups, or toward individuals belonging to more than one group that is being socially excluded by the dominant power structures in a society (for a full discussion of intersectionality, see Crenshaw, 2003, and Chapter 9 of this volume). Several chapters in this book examine resilience and vulnerability through this complex social justice framework.

BOOK SECTIONS AND CHAPTERS

This volume is divided into four parts in order to emphasize the primary foci and contributions: (I) theoretical frameworks, (II) disaster vulnerability, (III) resilience, and (IV) conclusions and lessons learned. Part I presents initial context (Chapter 2), and introduces the different lenses and theoretical perspectives for examining Katrina. These perspectives include an integrated theory of vulnerability and resilience (Chapter 3), complex systems dynamics (Chapter 4), and concepts of agency and structural violence (Chapter 5).

In Chapter 2, Campanella examines the attempts made to shrink the footprint of post-Katrina New Orleans to avoid future flooding. His chapter illustrates how much of the population movement toward land below sea level resulted from an ideology that has emphasized human technology as a way to conquer nature, such as building on former wetlands. The terrible consequence of this migration during the 20th century was the inundation of 80% of New Orleans structures by the hurricane surge of Katrina and the failure of technology, the levee system, to compensate.

Chapter 3, by Zakour and Swager, summarizes Vulnerability-plus (V +) theory. This theory is an integration of vulnerability and resilience concepts, theoretical models, and research. In the predisaster progression to vulnerability, the balance of disaster susceptibility and general resilience is shown to determine the level of vulnerability. The authors discuss how root causes, structural constraints/risk drivers, and unsafe conditions both produce disaster vulnerability and provide the baseline for needed postdisaster recovery. Postdisaster, the quantity and quality of networked resources increase the probability of a resilient recovery and the level of community functioning and wellness.

In Chapter 4, Mock, Shigoda, and Kadetz describe vulnerability and resilience in New Orleans through a systems lens. The concept of Complex Adaptive Social Systems (CASS) is used to illustrate the relationships and continuity between vulnerability and resilience, and to facilitate their integration. The *New Orleans Index* exemplifies a systems approach to recovery in New Orleans.

In Chapter 5, Laska, Howell, and Jerollerman examine how structural violence disrupts the networked agency of survivors; a necessary aspect of a resilient recovery. The chapter discusses the destructive outcomes of disregarding survivor agency and inflicting structural violence during and after disasters. The authors suggest reasons why agency is often ignored, and what can be done to resolve this.

Part II emphasizes the progression of disaster vulnerability that led to the Katrina disaster. This section examines the high level of vulnerability of the New Orleans area that resulted in the marked loss of life and property. This description of the progression to vulnerability applies both to New Orleans and its constituent populations. In Chapter 6, van Heerden warns of the ongoing increase in vulnerability for coastal and Southeast Louisiana. The author traces the historical loss of the protective wetlands, the rapidly subsiding coast, accelerating relative sea level rise, and the development of a major navigation channel that opened a surge conduit directly into the heart of New Orleans. The human errors and miscalculations that resulted in the catastrophic failure of the hurricane protection system are delineated. The author argues that a sustainable surge protection system, and aggressive coastal wetland and barrier island restoration, are keys to reducing future vulnerability.

Chapter 7, by Zakour and Grogg, complements the previous chapter by examining the historical socio-cultural causes of disaster susceptibility in New Orleans. The hazards of Katrina interacted with a high vulnerability level to trigger both a disaster and a humanitarian catastrophe. A lack of adequate resilience resources led to unsafe conditions and compromised a resilient post-Katrina recovery. The authors demonstrate that inequitably distributed vulnerability and resilience in New Orleans represent environmental injustice. Although vulnerable populations lacked resilience resources, social movements emerged post-Katrina that contributed to a more just recovery.

In Chapter 8, Ferreira and Figley discuss the reduction of psychosocial costs of disaster for vulnerable populations. The authors argue that a lack of a comprehensive understanding of disaster vulnerability accounts for the failure to identify communities susceptible to poor disaster mental health. The chapter offers improved, asset-informed, strategies for reducing the level of social vulnerability to disasters.

Chapter 9, by Kadetz and Mock, illustrates how vulnerability assessments must address the intersectionalities of inequality and social exclusion that are specific to any given context. In the context of New Orleans the intersectionality of gender, race, income, and age provides a more accurate depiction of the vulnerability of African-Americans, single mothers, and poor women of color over 65 years of age.

Part III concerning disaster resilience defines and discusses the nature of resilience, and demonstrates the importance of both a superior recovery and a community scope of analysis. The chapters in this section illustrate the relationships between disaster vulnerability and resilience. In the first chapter in this section, Chapter 10, Morris and Kadetz unpack the factors that facilitate resilience among musicians and the music community in New Orleans after the Katrina disaster. Both capabilities (protective factors) and liabilities (risk factors) in the environment are measured by an instrument for assessing postdisaster causal factors for resilient recovery. The chapter illustrates the central importance of a community's generation of cultural outputs, particularly music, for effecting community resilience.

Chapter 11, by VanLandingham, reveals the complex relationships between vulnerability and resilience. The chapter illustrates how a community can be both vulnerable and resilient in a disaster such as Katrina. The Vietnamese community in New Orleans East recovered resiliently due both to cultural factors, and what the author calls culture confounders (noncultural factors related to immigration and membership in an ethnic community). Cultural causes of resilience are shared narratives of survival, and shared perspectives or frames that perceive social relations as hierarchical.

In Chapter 12, Kreutziger, Blue, and Zakour examine the role of faith-based organizations in fostering resilience during the Katrina recovery and reconstruction periods. Through the provision of volunteers, faith-based organizations were leaders in a disaster-related utopian movement after Katrina. The United Methodist Church in post-Katrina New Orleans offers an illustrative case example of the role of faith-based organizations in fostering resilience.

In Chapter 13, Kadetz identifies how the complex relationships between social cohesion, social capital, and its elements of social trust and norms of reciprocity and collective efficacy, are foundational for a community's resilience to effectively respond to social change and catastrophic events. Although much of the literature regarding resilience focuses on the one dimension of social capital, the chapter demonstrates the multidimensionality of developing social infrastructure that also requires the inclusion of social capital and collective efficacy.

In Chapter 14, Mock and others demonstrate how the neighborhood-specific cultures that have evolved in New Orleans, and their needs during the reconstruction phase of Katrina, differ by community and subcommunity levels. The chapter describes the Recovery Action Learning Laboratory (RALLY) project. RALLY conducted participatory action research in support of recovery efforts, such as community action. The authors conclude that assessing recovery at the appropriate level of analysis allows for understanding populations' needs, and for design of the most appropriate interventions for that population.

Part IV concerns conclusions and lessons learned. This section of the book identifies the overriding lessons we can derive from Hurricane Katrina and a dozen years of disaster recovery and reconstruction. In Chapter 15, Laska examines higher education faculty from the New Orleans metropolitan area who experienced and survived the Katrina disaster. These researchers were both inside experts and barefoot scholars who were transformed by their experiences, and were ultimately motivated to achieve the highest quality professional work. Immersion in the context of Katrina has allowed these scholars to produce knowledge important both inside and outside of the academy.

In Chapter 16, Mock, a barefoot scholar, presents her own experiences in organizing and supporting others in responding and studying Katrina. All of the participants in the chapter engaged in experiential learning concerning disaster recovery. The chapter describes the evolution of a nonprofit organization that provided support for the barefoot scholars, and offers lesson about the deployment and organization of these scholars.

In Chapter 17, Zakour examines the lessons that can be drawn from applying Katrina-based research findings to a model integrating vulnerability and resilience theory. Evidence supporting Vulnerability-plus (V+) theory and its assumptions is reviewed. The authors identify substantial research support for V+ theory from the studies in this volume, and from other research on Katrina.

In the closing chapter of this volume, Kadetz, Mock, and Zakour argue that the lessons of Hurricane Katrina, particularly in terms of vulnerability and resilience, must be heeded and implemented in directing future disaster research, management, and interventions. Toward this goal, the offerings of this volume provide a guide to a more resilient future.

REFERENCES

Adams, V. (2013). *Markets of sorrow, labors of faith: New Orleans in the wake of Katrina.* Durham, NC: Duke University Press.

Aldrich, D. P. (2012). *Building resilience: Social capital in post-disaster recovery.* Chicago, IL: University of Chicago Press.

Ashcroft, B., Griffiths, G., & Tiffin, H. (1998). *Key concepts in post-colonial studies.* New York: Routledge.

Bourdieu, P. (1989). *Distinction: A social critique of the judgement of taste.* London: Routledge.

Bullard, R. D., & Wright, B. (Eds.), (2009). *Race, place, and environmental justice after Hurricane Katrina: Struggles to reclaim, rebuild, and revitalize New Orleans and the Gulf Coast.* Boulder, CO: Westview Press.

Button, G. (2010). *Disaster culture: Knowledge and uncertainty in the Wake of Human and Environmental Catastrophe.* Walnut Creek CA: Left Coast Press.

CDC. *Interim findings on formaldehyde levels in FEMA-supplied travel trailers, park models, and mobile homes.* (2008). Retrieved from <https://www.cdc.gov/nceh/ehhe/trailerstudy/pdfs/interim-findings-on-formaldehyde-levels-in-fema-supplied---final---2-29.pdf>. Accessed May 15, 2017.

Cigler, B. A. (2007). The "big questions" of Katrina and the 2005 great flood of New Orleans. *Public Administration Review, 67*(1), 64–76.

Crenshaw, K. W. (2003). Traffic at the crossroads: Multiple oppressions. In R. Morgan (Ed.), *Sisterhood is forever: The women's anthology for a new millennium* (pp. 43–57). New York: Washington Square Press.

Cutter, S. L. (1995). Race, class and environmental justice. *Progress in Human Geography*, *19*(1), 107–118.

Cutter S.L., Emrich C.T., Mitchell J.T., Boruff B.J., Gall M., Schmidtlein M.C., Burton, C. G., & Melton, G. The long road home: Race, class, and recovery from Hurricane Katrina, *Environment* 48(2), 2006, 9–20.

Echterling, L. G., & Wylie, M. (2013). In the public arena: Disaster as a socially constructed problem. In R. Gist, & B. Lubin (Eds.), *Response to disaster: Psychosocial, community, and ecological approaches.* (pp. 327–346). Philadelphia, PA: Routledge.

EPA. *Environmental justice.* (2017). Retrieved from <https://www.epa.gov/environmentaljustice>. Accessed June 20, 2017.

Farmer, P. (2004). An anthropology of structural violence. *Current Anthropology*, *45*(3), 305–325.

Freudenburg, W. R., Gramling, R., Laska, S., & Erikson, K. T. (2009). *Catastrophe in the making. The engineering of Katrina and the disasters of tomorrow.* Washington, DC: Island Press.

Gramsci, A. (1994). In F. Rosengarten (Ed.), *R. Rosenthal (Trans.). Letters from prison.* New York: Columbia University Press.

Headland, T., Pike, K., & Harris, M. (Eds.), (1990). *Emics and etics: The insider/outsider debate.* New York: Sage.

Hoffman, S. M. (1999). After Atlas shrugs: Cultural change or persistence after a disaster. In A. Oliver-Smith, & S. M. Hoffman (Eds.), *The angry earth. Disaster in anthropological perspective* (pp. 302–325). New York: Routledge.

IPCC. (2012). *Managing the risks of extreme events and disasters to advance climate change adaptation* (A special report of Working Groups I and II of the Intergovernmental Panel on Climate Change). Cambridge: Cambridge University Press.

IPCC. (2014). Summary for policy makers. In *Climate change 2014: Impacts, adaptation, and vulnerability Part A: Global and sectoral aspects, Contribution of the Working Group II to the Fifth Assessment Report of the Intergovernmental Panel on Climate Change.* Cambridge: Cambridge University Press.

Jordan, B. (1997). Authoritative knowledge and its construction. In R. Davis-Floyd, & C. Sargent (Eds.), *Childbirth and authoritative knowledge.* Berkeley, CA: University of California Press.

Kadetz, P. (2016). Safety net: The construction of biomedical safety in the global health discourse. *Asian Medicine: Tradition and Modernity*, *10*, 1–31.

Klein, N. (2007). *The shock doctrine: The rise of disaster capitalism.* New York: Henry Holt.

Knabb, R. D., Rhome, J. R., & Brown, D. P. (2005). *National Hurricane Center. Hurricane Katrina: August 23–30, 2005.* United States National Oceanic and Atmospheric Administration's National Weather Service.

Laska, S. (2012). Dimensions of resiliency: Essential resiliency, exceptional recovery and scale. *International Journal of Critical Infrastructures*, *8*(1), 47–62.

Mercer, J. (2012). Knowledge and disaster risk reduction. In B. Wisner, J. C. Gailard, & I. Kelman (Eds.), *The Routledge handbook of hazards and disaster risk reduction.* New York: Routledge. (Chapter 9).

Noji, E. K. (2005). Disasters: Introduction and state of the art. *Epidemiologic Reviews, 27*(1), 3−8.

Norris, F. H., Stevens, S. P., Pfefferbaum, B., Wyche, K. F., & Pfefferbaum, R. L. (2008). Community resilience as a metaphor, theory, set of capacities, and strategy for disaster readiness. *American Journal of Community Psychology, 41*(1/2), 127−150.

Oliver-Smith, A. (2004). Theorizing vulnerability in a globalized world: A political ecological perspective. In G. Bankoff, G. Frerks, & D. Hilhorst (Eds.), *Mapping vulnerability: Disasters, development and people* (pp. 10−24). London: Earthscan.

Oliver-Smith, A., & Hoffman, S. (2002). Introduction: Why anthropologists should study disasters. In A. Oliver-Smith, & S. M. Hoffman (Eds.), *Catastrophe & culture. The anthropology of disaster* (pp. 3−22). Santa Fe, NM: School of American Research Press.

Osofsky, H. J., Osofsky, J. D., Kronenberg, M., Brennan, A., & Hansel, T. C. (2009). Posttraumatic stress symptoms in children after Hurricane Katrina: Predicting the need for mental health services. *American Journal of Orthopsychiatry, 79*(2), 212.

Pike, K. (1967). *Language in relation to a unified theory of the structure of human behavior*. The Hague: Mouton.

Schweizer, T. (1998). Epistemology: The nature and validation of anthropological knowledge. In H. R. Bernard (Ed.), *Handbook of methods in cultural anthropology* (pp. 39−87). Walnut Creek, CA: Altamira Press.

Scott, J. C. (1985). *Weapons of the weak: Everyday forms of peasant resistance*. New Haven, CT: Yale University Press.

Stephens, K. U., Grew, D., Chin, K., Kadetz, P., Greenough, P. G., Burkle, F. M., et al. (2007). Excess mortality in the aftermath of Hurricane Katrina: A preliminary report. *Disaster Medicine and Public Health Preparedness, 1*(01), 15−20.

Tierney, K. (2014). *The social roots of risk: Producing disasters, promoting resilience (High reliability and crisis management series)*. Stanford, CA: Stanford University Press.

Wisner, B., Blaikie, P., Cannon, T., & Davis, I. (2004). *At risk. Natural hazards, people's vulnerability and disasters* (2nd ed.). New York: Routledge.

Wisner, B., Gailard, J. C., & Kelman, I. (Eds.), (2012). *The Routledge handbook of hazards and disaster risk reduction.*. New York: Routledge.

Zakour, M. J., & Gillespie, D. F. (2013). *Community disaster vulnerability: Theory, research, and practice*. New York: Springer.

CHAPTER 2

Settlement shifts in the wake of catastrophe

Richard Campanella
Tulane School of Architecture, New Orleans, LA, United States

CHAPTER OUTLINE

Introduction	25
Prologue: New Orleans' Historical Settlement Patterns	26
Post-Katrina Residential Settlement Patterns	28
Resettlement in Vertical Space	28
Resettlement by FEMA Flood Zones	32
Resettlement in Horizontal Space	36
Conclusions	40
Epilogue	42
Acknowledgments	43
References	43

INTRODUCTION

Urban risk may be understood as a function of hazard, exposure, and vulnerability (Crichton, 1999). In metro New Orleans, hurricane storm surge constitutes the premier hazard (threat); the exposure variable entails human occupancy of hazard-prone spaces; and vulnerability implies people's ability to respond resiliently and adaptively—which itself is a function of education, income, age, social capital, and other factors—after having been exposed to the hazard.

This chapter focuses on the second of these three variables, exposure, by analyzing how postdiluvian settlement patterns have shifted in vertical and horizontal (topographic, latitudinal, and longitudinal) space, and frame the findings in the context of urban resilience and sustainability (Campanella, 2015). Specifically, the research both maps and measures how many people lived above and below sea level (vertical space) 5 years prior to and after the 2005 flood, broken down by racial and ethnic groups. It does the same for those areas that actually flooded, as well as those areas predicted to flood in the event of a 100-year storm striking the metropolis, now that it is fortified with an improved Hurricane & Storm

Damage Risk Reduction System (HSDRRS). Next we track horizontal shifts in the population centroids of the metro area and assess their meaning. The results shed light on how these movements have affected denizens' exposure to future hazards and thus their overall urban risk—or, inversely, their sustainability.

PROLOGUE: NEW ORLEANS' HISTORICAL SETTLEMENT PATTERNS

Recent residential shifts in greater New Orleans are not the products from a specific reform or policy enacted after the Hurricane Katrina deluge of 2005, but rather from the *absence* of one. Because no decision was made to expropriate land in heavily damaged areas and "shrink the urban footprint," officials abrogated the opportunity to spatially guide the recovery and relinquished it to homeowners and landlords, who were permitted to return and rebuild where they wished. Viewed by some as a failure of planning, this outcome was hardly inexplicable: any political representative who advocated for neighborhood closure and "greenspacing" would have been deposed by voters, while at the state and federal level, no cash fund existed to compensate expropriated landholders fairly and swiftly. In addition, the Federal Emergency Management Agency's (FEMA's) updated Advisory Base Flood Elevation maps seemingly communicated federal endorsement (as well as actuarial encouragement) for status quo resettlement, and the Road Home Program's Option 1—to rebuild in place, by far the most popular of the three choices—provided grant money to do exactly that.

"Shrinking the urban footprint" became heresy; "greenspacing" took on sinister connotations; and rebuilding in flooded areas came to be valorized as a heroic civic statement. The Make It Right Foundation, for example, pointedly positioned its sustainable housing initiative below sea level, and near two surge-prone water bodies including the worst of the 2005 levee breaches, to illustrate that if the foundation "could build safe, sustainable homes in the most devastated part of New Orleans, [then it] would prove that high-quality, green housing could be built affordably everywhere" (History-Make It Right). Ignoring topography and hydrology gained currency in the discourse of community sustainability, even as it flew in the face of environmental sustainability.

In fact, topography and hydrology played fundamental roles in driving where New Orleanians settled for most of the city's history. The entire region originally lay above sea level, ranging from a few inches along the marshy perimeter, to a few feet along the Metairie-Gentilly Ridge, to 8–12 feet along the river's natural levee. Into the early 1900s, the vast majority of New Orleanians lived on the higher ground closer to the Mississippi. Uninhabited low-lying backswamps, while reviled for their (largely apocryphal) association with disease, nonetheless stored excess water and made the city surprisingly safeguarded from storm surges. Even the worst of the Mississippi River floods, in 1816, 1849, and 1871, mostly accumulated harmlessly in empty swamplands and, in retrospect, bore more benefits than costs. New Orleanians during the 1700s—1900s were less exposed to the hazard of flooding because the limitations of their technology forced them to live on higher ground (Campanella, 2008).

Circumstances changed in the 1890s, when engineers began designing and installing a sophisticated municipal drainage system to enable urbanization to finally spread across the backswamp to the lake. A resounding success from a developmental standpoint, the system came at a largely unforeseen cost (Campanella, 2010). What happened was that the pumps removed a major component of the local soil body—water—which opened up cavities, which in turn allowed organic matter (peat) to oxidize, shrink, and open up more cavities. Into those spaces settled finely textured clay, silt, and sand particles, which compacted and dropped below sea level. Over the course of the 20th century, former swamps and marshes in places like Lakeview, Gentilly, and New Orleans East sunk by as much as 8—12 feet, while interior basins such as Broadmoor dropped to 5 feet below sea level (Campanella, 2014a, 2014b).

Convinced that the natural factors that previously constrained their residential options had now been neutralized, New Orleanians migrated enthusiastically out of older higher neighborhoods and into lower modern subdivisions. Between 1920 and 1930, nearly every lakeside census tract at least doubled in population; low-lying Lakeview increased by 350%, while parts of equally low Gentilly grew by 636%. Older neighborhoods on higher ground, meanwhile, lost residents: Tremé and Marigny dropped by 10%—15%, and the French Quarter declined by one-quarter. The high-elevation Lee Circle area lost 43% of its residents, while low-elevation Gerttown increased by a whopping 1512% (Gilmore, 1937).

The 1960 census recorded the city's peak population of 627,525, double the number from the beginning of the 20th century. But while nearly all New Orleanians lived above sea level in the early 1900s, only 48% remained above sea level by 1960; fully 321,000 New Orleanians had vertically migrated from higher to lower ground (Campanella, 2008).

Subsequent years saw tens of thousands of New Orleanians migrate horizontally as well. They departed Orleans Parish neighborhoods primarily for social and economic reasons, namely school integration, rather than for any sense of environmental hazard. By 2000, the Crescent City's population had dropped 23% since 1960, representing a net loss of 143,000 mostly middle-class white families to adjacent parishes. Testifying to the level of unimportance ascribed to topographic elevation—and implicitly, the faith placed in drainage and flood control—most of those fleeing New Orleans proper unwittingly moved vertically to lower (and lowering) ground even as they sprawled out horizontally. By 2000, only 38% of Orleans Parish residents (and coincidentally 38% of all residents of the contiguous metropolis south of Lake Pontchartrain) lived above sea level (all figures were calculated by the author using highest-grain available historical demographic data, usually from the US Census, and LIDAR-based high-resolution elevation data captured in 1999—2000 by FEMA and the State of Louisiana).

Meanwhile, beyond the metropolis, coastal wetlands eroded at a pace that would reach 10–35 square miles per year, mostly on account of extensive canal excavation for navigation and oil and gas extraction, as well as the leveeing of the Mississippi River. Gulf waters crept closer to the metropolis' lateral floodwalls and levees, while inside that artificial perimeter of protection, land surfaces had subsided into a series of topographic bowls, half above and half below sea level.

The Katrina deluge served as a deadly reminder that topographic elevation, in fact, does matter, and the devices installed to neutralize it may prove unreliable. Satellite images of the flood footprint eerily matched the shape of the undeveloped backswamp in 19th-century maps, while higher areas quite naturally remained dry. But because so much of the public perception of flood protection had shifted away from living on higher ground in favor of building higher and stronger levees, most post-Katrina advocacy focused on the latter. "Make Levees, Not War" and "Category-5 Levees Now!" became popular bumper sticker slogans, and for good reason: it was, after all, federal levees that failed and ushered in Katrina's surge.

What few people seemed to appreciate was that it was those same levees— and the affiliated outfall canals and pumps—that lured development into flood zones in the first place, and it was the sunken landscape they helped create that impounded Katrina's surge and retained it for weeks. But calls to rethink resettling the bottoms of those bowls were effectively yelled off the table, for reasons aforementioned. One would be hard-pressed to find a "Move to Higher Ground!" bumper sticker in New Orleans today.

POST-KATRINA RESIDENTIAL SETTLEMENT PATTERNS

Resettlement patterns in 2006–07 initially suggested that New Orleanians were shifting to higher ground. The portion of the Orleans Parish population residing above sea level in 2007 increased to 50%, compared to 38% in 2000; by another approximated measure from February 2008, 55% of the city's 143,825 households receiving mail, a fair but imperfect indicator of repopulation, lay above sea level (Greater New Orleans Community Data Center, 2008). However, what accounted for the empirical rise in the percentage living above sea level was not people "voting with their feet" and moving to higher ground, but simply the fact that below-sea-level areas, because of their more severe damage, saw slower paces of repopulation.

RESETTLEMENT IN VERTICAL SPACE

The 2010 Census, unlike the surrogate datasets used for the above figures, finally allowed for more precise measurements of resettlement in vertical space (Fig. 2.1). These data show that residents of metro New Orleans shifted to higher ground by only 1%. Whereas 38% of metro-area residents lived above sea level in 2000, 39% did so by 2010, and that differentiation generally held true for each

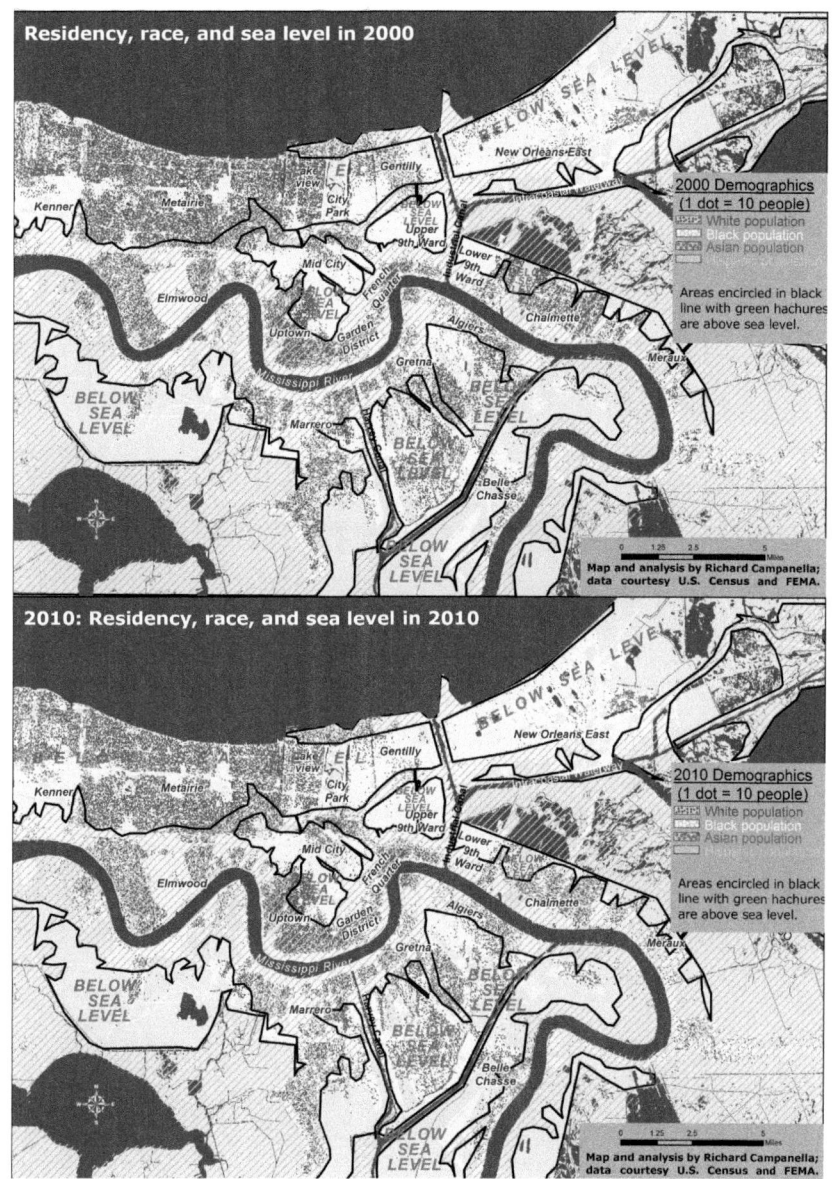

FIGURE 2.1

Resettlement by elevation above and below sea level. Analysis and maps by author.

racial and ethnic group. Whites shifted from 42% to 44% living above sea level; African-Americans from 33% to 34%, Hispanics from 30% to 29%, and Asians from 20% to 22% (Fig. 2.2). Clearly, elevation was not a major influence in resettlement decisions.

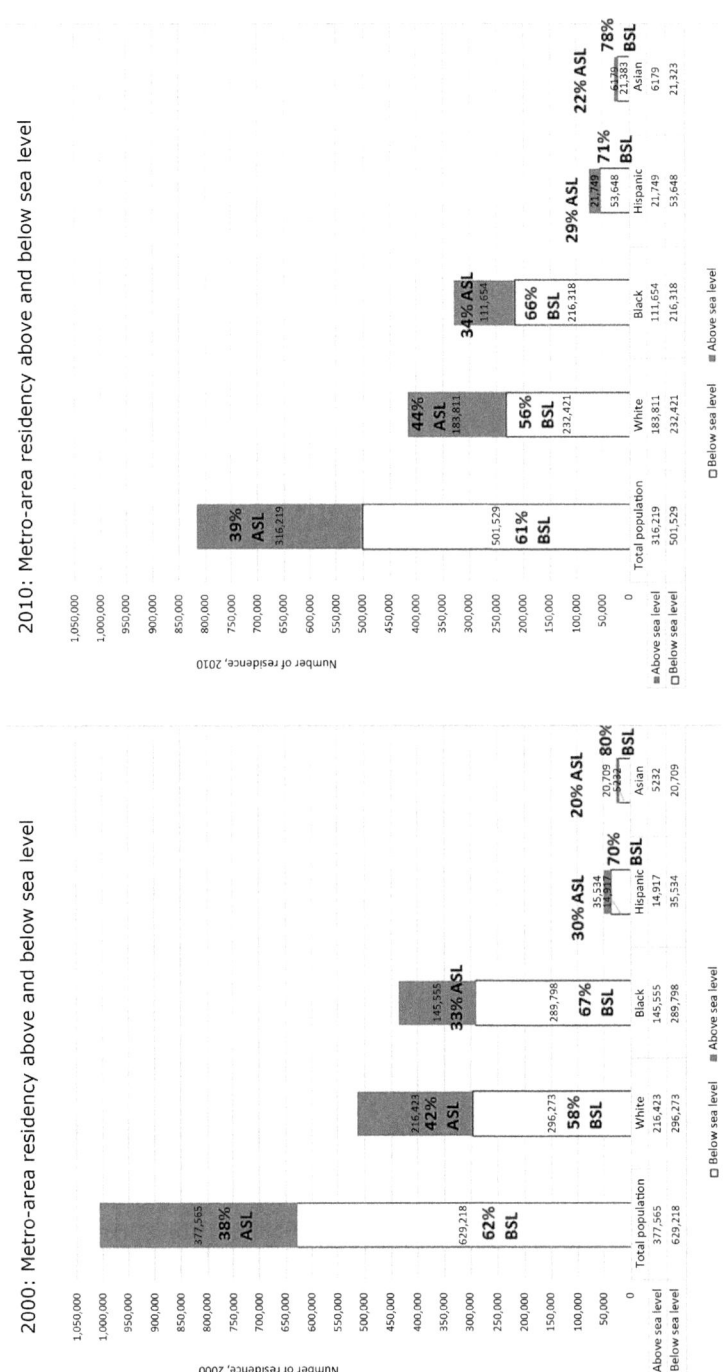

FIGURE 2.2

Occupancy of above- and below-sea-level areas. Analysis and graphs by author.

What impact did the experience of flooding have on resettlement patterns? The spatial extent of below-sea-level areas is not coterminous with the footprint of the Katrina flood. That is, not all inundated areas lay below sea level and not all above-sea-level areas remained dry; rather, various polders (hydrological sub-basins) flooded to varied levels. Thus, it is of interest to see how residency reconstituted within the flooded zone, regardless of elevation. [Only the main impounded deluge caused by the levee breaches is considered here, and not the brief shallow floodwaters seen in much of Metairie (outside Hoey's Basin) and Kenner, which were the result of splash-over and accumulated rain water in the absence of manned pump stations.]

The numbers are strikingly different here (Fig. 2.3). Whereas people shifted only slightly out of low-lying areas regardless of flooding, they moved significantly out of areas that actually flooded, regardless of elevation. Inundated areas lost 37% of their population between 2000 and 2010, with the vast majority departing after 2005. They lost 37% of their white populations, 40% of their black populations, and 10% of their Asian populations. Only Hispanics increased in the flooded zone, by 10%, in part because this population had grown dramatically region-wide, and because members of this population sometimes settled in the new housing opportunities found in neighborhoods they themselves helped rebuild.

The differing figures suggest that while low-lying elevation potentially exposes residents to the hazard of flooding, the trauma of actually flooding serves as a much more resounding—and costly—piece of evidence.

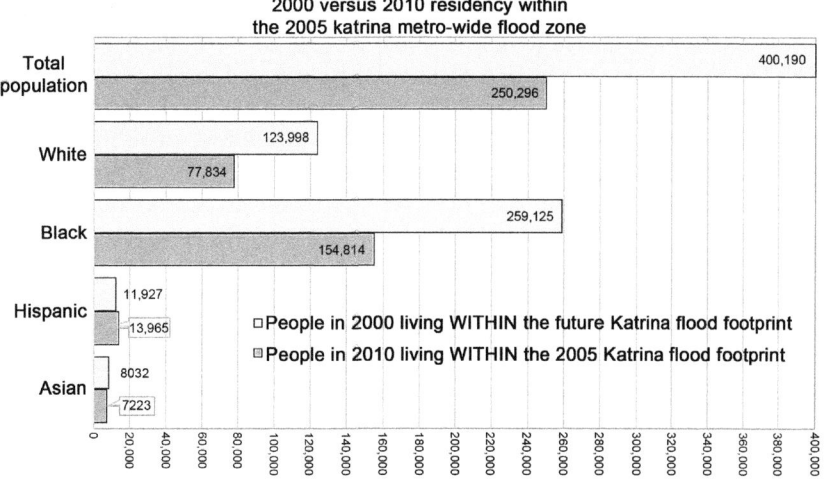

FIGURE 2.3

Occupancy of the area that actually flooded, 5 years before and after the fact. Analysis and graph by author.

What this research does not take into consideration is the fact that many houses are raised on piers above grade elevation, while others have been elevated in the past few years courtesy of Road Home Elevation incentive awards, Office of Community Development Hazard Mitigation grants, Increased Cost of Compliance grants from the National Flood Insurance Program (NFIP), or the Federal Emergency Management Agency's (FEMA's) Severe Repetitive Loss Mitigation Program. But a raised house does little to reduce the exposure of the neighborhood and its infrastructure, and an unflooded house in a flooded community hardly represents urban sustainability. Nevertheless, living above the grade is always a wise idea in this deltaic environment, and we should look positively on the new codes requiring residential structures to be raised (more on mitigation later).

What is less encouraging is that New Orleanians have not significantly reduced their exposure to flooding hazards by taking advantage of natural topography. Yet there is one positive angle to the fact that the above-sea-level percentage has risen, albeit barely (38%−39%): it marks the first time in New Orleans history that the percentage of people living *below* sea level has actually *dropped*.

RESETTLEMENT BY FEMA FLOOD ZONES

The above section tracked how New Orleanians repositioned themselves with respect to topographic elevation, above and below sea level. Next, we track resettlement by the related but not identical notion of actuarial flood zones as mapped by FEMA.

During 2006−11, the U.S. Army Corps of Engineers and its contractors designed and built a $14.5 billion HSDRRS; an integrated network of raised levees, strengthened floodwalls, barriers, gates, and pumps (Greater New Orleans Hurricane & Storm Damage Risk Reduction System, 2013). Combined with internal and nonfederal levees, as well as increased municipal drainage capacity, the HSDRRS has fundamentally altered the geography of potential flooding. These new flood zones are communicated to the public via FEMA flood maps, which drive the availability and cost of coverage in the NFIP.

To determine NFIP zones and rates, FEMA scientists analyze topographical and hydrological conditions and collaborate with Army Corps and local authorities to incorporate the influences of the latest levees, floodwalls, pumps, and storm water drainage capacity. With the aid of LIDAR elevation data and geographic information systems, FEMA delineates various levels of flood risk and releases the results in the form of Flood Insurance Rate Maps (FIRMs, their digital counterparts being called D-FIRMs). Because real estate transactions cannot be put on hold while these complex data are being processed and certified, the FEMA issues "preliminary D-FIRMs" as swiftly as possible. These draft maps then go through 3 months of public hearings, comments, and adjustments, followed by 6 months for local governments to approve and embed them into their zoning and codes. Jurisdictions that reject preliminary D-FIRMs run the risk of exclusion from the NFIP; those that accept can obtain coverage with premiums dependent on the flood zones mapped. Once fully approved, "preliminary D-FIRMs" are called "effective D-FIRMs." For most prospective homebuyers, however, there is usually little appreciable difference between the preliminary and effective maps. [This explanation reflects various FEMA and FIRM government literature, websites, FAQs, and brochures, some of which may be found at https://www.fema.gov/floodplain-management/flood-insurance-rate-map-firm, https://www.fema.gov/national-flood-insurance-program, and https://www.fema.gov/fema-region-vi-updating-flood-maps-greater-new-orleans-area.]

D-FIRMs delineate areas from highest to lowest flood risk and label them with the letters "V," "A," "B," or "C," respectively. Unprotected coasts may earn "V" labels (Velocity), indicating the highest risk in the face of the wave action of fast-moving surges. Some areas in the eastern Rigolets and Lake Catherine region of Orleans Parish, outside the HSDRRS, are V zones. Nearly all homes on this narrow land bridge were obliterated by Katrina's surge, and those that have been rebuilt are all raised by well over 10 feet.

The vast majority of lands inside the HSDRRS, where 98% of metro-area residents live, fall either within FEMA flood zones "A," which have a 1.0% chance of seeing standing water in any given year (the so-called 100-year flood, not necessarily resulting from hurricanes or their surges), or "B zones," which are relatively lower-risk areas with a 0.2% chance of flooding (ie., a 500-year flood).

Other areas elsewhere in the nation are "C" zones, which have minimal risk of flooding. Alas, there are no C zones in New Orleans. Even with the HSDRRS, flooding remains a statistical likelihood for roughly half of metro-area residents over the next few decades—not necessarily Katrina-like 6–10 feet of inundation, more likely 1–2 feet, but enough to inflict damage nonetheless. This is, after all, a deltaic plain, and before levees were erected, it constituted one vast flood plain.

"A" and "B" zones in metro New Orleans generally, though not perfectly, correlate to natural elevation (Fig. 2.4): areas higher in topographic elevation are usually, but not always, "B" zones, and lower areas tend to be "A" zones. Deviations occur because urbanization and municipal drainage have reworked how water would otherwise flow across the landscape. For example, there are some blocks riverside of St. Charles Avenue in uptown that are 2–3 feet higher than those across the avenue, yet are counterintuitively labeled as higher-risk "A" zones, while the lower-elevation blocks are in drier "B" zones. The reason: the berm beneath the streetcar tracks, which run along the avenue, prevents runoff from flowing from higher to lower ground, and tends to trap it like a dam on the higher side. Despite that this particular "A" zone is safer from macroscopic Katrina-style flooding, it is more prone to the minor street flooding that commonly follows heavy rainfalls. The ongoing Southeast Louisiana Urban Flood Control Project will likely remedy this situation by increasing capacity of below-ground drainage canals on Louisiana, Jefferson, and Napoleon avenues. If so, the reduced flood risk would be reflected in updated D-FIRMs.

The height of the top of the water level in the event of a 1% flood is called Base Flood Elevation (BFE). After FEMA computes the BFE for its "A" zones, it relabels

34 CHAPTER 2 Settlement shifts in the wake of catastrophe

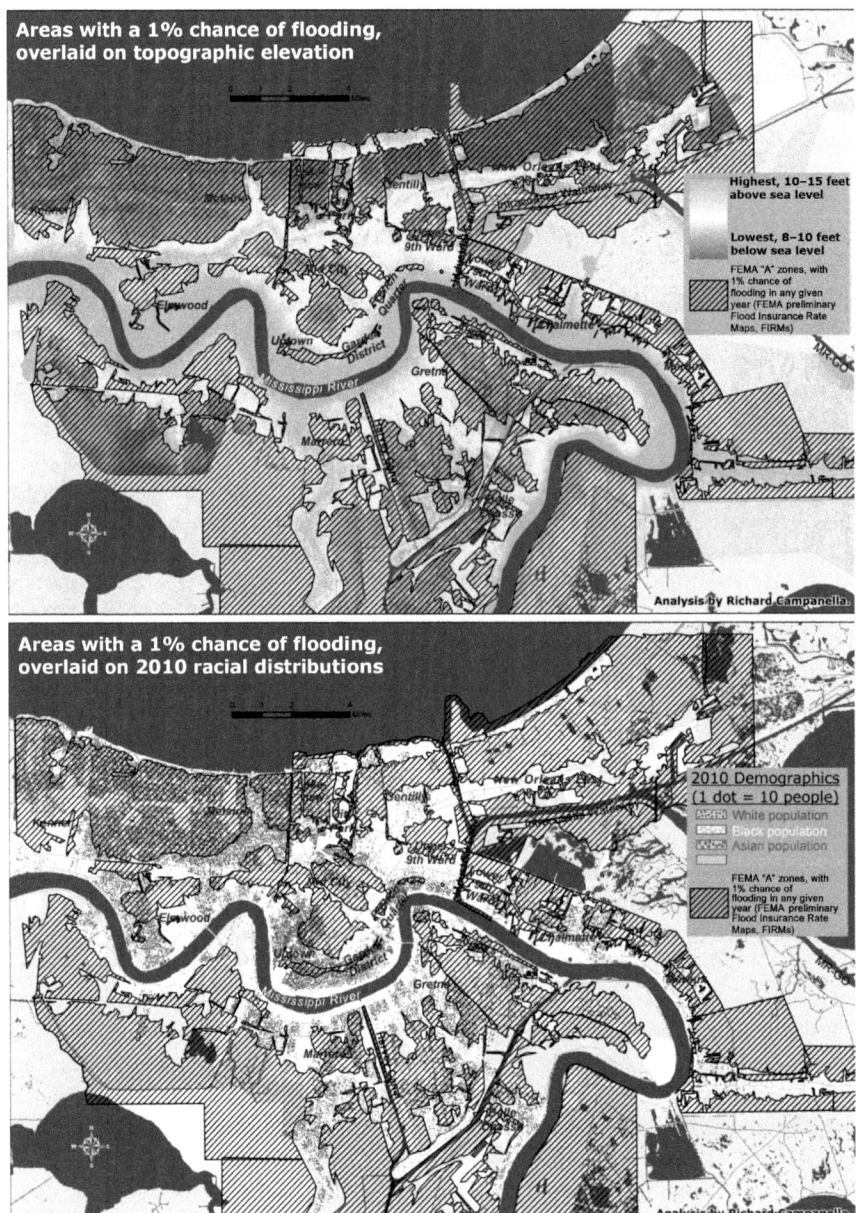

FIGURE 2.4

FEMA flood zones and their relationship to elevation (*top*) and 2010 settlement patterns (*bottom*). Analysis and maps by author.

them "AE" zones and posts the BFE figure adjacently in its D-FIRMs. Thus, in a section of Lakeview labeled "ZONE AE EL-5," a homeowner would have to raise a house at least 3 feet above the five-foot-below-sea-level (-5) elevation contour, *or* 3 feet above the grade level, whichever is higher. Once raised and certified by a surveyor, the homeowner receives a NFIP Elevation Certificate, and only then can the house qualify for federal flood insurance. A noncertified house positioned in an "A" zone (1% annual chance of flooding) below BFE is a tough sell on the real estate market. Houses in "B" zones (0.2% annual chance of flooding) do not need an Elevation Certificate to qualify for flood insurance, which adds not only to their security but their value and marketability. Thus, the value of natural topography.

It is worth noting that 1% flood risk equates to a roughly 1-in-3 chance of flooding during the span of a typical 30-year mortgage. Insurance is imperative to a homeowner's equity, not to mention the economic viability of the metropolis, and FIRM zones, BFEs, and NFIP Elevation Certificates determine whether flood insurance will be available and affordable. They represent the "moment of truth" for anyone owning property in this region, the point at which geophysical abstractions convert to lines on maps and bills in the mail.

Who lives in "A" zone (1.0% annual chance of flooding) and "B" zone (0.2% annual chance of flooding)? The demographics of flood risk are remarkably ecumenical (Fig. 2.5). For nearly every socio-economic segmentation analyzed, from race to poverty to education levels, local society was roughly evenly distributed across "A" and "B" flood zones, with only slight disproportions.

Of the total 817,776 people living within the contiguous metropolis south of Lake Pontchartrain as of the 2010 Census, slightly more than half (51.4%) live in "A" flood zones and 48.6% live in "B" Zones. White and black populations are similarly split, with 49.9% of whites and 51.5% of African-Americans residing in "A" zones. The percentages of "A"-zone occupancy are higher for Hispanic and Asian-ancestry populations (61.3% and 62.2%), although their absolute numbers are much lower.

When compared to their proportions of the total metro-area population, flood-zone residency becomes even more transcendent of race and ethnicity. Whites make up 50.9% of the metropolis and 49.4% of those living in "A" zones; blacks 40.1% of the metropolis and 40.2% of those living in "A" zones; Hispanics 9.2% and 11.0% respectively; and Asians 3.4% and 4.1%, respectively.

No one group has a monopoly on either flood risk or flood safety. Even when broken down by metrics of social vulnerability—namely, poverty, elders living alone, children living in households without a father, and educational attainment—households straddled "A" and "B" zones fairly evenly. (This particular analysis of social vulnerability factors was based not on the 2010 Census, but rather the 2011 American Community Survey estimates mapped at the coarser block-group level. The Census does not collect and plot responses to these metrics at the block level.)

A society living and working at its own actuarial risk will likely divest if hazard and exposure surpass a certain threshold. These data show that segments of

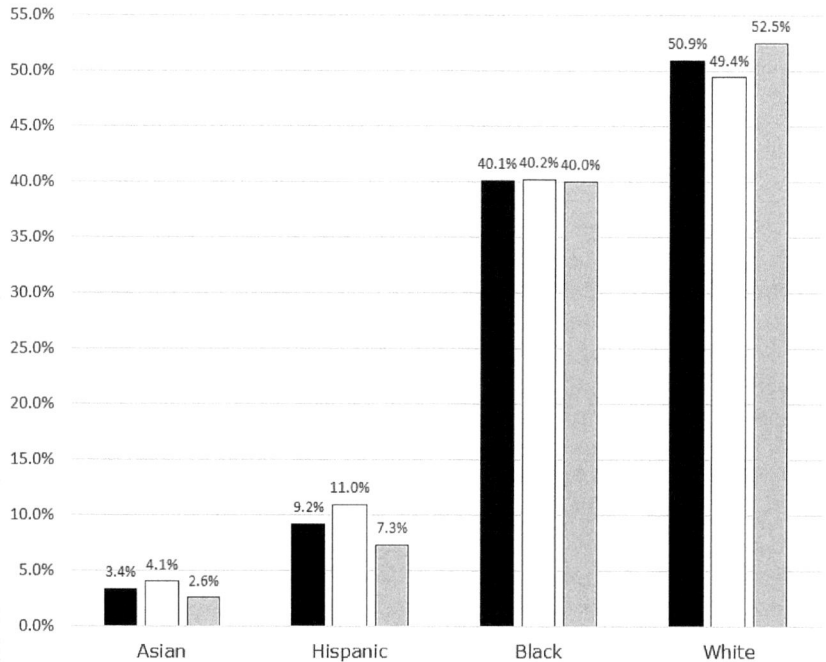

FIGURE 2.5

Comparison of demographic composition of metro area, higher-risk "A" flood zones, and lower-risk "B" flood zones. Analysis and graph by author.

the metro New Orleans population experience this risk, in so far as FEMA flood zones indicate, in proportions of remarkable parity. People of all backgrounds, ages, education levels, and income levels may be found in comparable numbers inside and outside of flood zones. This is a shared, society-wide problem.

RESETTLEMENT IN HORIZONTAL SPACE

The early repopulation of post-Katrina New Orleans defied easy measurement. Residents living "between" places as they rebuilt, plus temporarily broken-up families, peripatetic workers, and transient populations, all conspired to make the city's 2006–09 demographics difficult to estimate. The 2010 Census finally provided a precise figure: 343,829. By 2015 over 389,000 lived in Orleans Parish, or 8 out of every 10 of the pre-Katrina figure. Of course, not all of the current population resided here pre-Katrina; a 2010 study found that around 10% of the city's population had settled here after the storm (The Henry J. Kaiser Foundation, 2010).

When mapped, the total postdiluvian population, including both returnees and newcomers, dispersed themselves in patterns somewhat ajar from times prior. The shifts invoke the exposure variable in our urban risk formulation, this time from a horizontal (latitudinal−longitudinal) dimension. Did populations nudge inland after the storm? Did they move away from gulf or canal water? If so, was it because of increased settlement in less-exposed unflooded areas—or simply slow return rates in more-exposed flooded areas?

Contrasting before-and-after residential patterns may be done through traditional methods, such as comparative maps and demographic tables. What this investigation offers is a more singular and synoptical depiction of spatial shifts: by computing and comparing spatial central tendencies, or centroids.

A population centroid is a theoretical point around which people within a delimited area are evenly distributed. [Defining the study area is essential when reporting centroids. New Orleans proper, the contiguous metro area, and the Metropolitan Statistical Area, which includes St. Tammany and other outlying parishes, would all have different population centroids. This study uses the metro area south of the lake shown in the accompanying maps. It is also important to use the finest-grain—that is, highest spatial resolution—demographic data to compute centroids, as coarsely aggregated data carries with it a wider margin of error. This study uses block-level data from the decennial US Census, the finest available.] Centroids are useful in that they capture complex shifts of millions of data with a single point. But they do not tell the entire story. A centroid for a high-risk coastal parish, for example, may shift inland not because people have moved away from the seashore, but because prior residents decided not to return. It is also worth noting it takes a lot to move a centroid, as microscale shifts in one area are usually offset by similar shifts elsewhere. Thus, apparent minor centroid movements can actually be significant.

In 2000, 5 years before the flood, there were 1,006,783 people living within the metro area as delineated for this particular study, of whom 512,696 identified their race as white; 435,353 as black; 25,941 as Asian; and 50,451 as Hispanic in ethnicity. Five years after the flood, these figures had changed to 817,748 total population, of whom 416,232 were white; 327,972 were black; 27,562 were Asian, and 75,397 were Hispanic. [Figures do not sum to totals because some people chose two or more racial categories while others declined the question, and because Hispanicism is viewed by the Census Bureau as an ethnicity and not a race.] When their centroids are plotted (Figs. 2.6 and 2.7, "2000−2010 Population Centroid Shifts by Race for Contiguous New Orleans Metropolitan Area"), they illustrate that metro residents as a whole, and each racial/ethnic subgroup, universally shifted westward and southward between 2000 and 2010. What accounts for these consistent shifts is the fact that the eastern half of the metropolis bore the brunt of the Katrina flooding, for reasons explained later. The ensuing deluge meant populations here were less likely to completely reconstitute, which thus nudged centroids westward. Robust return rates in unflooded parts of uptown also aided the southwestward shift of centroids.

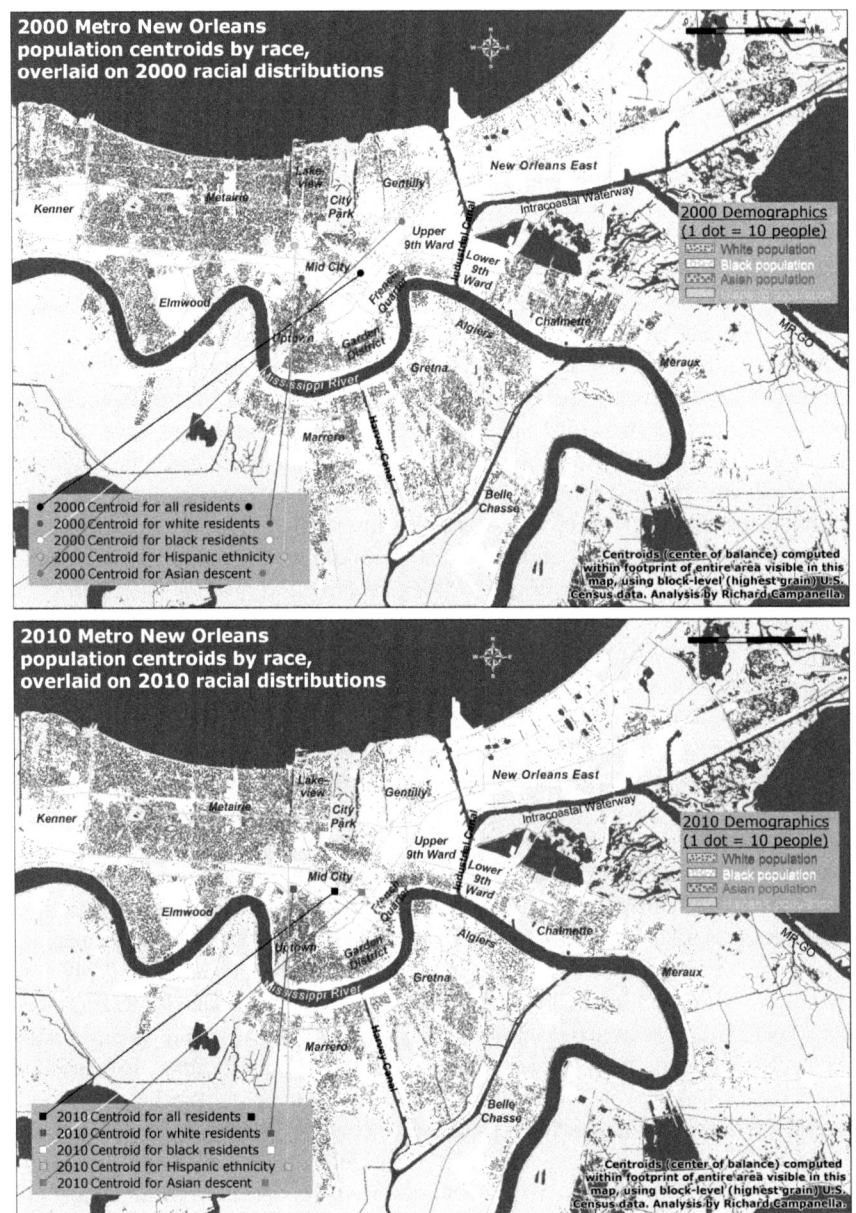

FIGURE 2.6

Population centroids by race, metro New Orleans, 2000–10. Analysis and maps by author.

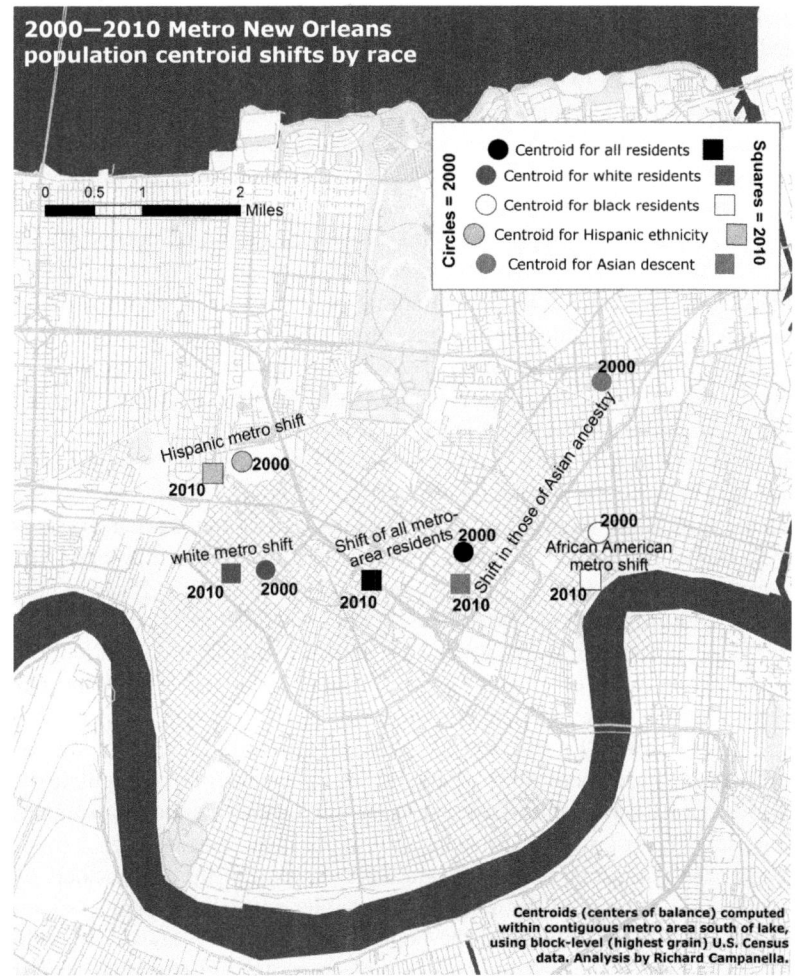

FIGURE 2.7

A closer look at the metro-area population centroid shifts by race, 2000–10. Analysis and map by author.

The yawning gap between the white and black centroids for both 2000 and 2010 shines light on an important but rarely observed fact: African-American populations have, since the late 20th century (and despite some major exceptions, such as St. Bernard Parish) predominated on the eastern half of the metropolis, the same area that suffered the lion's share of Katrina flooding. This area is generally at more risk of hurricane-induced surges because (1) it is scored and scoured by three major manmade navigation canals (the Industrial Canal, Intracoastal Waterway, and Mississippi River-Gulf Outlet Canal), all of which communicate with the Gulf of Mexico; (2) it abuts highly degraded and eroded wetlands where storm-driven Gulf of Mexico surges incur minimum friction as they move inland; and (3) it lies lower in elevation, being on the farther side of the Mississippi River's downsloping, prograding fluvial deltaic plain. The western half of the metropolis, meanwhile, while by no means risk-free, has no major gulf-access navigation canals and enjoys a greater terrestrial buffer, farther geographical distance to the open gulf, and slightly higher terrestrial elevations vis-à-vis the most dangerous surge paths. [Subsided soils and outfall canals for municipal drainage, which also increase the risk of surge flooding, are found throughout the metropolis.] This longitudinal pattern explains more of the racial variation in the Katrina flooding statistics than elevation-based patterns. In other words, the fact that blacks generally lived on the higher-risk eastern side of the metropolis, led to more black flood victims than did their settlement patterns in terms of topographic elevation (Campanella, 2014a, 2014b).

Similarly, the shift in the Asian centroid likely reflects the concentrated nature of most the Vietnamese population, which has its largest cluster on the eastern side of the East Bank (Versailles and Village de L'est) and smaller enclaves on the West Bank. The former flooded; the latter did not. The southwestwardly tilt of the centroids, thus represents population changes on the East Bank in the face of stability and growth on the West Bank (which, of course, is actually south of the city proper). The southward shift of the African-American population reflects a similar phenomenon, of greater stability on the unflooded West Bank pulling the centroid in that direction. The Hispanic shift is the only one mapped here that is not a direct result of Katrina and the population loss on the eastern side of the metropolis (few lived there prior to 2005), but rather a result of population gain on the western half. Hispanics have settled in substantial numbers in Kenner and Metairie, areas that mostly experienced only minor nuisance flooding.

In sum, the consistent southwestward shift shows that the eastern half of the metropolis bore the brunt of the deluge, such that its populations were less likely to reconstitute by 2010. This nudged the population centroids westward. In addition, flooding from Lake Pontchartrain, through ruptures in the outfalls canals, disproportionally damaged the northern tier of the city. Combined with robust return rates in unflooded parts of uptown as well as the West Bank, which sit to the south and west of the worst-damaged areas, they abetted a southwestward shift of the centroids. In a purely empirical sense, this change means more people now live in less-exposed areas. But, as we saw with the vertical shifts, the shifts reflect passive responses to flood damage, more so than active decisions to avoid future floods.

CONCLUSIONS

This research finds that resettlement patterns have only marginally reduced residential exposure to the hazard of storm surge. Metro-area residents today occupy below-sea-level areas at only a slightly lower rate than before the deluge, 61% as opposed to 62%, although the change represents the first-ever reverse (decline) of the century-long drift into below-sea-level areas. Likewise, residents' horizontal shifts, which were in southwestward directions, seemed to suggest a movement away from hazard. But these shifts were more a product of passive than active processes; that is, not returning to damaged areas rather than proactively moving to alternatives. Residents also occupied actuarial flood zones in demographic patterns, largely representative of the larger metropolitan society with all its vulnerabilities and resilience.

Metro New Orleans, we should note, *has* reduced its overall risk—but largely on account of its improved federal HSDRRS. All other risk drivers—the condition of the coastal wetlands, subsidence and sea level rise, social vulnerability, and, as evidenced in this paper, exposure—have either slightly worsened, only marginally improved, or generally remained constant.

In many ways, the exposure-related patterns reported here are outcomes of the "Great Footprint Debate" of 10 years ago (Campanella, 2008). Months after Katrina, when it became clear that no neighborhoods would be closed and the urban footprint would not be shrunk, decisions driving resettlement patterns in the flooded region effectively transferred from leaders to homeowners. Rather inevitably, the *laissez faire* recovery strategy proved to be exactly that.

Ten years later the resulting patterns are a veritable Rorschach Test. Some observers may point to the 75%–90% repopulation rates of certain flooded neighborhoods and view them as heroically high, proof of New Orleanians' resilience and "topophilia" (Tuan, 1974). Others might look at the 25%–50% rates of other areas and call them scandalously low, evidence of corruption and ineptitude. Still others might point to the thousands of scattered blighted properties and weedy lots and concede—as St. Bernard Parish President David Peralta admitted on the ninth anniversary of Hurricane Katrina—that "we probably should have shrunk the footprint of the parish at the very beginning" (Alexander-Bloch, 2014).

New Orleanians should be relieved that their new HSDRRS has generally achieved its eponymous task in reducing urban risk. In this light, perhaps it should neither surprise nor disappoint us that residents did not put too much of a premium on natural advantages as they resettled after the flood.

Unfortunately, continual subsidence and erosion vis-à-vis rising seas, coupled with costly and as-yet undetermined maintenance and certification responsibilities, will gradually diminish the safety dividend that came with the completion of the HSDRRS. The nation's willingness to pay for continued upkeep, meanwhile, may grow tenuous; indeed, it is not even a safe bet locally. Voters in St. Bernard, which suffered near-total inundation from Katrina, defeated not once but twice a tax millage needed to pay for levee and drainage maintenance, a move that may well increase their flood insurance (Schleifstein, 2015).

Residents throughout the metropolis may be in danger of repeating the same mistakes they made during the 20th century: of dismissing the importance of natural elevation, of over-relying on engineering solutions, of under-maintaining these structures in a milieu of inadequate federal funding, and of developing a false sense of security about flood "protection."

This research suggests that residents ought to recognize the limits of our ability to neutralize hazards—that is, to presume that levees will completely protect us from storm surges—while appreciating the benefits of reducing our exposure to this hazard. Beyond the metropolis, this means aggressive coastal restoration using every means available as soon as possible, an effort that may well require some expropriations. Within the metropolis, it means living on higher ground and/or raising structures above the grade. In the words of University of New Orleans disaster expert Dr. Shirley Laska, "mitigation, primarily elevating houses, is [one] way to achieve the affordable flood insurance [....] It is possible to remain in moderately at-risk areas using engineered mitigation efforts, combined with land use planning that restricts development in high-risk areas" (S. Laska, Personal Communication, April 12, 2015). Planners are also encouraged to maximize the residential utilization of above-sea-level areas, through land use and density changes, coupled with mitigation strategies in lower-lying areas and inclusionary zoning of affordable housing to counter the deleterious effects of gentrification. While it is far too late to revisit the bitter issue of footprint shrinkage, now is the time to ensure residents understand where flood risk is located, what it costs, and how to mitigate it.

EPILOGUE

This brings the discussion back to flood zones, flood insurance premiums, and the NFIP. The federal government got into the flood insurance business largely because the private sector had become wary of the uniquely destructive and replicating nature of inundation. Worse, the phenomenon known as adverse selection incentivized only those homeowners at highest flood risk to acquire insurance, while those with little or no risk did not bother enrolling. This resulted in a potentially large number of costly claims paid from a relatively small pool of premium-payers, which left little room for profit and thus no incentive to get into the marketplace (National Academy of Sciences, 2015).

Seeking to fill the gap, Congress in 1968 created the NFIP. NFIP aimed to protect federally backed mortgages for homes in flood zones and reduce the costs of disaster recovery by pooling and managing risk before disaster struck. But NFIP also lured development into flood-prone areas. When catastrophes occurred—namely hurricanes Katrina, Ike in Texas, and Sandy in New York—claims surged, and NFIP fell into debt by over $20 billion (National Academy of Sciences, 2015).

An attempt was made by Congress in 2012 to make NFIP solvent by charging premiums reflective of the mounting risk. Proving just how risky things had become, the resulting Biggert-Waters Flood Insurance Reform Act had the effect of sky-rocketing federal flood insurance premiums in many cases by 1000%–3000%, a consequence that, incredibly, seemed to catch its backers by surprise even after the bipartisan bill was signed into law by the president.

Biggert-Waters was itself "reformed" in 2014 with the Homeowner Flood Insurance Affordability Act, which implemented gradual rate increases and surcharges rather than one traumatic jump. But the writing is on the wall: flood hazard is increasing, and if we insist on exposing ourselves to it, someone will have to bear that cost. During 1980–2013 the nation suffered over $260 billion in flood damages, and floods accounted for 85% of all declared disasters. Furthermore, states FEMA, "the costs borne by the Federal government are more than any other hazard, [and] with climate change, we anticipate that flooding risks will increase over time" (Federal Emergency Management Agency, 2015).

In the wake of Hurricane Sandy, the federal government's latest attempt to confront rising seas came in the form of Executive Order 13690, otherwise known as the Federal Flood Risk Management Standard (FFRMS). In the words of RAND Corporation, the Standard "ask[s] project planners to add two feet to currently estimated flood levels for non-critical projects and three feet for critical ones," which it defines as "any activity for which even a slight chance of flooding would be too great" (Knopman, Lempert, & Fischbach, 2015). Currently open for public comments, the Standard, in FEMA's words, "specifically requires agencies to consider current and future risk when taxpayer dollars are used to build *or rebuild* floodplains" (emphasis added) (Federal Emergency Management Agency, 2015). Rebuilding floodplains with taxpayer dollars is essentially what happened 10 years ago, after suggestions to do otherwise—that is, to shrink the urban footprint and build in higher density on higher ground—were spurned.

Next time, we might not have the choice.

ACKNOWLEDGMENTS

The author wishes to thank Gulf of Mexico Program Officer, Kristin Tracz of the Walton Family Foundation, and the Gulf Coast Restoration Fund at New Venture Fund for their support of this research.

REFERENCES

Alexander-Bloch, B. (2014). *Hurricane Katrina +9: Smaller St. Bernard Parish grappling with costs of coming back. New Orleans Times-Picayune, August 29, 2014.*

Campanella, R. (2008). *Bienville's dilemma: A historical geography of New Orleans and geographies of New Orleans.* Baton Rouge, LA: University of Louisiana Press.

Campanella, R. (2010). *Delta urbanism: New Orleans.* Chicago, IL: American Planning Association.

Campanella, R. (2014a). *The Katrina of the 1800's was called suave's crevasse*. New Orleans Times-Picayune, "Cityscapes" column, June 13, 2014.

Campanella, R. (2014b). *Two centuries of paradox: The geography of New Orleans' African-American population. Hurricane Katrina in transatlantic perspective*. Baton Rouge, LA: Louisiana State University Press.

Campanella, R. (2015). Special report: The laissez faire New Orleans rebuilding strategy was exactly that. *New Geography Journal*. Retrieved from <http://www.newgeography.com/content/004995-special-report-the-laissez-faire-new-orleans-rebuilding-strategy-was-exactly-that>.

Crichton, D. (1999). *The risk triangle. Natural disaster management: A presentation to commemorate the International Decade for Natural Disaster Reduction (IDNDR), 1990–2000* (pp. 102–103). Leicester: Tudor Rose.

Federal Emergency Management Agency. (2015). *Federal flood risk management standard (FFMRS)*. Retrieved from <https://www.fema.gov/federal-flood-riskmanagement-standard-ffrms>.

Gilmore, H. W. (1937). Some basic census tract maps of New Orleans. New Orleans, LA: Tulane University.

Greater New Orleans Community Data Center. (2008). Retrieved from <http://www.datacenterresearch.org>.

Greater New Orleans Hurricane and Storm Damage Risk Reduction System (HSDRRS). (2013). Retrieved from <http://www.mvn.usace.army.mil/Missions/HSDRRS.aspx>.

History-Make It Right. Retrieved from <http://makeitright.org/about/history/>. Accessed 13.02.05.

Knopman, D., Lempert, R., & Fischbach, J. (2015). *Future of coastal flooding*. Retrieved from <http://www.rand.org/blog/2015/02/future-of-coastal-flooding.html>.

National Academy of Sciences Committee on the Affordability of National Flood Insurance Program Premiums. (2015). *Affordability of national flood insurance program premiums: Report 1*. The National Academies Press (pp. 21–25).

Schleifstein, M. (2015). *St. Bernard tax defeat means higher flood risk, flood insurance rates, levee leaders warn. Times-Picayune*. Retrieved from <http://www.nola.com/environment/index.ssf/2015/05/st_bernard_tax_defeat_means_hi.html>.

The Henry J. Kaiser Family Foundation. (2010). *New Orleans five years after the storm: A new disaster amid recovery*. Retrieved from http://kaiserfamilyfoundation.files.wordpress.com/2013/02/8089.pdf.

Tuan, Y. (1974). *Topophilia: A study of environmental perception, attitudes and values*. Upper Sadle River, NJ: Prentice-Hall.

CHAPTER 3

Vulnerability-plus theory: The integration of community disaster vulnerability and resiliency theories

Michael J. Zakour[1] and Charles M. Swager[2]
[1]West Virginia University, Morgantown, WV, United States
[2]Essential Foundations PLLC, Morgantown, WV, United States

CHAPTER OUTLINE

Vulnerability-plus Theory	46
The Development Perspective	47
Chain of Causality: Root Causes, Structural Constraints, and Unsafe Conditions	48
Access to Resources Model	50
The Resilience Perspective	51
Attributes of Resilience Resources	52
Types of Resources	53
Comparing and Contrasting Vulnerability and Resiliency	56
Vulnerability and Resilience Theories Are Complementary	57
Continuity Between Vulnerability and Resilience Theories	59
Vulnerability-plus Theory as Integration of Theories	61
Assumptions of Vulnerability-plus Theory	62
Empirical Support	64
Causal Chains in V+ Theory	68
Root Causes	68
Structural Constraints	69
Unsafe Conditions	69
Hazard Types	70
Disaster Characteristics	71
Resources	71
Summary and Conclusions	72
References	73

Over the past two decades, disasters have affected increasing numbers of people and property losses have escalated. The Indian Ocean Tsunami and the 2010 Haiti earthquake together killed more than 500,000 people. In 2005 Hurricane Katrina became the costliest natural disaster in the history of the United States, and left over 1800 dead. It was followed by the 2010 Gulf Oil Spill, a disaster that will have unknown consequences for the ecology of the Central Gulf Coast for years or decades. The triple disasters (earthquake, tsunami, and nuclear accident) in Japan killed 10,000 people, while the area surrounding the meltdown at the Fukushima Daiichi plant will be uninhabitable for generations.

As disasters have increased in scope and impact, the need for loss reduction has become more urgent. For emergency managers, it has become increasingly apparent that the hazards triggering disasters are very difficult to eliminate. Because hazards, especially natural hazards, are recognized as intractable, there has been a renewed interest among disaster researchers and managers to understand the social causes of vulnerability. Related to this interest in social causes, there is an expanding interest in the development of resilience resources to help communities recover rapidly and well following disasters. The increased focus on both vulnerability and resiliency is part of a trend in emergency management toward a more proactive approach to disasters (Zakour & Gillespie, 2013).

In this movement toward proactive disaster management, numerous disciplines and professions have incorporated a focus on vulnerability and resilience in their disaster and emergency management research. These disciplines and professions use different approaches to disaster research, as well as diverse concepts, and they study different aspects of disaster (McEntire, 2004a, 2004b). As we have noted earlier, because of this diversity of disaster research, vulnerability is being increasingly used as a bridging concept linking findings from disaster research (Zakour & Gillespie, 2013). Vulnerability encompasses each kind of hazard and disaster stages, including mitigation, planning, response, and recovery. The concept of vulnerability has theoretical and practical significance in many disciplines and professions, such as sociology, social work, public health, international development, anthropology, psychology, and political science (Gillespie, 2008a).

VULNERABILITY-PLUS THEORY

From the usefulness of vulnerability and resilience concepts in these disciplines and professions, it has become clear that disaster vulnerability and resilience perspectives are complementary rather than being unrelated or competing approaches. As shown in this chapter the similarity of the two concepts is used to integrate vulnerability and resilience theory, resulting in Vulnerability-plus (V+) theory. This chapter identifies the empirical support for V+ theory and its major assumptions. The discussion focuses on root causes, sustainable human development, and resources, along with liabilities and capabilities (risk and protective factors). As we previously noted (Zakour & Gillespie, 2013) the applicability of this new theory for disaster management is described.

As an integration of vulnerability and resilience theories, V+ theory is consistent with conceptualizations of vulnerability (B. L. Turner et al., 2003). For these disaster and political ecology researchers, the vulnerability of socio-ecological systems consists of a combination of exposure to hazards, the susceptibility of these systems, and their resilience. For most vulnerability theorists, susceptibility is defined as both the likelihood of people and communities suffering damage from a hazard, and the likely degree of that damage. Conversely, resilience is the capacity of a system to regain its level of functioning following disaster, at the same or higher level compared with predisaster functioning. Following McEntire (2004a, 2004b), in V+ theory disaster vulnerability is defined as the balance of susceptibility and resilience (Zakour & Gillespie, 2013). Resilience resources can buffer and/or counterbalance the susceptibility of communities to disaster, leading to a reduction of disaster loss and vulnerability.

THE DEVELOPMENT PERSPECTIVE

The social development perspective has contributed substantially to vulnerability theory and is an important foundation for V+ theory. The political ecology framework vulnerability of B. L. Turner et al. (2003) is used to examine the political economy, institutional, and global trends influencing the vulnerability of communities. As stated in our earlier book (Zakour & Gillespie, 2013), we extensively use the research of Blaikie, Cannon, Davis, & Wisner (1994) who helped establish the foundations of vulnerability theory. Wisner and his colleagues conducted a qualitative meta-analysis of a range of disaster types in a number of countries to document the progression of vulnerability (Wisner, Blaikie, Cannon, & Davis, 2004).

In their development of the vulnerability concept and theory, Wisner et al. (2004) emphasized social processes influencing the effects of hazards and disasters on communities and their populations. They theorized that social processes occur in the socio-cultural and political—economic environments of communities, conceptualized as socio-ecological systems. Wisner, Gaillard, and Kelman (2012) have emphasized the close relationships between society and the environment by identifying the lack of biodiversity resources, arable land, and water as causes of vulnerability. Unsafe conditions (including fragile livelihoods and unsafe locations) combine with a hazard to result in a disaster, with its loss and trauma (Bolin, 2007; Wisner et al., 2004). Climate change and ecological issues such as plant disease, pests, invasive species, and erosion of biodiversity are included by Wisner et al. (2012) among hazard characteristics or types.

In this development approach to disaster, variables in the social, built, and natural environments can be classified as either liabilities or capabilities

(Gillespie, 2008b; McEntire, 2004b; Wisner, 2016; Wisner et al., 2004). Liabilities increase susceptibility to disaster, while capabilities decrease susceptibility. Both liabilities and capabilities help determine the level of disaster vulnerability within a community. While the concept of capacity refers to the presence of disaster-relevant resources, the capability concept additionally includes mobilization of these resources (Wisner, 2016). For this reason, we prefer the use of the capability concept (Zakour & Gillespie, 2013). In the present context the natural environment consists of natural capital, the characteristics of nature, and trends in the biosphere including global climate change (B. L. Turner et al., 2003). The physical environment includes the built environment and physical characteristics of communities.

CHAIN OF CAUSALITY: ROOT CAUSES, STRUCTURAL CONSTRAINTS, AND UNSAFE CONDITIONS

The Pressure and Release model

In the widely used theoretical model of Wisner et al. (2004), vulnerability is governed through social processes over time, beginning with root causes that set up dynamic pressures (structural constraints) and translate effects from the root causes into unsafe conditions. This three-level time-based framework for explaining vulnerability to disaster is referred to as the Pressure and Release (PAR) model (Wisner et al., 2004). The pressure part of the model represents the unsafe conditions (liabilities) that build up over time to make socio-ecological systems more and more vulnerable to disaster. The release portion represents a reduction in the unsafe conditions or promotion of safe conditions (capabilities).

Root causes

As noted by Zakour and Gillespie (2013) liabilities may be ubiquitous, yet nearly invisible, and these liabilities make up the root causes of vulnerability. In the social development perspective, root causes represent the theoretical point of origin in explaining vulnerability. Wisner et al. (2004) identify a number of root causes, including limited and inequitable access to resources. Root societal causes embedded deep in the social structure guide the distribution of resources. Many root causes are reflected in economic and political ideologies (Renfrew, 2009, 2012). Root causes are the most distal of the types of causes of vulnerability, acting over generations or even centuries. Over time, root causes interact with structural constraints, resulting in more or less unsafe conditions.

Structural constraints

More proximal liabilities are called structural constraints or dynamic pressures. As we explained earlier (Zakour & Gillespie, 2013), we prefer the term structural constraints because of the societal and macrolevel nature of these variables, and because they usually act over decades or generations. In the social development perspective, structural constraints are critical points for explanation of disaster vulnerability. The structural constraints causing disaster include societal deficiencies such as lack of local institutions, social capital, investments, and markets (Wisner et al., 2004, 2012). Structural constraints include a number of macroforces affecting social-ecological systems. For Wisner et al. (2012), about half of these macroforces deal with damage to natural capital in social-ecological systems. These include land grabbing, mining, overfishing, decline in soil productivity, and decline in biodiversity. Societal institutions including governmental, private nonprofit, and private for-profit organizations often cause unsafe conditions by failing to redistribute resources through disaster-relevant networks.

Unsafe conditions

Structural constrains combine with root causes to create unsafe conditions. In the social development perspective, unsafe conditions represent the surface manifestation of vulnerability. Unsafe conditions include both unsafe locations and fragile livelihoods. These unsafe conditions are observed on a day-to-day basis, and are a primary focal point for vulnerability theorists (Zakour & Gillespie, 2013). Among unsafe conditions Wisner et al. (2004, 2012) identified dangerous locations, unprotected buildings and infrastructure, fragile health of populations, lack of disaster preparedness and social protection, and a lack of economic resources resulting in low socioeconomic status for populations. Lower socioeconomic status individuals are less able to protect themselves from hazards in large part because they cannot afford to pay for protection such as home and property insurance. Lower income individuals may additionally reside in communities with poor tax bases and inadequate levees. As an example, Zakour and Harrell (2003) showed that the urban ecology of New Orleans reinforces low access to disaster mitigation and response services, especially for vulnerable populations such as the very young, the very old, and people of color.

Livelihoods

A livelihood includes people, their capabilities, and their arrangements for making a living. Means of living includes food, income, and assets. Livelihoods encompass all resources (capacities) to sustain basic needs, including food, shelter, clothing, cultural values, and social relationships. Sustainable livelihoods maintain and enhance assets on which livelihoods depend. Sustainable livelihoods also have beneficial effects on other livelihoods, including those of future generations. Natural hazards can be a serious threat to livelihoods, but socially sustainable livelihoods can cope with and recover from stress and shocks. Sustainable livelihoods mean that basic needs are met on a daily basis and in the long term (Wisner et al., 2004).

Leverage points in the PAR model

The complex social structures that cause vulnerability are potential points for intervention. Root causes are the most powerful leverage points for reducing the level of vulnerability. Because root causes are distal, they may be resistant to change, especially in the short term (Blaikie & Brookfield, 1987). They are often long-standing in communities or societies, and changing them may involve major social and cultural change. Unsafe conditions are the end-point in the progression to vulnerability. As noted earlier (Zakour & Gillespie, 2013), focusing on the safety of conditions is probably the best leverage point, because intervention here can rapidly produce positive outcomes. Making community conditions safer represents the release part of the PAR model, and is conceptualized as the progression to safety (Wisner et al., 2004, 2012).

ACCESS TO RESOURCES MODEL

The PAR model explains the progression to disaster vulnerability, but this model cannot account for the changes that occur during and after a disaster. To explain the changes occurring during the transition to disaster, the Access to resources model of vulnerability looks at "normal life" before disaster strikes, and places the transition to disaster conditions between unsafe conditions and the natural hazard. The Access model "sets out to examine at a micro-level the establishment and trajectory of vulnerability and its variation between individuals and households. It deals with the impact of a disaster as it unfolds, the role and agency of people involved, what the impacts are on them, how they cope, develop recovery strategies and interact with other actors" (Wisner et al., 2004, p. 88).

Capabilities, referred to as assets or resources in the Access model, include five types of capital:

- human capital, such as skills, knowledge, health, and energy;
- social capital, the resources available through networks;
- physical capital such as infrastructure, technology, and equipment;
- financial capital including credit; and
- natural capital, made up of natural resources, land, water, fauna, and flora.

These capabilities are an important part of the Access to resources model, and they are prominent as safety of conditions variables in the PAR model (Wisner et al., 2012). The lack of these capabilities leads to unsafe locations and fragile livelihoods. As discussed later in this chapter, resilience models focus on the capabilities included in vulnerability models, especially financial, social, human, and political resources (Norris, Stevens, Pfefferbaum, Wyche, & Pfefferbaum, 2008).

Unlike the PAR model the Access model is essentially dynamic, and iterates through time to provide a precise understanding of how people are impacted by a hazard event and their trajectories through that event. The model is economistic, implicitly quantitative, and structuralist. It isolates important political—economic processes of normal life. The model does not explicitly deal with agency (e.g., creativity, ingenuity), which is believed to be much harder to measure quantitatively. Also, coping mechanisms are not central in this model, and are only measured qualitatively (Wisner et al., 2004). As we discuss later, agency and coping skills are explicitly taken up by resilience models.

The influence of root causes on access to resources during a disaster connects the PAR model with the Access to resources model, as well as with resilience models. The political economy of normal life, as detailed in the PAR model, shapes social relations and structures of domination in a community or society (Wisner et al, 2004). Both social relations and structures of domination are root causes in the PAR model. Social relations refer to the flow of financial and other resources within a community or society, and among populations. Structures of domination refer to the unequal relationships among categories of people, defined by class, age, gender, and other socio-demographic variables. Structures of domination also refer to the relationships of individuals or households with government at the local to national levels. During a disaster, social relations and structures of domination determine access to both resources and opportunities for the populations that make up a community or society. Both of these root causes influence how well people cope, adapt, and change in response to disasters.

In the following sections of this chapter, resilience theory is described and then combined with the Access and PAR models to integrate vulnerability and resiliency theories. The Access to resources model is a systems framework and can accommodate or be combined with other theories focusing on access to resources (Wisner et al., 2004). Both the Access model and resilience theory focus on access to resources as a means of coping with disaster. As explained later, vulnerability and resilience models focus on different resources and capabilities, as well as on different dimension of these resources.

THE RESILIENCE PERSPECTIVE

Community disaster resilience is "a process linking a set of networked adaptive capacities to a positive trajectory of functioning and adaptation in constituent populations after a disaster" (Norris et al., 2008, p. 131). A resilient recovery effectively networks and mobilizes community resources or capabilities. As we noted earlier, resilience can also be conceptualized as "the ability of an individual or community to return to a normal or improved state of functioning, or to recover more quickly than expected" (Zakour & Gillespie, 2013). Although resilience is a process and not an outcome, wellness is an important outcome implied by both of these resilience definitions.

Wellness as an outcome of disaster resilience goes beyond an absence of illness and includes a high level of functioning. For resilience researchers the components of individual wellness include (1) a lack of psychopathology and disease; (2) adequate role functioning at home, school, or work; (3) a lack of generalized distress; and (4) a high quality of life. In disasters individuals can suffer from generalized distress, a low quality of life, and the disruption of health behaviors and role functioning; yet they do not suffer from psychopathology (Norris et al., 2008). Even without health or mental health pathology, these individuals and populations can experience considerable suffering that needs to be addressed by disaster managers.

Resilience researchers conceptualize those resources or capabilities, that have already been developed in a system predisaster, as general or inherent resilience (Tierney, 2014). During the progression to vulnerability, resilience capabilities may outweigh the susceptibility of socio-ecological systems. Such a balance results in a reduced level of vulnerability, and absorption of some of the otherwise disruptive forces of a hazard. This is consistent with defining vulnerability as the balance of susceptibility and resiliency.

ATTRIBUTES OF RESILIENCE RESOURCES

Theorists generally view resilience as a postdisaster process of resource mobilization (Norris et al., 2008). One approach to measuring resilience is identifying resources necessary for its implementation, and then focusing on key attributes of those resources. Enhancing the resources that promote resilience can reduce vulnerability. These resources, or capabilities, can be thought of as capabilities similar to those in vulnerability theory (Wisner, 2016). Community resources, or networked adaptive capacities, vary in their robustness, redundancy, and rapidity. Resilience is facilitated to the degree that resources are robust, redundant, or rapidly deployed. When planning and acting to enhance resilience, communities need to consider what resources are needed and available (Cutter, Boruff, & Shirley, 2003).

Robustness

Both the adequacy and the quality of resources are essential for community disaster resilience (Tierney, 2014). Robustness is the strength of a resource and its accessibility throughout the recovery process (its ability to withstand stress without deterioration). Resource robustness is particularly important in disasters that require a long recovery period. The provision of social resources and the altruistic community typically diminish over time and come to an end before the recovery period is completed (Kaniasty & Norris, 2009).

Redundancy

Redundancy means elements of a system can be substituted for one another (Streeter, 1992). When one element is incapacitated, another element of the same or similar type can replace it and the system can continue to operate (Norris et al., 2008). Redundancy is especially important when a disaster is severe and damages both social support networks and the disaster response and recovery system of governmental and community-based organizations.

Rapidity

The rapidity of a resource refers to how quickly the resource can be mobilized during and after disasters. Rapidity is most important when marginalized populations are otherwise unable to access recovery resources in a timely manner. Timely resource provision is critical during the immediate aftermath of disaster (Meichenbaum, 1997; Norris, Murphy, Kaniasty, Perilla, & Ortis, 2001). Many needs in a disaster must be met rapidly, including medical treatment, safety precautions, and utility restoration.

TYPES OF RESOURCES

In addition to its focus on the quality of needed resource, resilience theory also specifies the types of community adaptive resources or capabilities. These resources types are (1) economic development, (2) social capital, (3) information and communication, and (4) collective action (Norris et al., 2008; Zakour & Gillespie, 2013). Economic development is a prerequisite for social capital and information/communication adaptive capacities. In turn, social capital and information/communication are prerequisites for collective action (Norris et al., 2008). Unfortunately, damage to resources from disasters, and the interaction of particular characteristics of a disaster and community resources and their mobilization, almost always results in transient community dysfunction. After a disaster, communities tend to evolve on trajectories of either resilience or dysfunction (Kaniasty & Norris, 2009; Norris et al., 2008).

Economic development

Economic development, including the fairness or equity of vulnerability of constituent populations, is the foundation on which resilience is based. It is the prerequisite for all other adaptive capacities. Economic development is the level and diversity of economic resources, as well as the equitable distribution of resources. The overall level of economic development determines the level of available resources. The diversity of economic resources influences how rapidly the community is able to mobilize in a disaster. If some of the populations in a community or society are vulnerable because of the unfair and inequitable distribution of high-quality resources, then the community or society is also vulnerable to disaster (Norris et al., 2008).

Social capital

Although economic development is the foundation for resilience, social capital is the most important capability needed for resilience, and it is necessary for a resilient recovery (Norris et al., 2008). Social capital refers to resources individuals or collectives obtain from their social networks. The social resources embedded in networks are economically valuable and vital for community disaster resilience. In general, social capital consists of expected benefits from cooperation and preferential treatment among individuals and groups. Preferential treatment is closely related to access qualifications in the Access to resources model. Access qualifications refer to resources and social attributes (e.g., human and social capital) required for an individual to access social resources and income opportunities, as well as the payoffs of these opportunities. Access profiles are all of the resources, especially social resources embedded in networks, a household or population has available. Access qualifications and profiles determine the type and amount of postdisaster support provided victims with different social statuses.

Social capital includes received and perceived support (Kaniasty & Norris, 2009), citizen participation, a sense of community, attachment to place, social embeddedness (informal ties), and interorganizational linkages and cooperation. Citizen participation and/or leadership is comprised of volunteering, membership in voluntary or volunteer organizations, and the mass assault after a disaster (Barton, 1969, 2005). A sense of community includes shared values and mutual concerns among community members, as well as a perception of needs fulfillment. Place attachment refers to an emotional connection to one's community as a geographic place. At the community level, attachment to place encourages community revitalization, as well as altruism and community spirit (Solnit, 2009; Wallace, 1956/2003).

The two remaining aspects of social capital refer more explicitly to social networks. These variables are social embeddedness, or informal links, and interorganizational networks (Norris et al., 2008). Both informal (social embeddedness) and formal relationships (between individuals and organizations) facilitate information seeking, the process of actively gathering information concerning risks and their mitigation. Cooperative links and coordination of interorganizational networks allow disaster relief to be delivered in a timely fashion (Gillespie, Colignon, Banerjee, Murty, & Rogge, 1993). As we stated earlier (Zakour & Gillespie, 2013), preexisting interorganizational networks are the key to mobilizing and redistributing resources after disaster strikes.

Information and communication

The informal and formal network relationships within which social resources are embedded promote access to community information and communication capabilities. Information and communication adaptive capabilities are based in community skills and infrastructure, and include trusted sources of information, a responsible media, and shared community narratives. Multiple trusted sources of information can repeat and confirm each other's messages and promote disaster resilient recoveries. Electronic and print media are responsible when they accurately report on disasters, and disseminate information useful for a resilient disaster response. Irresponsible media outlets frame disasters in a sensationalist manner, blame the most vulnerable populations for their own losses, or help disaster management and other organizations avoid accountability for their response and relief failures (Solnit, 2009).

Community narratives are a critical part of information and communication resources in resilience. Narratives emphasize community efficacy in disasters, and express the belief that predisaster social customs and practices will be restored or improved by recovery (Norris et al., 2008). Community narratives about successful disaster recovery promote cognitive reprocessing, and these narratives create and reinforce shared meaning and purpose (Solnit, 2009). Cognitive reprocessing of the situation often forms the core of community revitalization movements that enhance the recovery. Cognitive and emotional processing facilitates resilience because they promote a strong sense of purpose related to recovery and reconstruction, including new meanings, new assumptions, and more realistic worldviews (Wallace, 1956/2003). The predominance of outsider and expert knowledge, especially if it is promoted by organizations and sectors that fear liability for a disaster, can damage communication and information adaptive capacities.

Collective action

Collective action is the fourth type of adaptive capability or resilience resource. Also called community competence (Norris et al., 2008), collective action is supported by both social capital and communication and information variables. Community narratives and citizen participation are especially important for collective action. Collective action manifests through

- collective efficacy,
- flexibility and creativity,
- critical reflection and problem-solving skills,
- community action, and
- political partnerships (Norris et al., 2008).

Identification and description of collective action capabilities represent an important contribution of resiliency theory. The variables in collective action go beyond the economic and structural variables in the Access to resources model proposed by vulnerability theorists. Collective action is the networked version of human agency, which is central to resilience. Collective efficacy is the capability of community members to organize and make decisions, and is a basis for exercising human agency. Collective efficacy empowerment is the ability of members of a community to organize and make decisions to improve quality of life (Zakour & Gillespie, 2013). Communities with collective efficacy tend to succeed in disaster recovery because collective efficacy increases the amount and/or quality of resources and access to resources (Norris et al., 2008). Flexibility and creativity are also at the heart of human agency, as emphasized by resilience theorists. Planning is always necessary, but because of unexpected and synergistic impacts of disasters, human agency is essential if plans are to be effective (Norris et al., 2008).

Two additional variables in collective action include critical reflection and problem solving, and community action. These variables, like the other variables contributing to collective action, are essential to collective agency in disaster. Through critical reflection and problem solving, communities identify major problems or opportunities in disaster recovery. In disaster problem solving, participants identify recovery problems, opportunities, and goals, and they address important community issues (Zakour & Gillespie, 2013). Community action in a disaster is the capability to organize for a specific purpose, and to take action to achieve goals shared by the community. Community action aims to gain access to additional resources or types of resources, as well as higher quality resources (Norris et al., 2008). Community action typically involves marginalized populations in a community challenging the access qualifications, benefiting high-status populations.

Political partnerships, the final variable in collective action, facilitate coordination and reduce conflict among governmental and other types of organizations. This aspect of collective action is closely related to a number of social capital resources, including formal and informal relationships. Political partnerships within the community, among different communities, and between local, regional, and national governments are an important type of resilience resource and capability. This capability may determine if a community survives and thrives after a disaster, or if a community is hollowed-out or even annihilated (Sundet & Mermelstein, 2000).

COMPARING AND CONTRASTING VULNERABILITY AND RESILIENCY

Vulnerability and resilience are different kinds of concepts, and they focus on different disaster phases. They also focus on different levels of analysis (Zakour & Gillespie, 2013). Vulnerability and resilience theorists work with separate models, although the models share a number of similar concepts (Norris et al., 2008; Wisner et al., 2012). The two theories focus on different dimensions of disaster resources. Vulnerability is descriptive of continuously changing predisaster conditions. During the progression to vulnerability, safe conditions can deteriorate. Conversely, unsafe conditions can be removed or mitigated, resulting in a progression to safety. General or inherent resiliency resources interact with disaster susceptibility to determine the safety of conditions. However, resilience refers especially to processes during disaster conditions, and especially disaster recovery. Resilience entails a return to predisaster functioning, or to an improved level of functioning.

Along with a focus on different disaster phases, vulnerability and resiliency theorists focus on different levels of analysis. Vulnerability theorists emphasize structural and economic variables, while resiliency theorists emphasize individual and population outcomes (Norris et al., 2008). Models of resilience, with their focus on wellness, directly complement vulnerability theory (Zakour & Gillespie, 2013).

Although vulnerability theorists emphasize household and political economies, they explicitly recognize the importance of wellness (Wisner et al., 2004).

In addition to a focus on different concepts, disaster phases, and levels of analysis, vulnerability and resiliency researchers work with separate but conceptually similar models using essentially the same central idea (Zakour & Gillespie, 2013). In vulnerability theory, adaptation to the natural environment is central to reducing vulnerability, and adaptation entails fair and equitable human development for all constituent populations in a community or other type of socioecological system (Taylor, 2015). This is conceptualized using a PAR model and an Access to resources model (Wisner et al., 2004). In resilience theory, adaptation is conceptualized as community coping. In this perspective, coping to an environment altered by disaster occurs through mobilization of networked adaptive capabilities (Norris, Galea, Friedman, & Watson, 2006). The proximal variables categorized as Unsafe Conditions (unsafe locations and fragile livelihoods) in the PAR model are capabilities and resource variables (Wisner, 2016; Wisner et al., 2012) similar to networked adaptive capabilities in resilience models (Norris et al., 2008).

Finally, in addition to the differences described earlier, vulnerability and resiliency researchers focus on different dimensions of disaster-relevant resources. These different dimensions are complementary, representing the various facets of community life (Zakour & Gillespie, 2013). Resource capabilities specified by vulnerability theorists are economic and structural, emphasizing how structural changes over time impact vulnerability. Resiliency theorists specify the dimensions of processes and resources necessary for a timely and socially just disaster recovery. If resources are robust, redundant, and rapidly accessible, they are thought to be adaptive capacities or capabilities. Community disaster resilience is the linkage of quality resources to outcomes of wellness (Norris et al., 2008). Resilience is ideally transformative so that communities learn to cope with and achieve a level of functioning higher than predisaster, and to apply lessons learned for future disasters.

VULNERABILITY AND RESILIENCE THEORIES ARE COMPLEMENTARY

While vulnerability and resiliency concepts differ from one another, they neatly complement each other (Zakour & Gillespie, 2013). Each theory follows a similar chain of causality (Blaikie & Brookfield, 1987). Overall, vulnerability theory subsumes the concepts, assumptions, and findings from resiliency theory (Norris et al., 2008). Consistent with the conceptualization of resilience in vulnerability theory, resilience occurs despite collective adversity in a system such as a household, organization, community, or society (Zakour, 2008, 2010). Disaster vulnerability and resiliency theorists recognize that every community has varying degrees of disaster vulnerability and the potential to recover resiliently (Queiro-Tajalli & Campbell, 2002).

Vulnerability theory concentrates on the progression to vulnerability during predisaster conditions. Resiliency theory concentrates on access to and use of resources after disaster strikes. Given they are complementary, we believe that the greatest potential of these theories will be realized by integrating the resilience concept and process with vulnerability theory (Zakour & Gillespie, 2013). We discuss the merits of this integration and designate it as V+ theory. Integrating the process of resilience into vulnerability theory to create V+ theory yields a more comprehensive, balanced, and potentially productive theory.

One additional way that vulnerability and resilience theories are complementary is the effect of resilience in countering vulnerability (Zakour & Gillespie, 2013). In the stress and coping approach of community psychologists (Norris et al., 2006, 2008), resilience is a postdisaster process that can counter the damaging effects of high levels of disaster vulnerability. Resilience and vulnerability are generally inversely related, so that the variables facilitating resilience tend to reduce vulnerability. Given this inverse relationship the theory of vulnerability can be expanded and enhanced by explicitly incorporating resilience as process. The level of community disaster vulnerability is reduced by diminishing liabilities while strengthening capabilities (McEntire, 2004b, 2005). Promoting opportunities for resilience to emerge during disaster response and recovery, and increasing the amount and quality of resilience resources, reduces vulnerability to future disasters.

The second additional area in which vulnerability complements resilience is their common focus on capabilities (Zakour & Gillespie, 2013). Environmental capabilities are hypothesized, in both theories, as associated with reduced levels of disaster vulnerability and an increased likelihood of resilient recovery. Capacities, resources, and capitals are all capabilities and are often used interchangeably by vulnerability and resiliency researchers and theorists (Norris et al., 2008; Wisner, 2016). For both theories, capabilities include economic, cultural, social, physical, natural, and political capital. In vulnerability theory the level of these capacities or resources is associated with the safety of living conditions (Wisner et al., 2012). In resiliency theory these resources are called adaptive capacities. The networking of these capacities during disasters is the essence of community disaster resiliency.

The third additional area in which vulnerability complements resilience is in their common focus on economic and social marginality (Zakour & Gillespie, 2013). In both resiliency and vulnerability theories, poverty and social isolation are important and related environmental liabilities (Pulwarty, Broad, & Finan, 2004; Zakour, 2010). Poverty, or economic marginality, means that some households and populations in a community do not have the resources needed for vulnerability reduction or for recovery. Economic inequality can benefit affluent households or populations, who may have a surplus of resources to absorb the negative effects of disasters and to rapidly recover. Social marginalization is isolation from neighbors, kin, and formal organizations. Socially marginalized individuals and households are unable to access social capital and associated social support for disaster recovery (Wisner, 2016; Wisner et al., 2012). This includes information from kin or neighbors to help make evacuation decisions or obtain relief services (Klinenberg, 2002).

For both vulnerability and resilience theories, economic and social marginality tend to co-occur in neighborhoods, particularly those affected by racial segregation. These neighborhoods have few voluntary or volunteer organizations to provide social services and to build social capital for organizations and their clients. The lack of volunteer and membership opportunities also limits the social capital of individuals in the community (Putnam, 2000). The neighborhoods have lower levels of financial, in-kind, and other donations to support voluntary organizations or volunteers. There is also a lack of a strong tax base (Zakour & Gillespie, 1998). These conditions lead to lower levels of social capital, fewer human services organizations, and especially fewer local disaster mitigation projects (Zakour & Harrell, 2003).

CONTINUITY BETWEEN VULNERABILITY AND RESILIENCE THEORIES

There is substantial continuity from the predisaster progression to vulnerability and the postdisaster process of resilience. The progression to vulnerability has a strong influence on the resilience process, which in turn affects vulnerability to future disasters. Resilience as process occurs largely during and after disaster conditions. However, resilience temporally follows and is influenced by the nature of safety of conditions, the last stage in the progression to vulnerability. A very important area of continuity is the baseline of community functioning occurring during the safety of conditions period, before a disaster occurs. When community vulnerability and dysfunction is high predisaster, the path to a resilient disaster recovery and a high level of community functioning is much longer and fraught with difficulty. The greater the inequity and unfairness in the distribution of vulnerability, the less likely it is that communities will be able to recover in a resilient fashion (Laska, 2012). When a disaster occurs, elite groups who may have benefited from inequality often attempt to reassert their relative advantages (Oliver-Smith, 2009; Taylor, 2015). Elites in business, government, and even in nonprofits such as universities, will quickly attempt to reestablish the status quo and regain power advantages in the aftermath of disaster (Solnit, 2009).

The second area of continuity between vulnerability and resilience is the influence of root causes and structural constraints on both the progression to vulnerability and the unfolding of resilience. Root causes and structural pressures in the PAR model determine the access qualifications and profiles (in the Access model) as well as the level of access to recovery resources. While root causes and structural constraints are distal causes determining vulnerability, in resiliency theory access qualifications and profiles help determine which individuals, households, and populations can access adaptive capacities. Individuals and households with greater levels of resources (access profiles) and higher social statuses (access qualifications) tend to receive more support from the altruistic community (Kaniasty & Norris, 2009).

The third area of continuity between vulnerability and resilience theories are the actions of social relations and structures of domination both as root causes of vulnerability, and as constraints on the resilience of populations. Social relations and structures of domination form the institutional framework for the workings of resiliency. Social relations as a root cause of vulnerability determine the flow of economic and other resources among populations and individuals in a community. Structures of domination are the unequal power relations between populations as categories in a stratified system, such as between managers and workers, men and women, the elite and low-SES (socioeconomic status) people, and between individuals and populations in a community and in national governments (Wisner et al., 2004). In the progression to vulnerability, the most proximate causes of vulnerability are the lack of safety of conditions and the absence of adequate types and levels of capabilities. These conditions are the result of social relations and structures of domination that have benefited economic and power elites in a community.

The fourth area of continuity between the progression to vulnerability and postdisaster resilience is the capital determining the safety of conditions, and forming the basis for the networked adaptive capacities and resilience (Norris et al., 2008). The capabilities determining the safety of conditions overlap with those needed for resiliency. Wisner et al. (2012) have elaborated the PAR model, revealing striking similarities between capabilities in the PAR model, and resiliency adaptive capacities. In this elaboration of the PAR model the Unsafe Conditions (Fragile Livelihoods and Unsafe Locations) variables include six types of resources, or capital, that make up the categories for specific variables in Unsafe Conditions. In both vulnerability and resiliency theories a lack of these capitals and capabilities leads to greater vulnerability and marginality of populations, as well as a greater likelihood of failed recovery and persistent dysfunction in socio-ecological systems.

Economic development is a fifth area of continuity between vulnerability and resilience. Economic development affects the progression to vulnerability, and development is a foundation for the networked capacities promoting a resilient recovery. The Pressure and Release (PAR) model of vulnerability is an economic model focused on development (Wisner et al., 2012). Failed or inadequate development increases the level of community disaster vulnerability. The economic development variables in resilience models include (1) fairness of vulnerability to hazards, (2) an adequate level and diversity of economic resources, and (3) equitable resource distribution (Norris et al., 2008).

A final area of continuity concerns ideologies, worldviews, and culture. During the progression to vulnerability, ideologies justify the root cause of structures of domination. Similar to predisaster conditions, after disasters the power elite continues to hold ideologies which justify their self-interest. Usually the media express ideologies supportive of the power elite and reinforce these beliefs. Ideologies both justify and inspire the elite behavior in disasters called "elite panic" (Solnit, 2009). The effects of media worldviews are particularly strong in terms of the information and communication component of resilience. If the media supports the status quo and community elites, this can prevent a community's development of narratives, skills and infrastructure, and trusted sources of information. Both community elites and an irresponsible media will also have an effect of dampening citizen participation and leadership roles, important aspects of the social capital component of resilience (Norris et al., 2008).

VULNERABILITY-PLUS THEORY AS INTEGRATION OF THEORIES

By identifying and using the various complementary aspects of the two theories, vulnerability and resilience theory can be integrated by adding resiliency resources to the PAR model (see Fig. 3.1). The continuities among vulnerability and resiliency theories show that this integration enriches the chain of causality from root causes of vulnerability to resilience. B. L. Turner et al. (2003) developed a model of vulnerability providing an additional framework for the integration of vulnerability and resiliency models. This integration creates V+ theory, a more complete theory of vulnerability. V+ theory guides the reduction of the level of vulnerability and promotes resilient recoveries. Changes in vulnerability and resilience lead to higher levels of functioning in the postdisaster environment, or the "new normal."

Resiliency theory adds a host of additional variables known to both foster resilient recoveries and enhance the progression to safety. In resilience theory, economic development is a basis for social capital as well as information and communication capabilities. Social capital includes variables that are not part of original vulnerability theory. These variables include received and perceived social support, social embeddedness (informal ties), attachment to place, and sense of community. Community narratives of successful recoveries, communication skills and infrastructure, and trusted sources of information are also new to vulnerability theory.

V+ theory contains new variables and causal relationships to explore, beyond those offered by traditional vulnerability or resiliency theory alone (Zakour & Gillespie, 2013). Like traditional vulnerability theory, V+ theory includes social and economic variables (Wisner et al., 2004, 2012). Similar to some vulnerability models and especially resilience theory (Norris et al., 2008), V+ theory also includes variables at different analytic levels, such as the physical, social, and human capital underlying collective action. Other tangible and intangible adaptive capacities (e.g., emotional support, information, and communication) are also included in V+ theory. The integration of traditional resiliency theory with vulnerability theory increases our understanding of the role of noneconomic factors in recovery, especially collective action and community empowerment. V+ theory holds the potential for achieving not only safer communities, but also collective human agency, resilient recoveries, and community wellness.

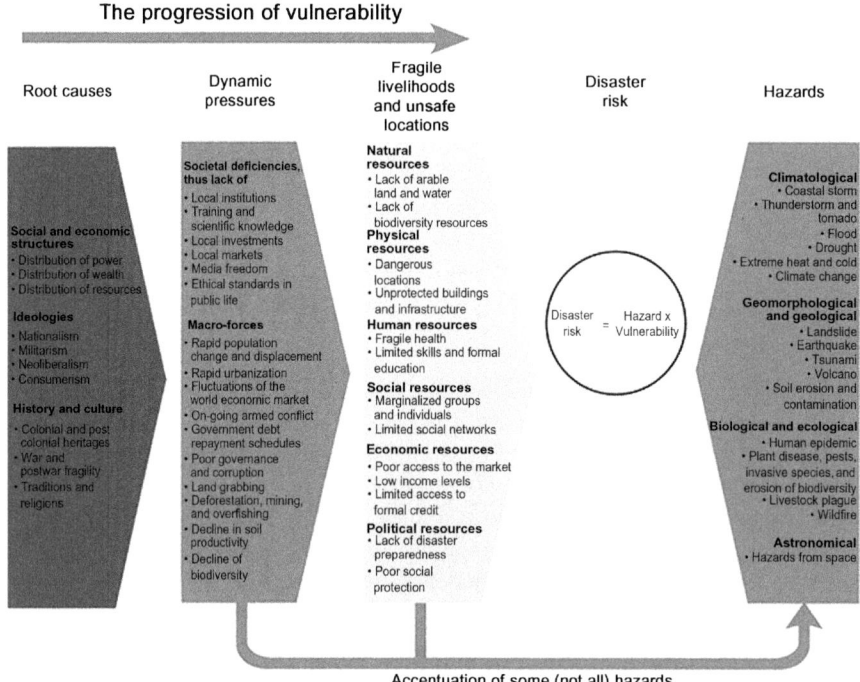

FIGURE 3.1

The Progression to Vulnerability

From Wisner, B., Gaillard, J. C., & Kelman, I. (2012). Framing disaster: Theories and stories seeking to understand hazards, vulnerability and risk. In B. Wisner, J. C. Gaillard, & I. Kelman (Eds.), The Routledge handbook of hazards and disaster risk reduction (pp. 18–31). New York: Routledge. Copyright 2012 by the authors and by Routledge Press. Reprinted with permission.

ASSUMPTIONS OF VULNERABILITY-PLUS THEORY

"Theories are created through the assumptions they make" (Zakour & Gillespie, 2013). V+ theory is based on 12 general assumptions. The first three assumptions cover the definition of vulnerability (#1), its distribution (#2), and its dimensionality (#3). The remaining nine assumptions set out a broad causal framework. This framework identifies the kind of variables that cause disaster vulnerability as well as their proximity to one another. Six of the assumptions are from vulnerability theorists and researchers (Assumptions 4, 5, 7, 8, 9, and 12). The remaining assumptions of V+ theory are based on theoretical ideas from the disaster literature or from related literatures (Assumptions 1, 2, 3, 6, 10, and 11). Box 3.1 lists and briefly discusses the 12 assumptions in summary form, along with citations of research that supports each assumption.

BOX 3.1 ASSUMPTIONS OF V+ THEORY AND THEIR EMPIRICAL SUPPORT

1st assumption: *The vulnerability of social systems is the reduced capacity to adapt to environmental circumstances* (Cardona, 2004). This assumption is based on political ecology and development ideas (Collins, 2008a; McEntire, 2004a). It suggests that the lack or failure of sustainable human development is a central cause in the progression to disaster vulnerability (Zakour, 2010). Development is understood as a process of increasing harmony between a community's social system and its physical and natural environment (Gillespie, 2008a, 2010).

2nd assumption: *Vulnerability is not evenly distributed among people or communities.* Communities with more and greater environmental liabilities have a higher level of vulnerability (Gillespie, 2008b, 2010). Disaster vulnerability is inequitably distributed across the world, regions, countries, communities, and populations within communities. This assumption is made on the basis that almost all communities are stratified along some dimension (Bottero, 2007).

3rd assumption: *The concept of disaster vulnerability is multidimensional.* This premise suggests that disasters are all-encompassing experiences that impact every dimension of life in a community (Wallace, 1956) and affect its systems at all levels of analysis. Assessing community vulnerability involves examining the vulnerability of each constituent subsystem, as well as assessing the community as a system.

4th assumption: *The availability and equitable distribution of resources in a community decreases disaster vulnerability and facilitates resilience.* This assumption proposes that capitalist production and the market approach to socio-ecological systems generally create inequality (Robbins, 2012) and that markets structure relationships of unequal power and wealth (Taylor, 2015). This leads to an unequal distribution of resources and ultimately to increased disaster susceptibility and vulnerability (Renfrew, 2009, 2012; Wisner et al., 2012).

5th assumption: *Vulnerability is largely the result of environmental capabilities and liabilities.* It is assumed that environmental liabilities increase the susceptibility of systems and their functioning to disasters. Alternatively, capabilities increase the general or inherent resilience of systems, and help to reduce vulnerability (McEntire, 2004b).

6th assumption: *Social and demographic attributes of people are associated with, but do not cause, disaster vulnerability.* The frequent association between demographic variables and environmental liabilities is the key to this assumption. External variables generally account for these relationships (see Rosenberg, 1968). Demographic characteristics such as age and household income are associated with different levels of vulnerability because of inequitable exposure to the environmental liabilities and capabilities of the community. People defined as vulnerable according to their socio-demographic attributes, such as the very young and very old, tend to be economically and socially marginalized, and have limited access to resources for protection and disaster recovery (Rosenfeld, Caye, Ayalon, & Lahad, 2005; Thomas & Soliman, 2002).

7th assumption: *Unsafe conditions in which people live and work are the most proximate and immediate societal causes of disaster.* This assumption advises that unsafe conditions include both dangerous locations and fragile livelihoods. Low levels of capital, including social and human capital, characterize unsafe conditions (Greene & Livingston, 2002; Norris et al., 2008; Wisner et al., 2004, 2012).

8th assumption: *Root causes, the socio-cultural characteristics of a community or society, historically and in the present, are the ultimate causes of disasters* (Blaikie, Cameron, & Seddon, 1979, 1980; Wisner et al., 2012). It is understood that root causes of disaster are thought of as distal, because they are so ingrained in society, over many years, that they are difficult to perceive. Root causes are also conceptualized as ultimate causes, set at the beginning of a chain of causality explaining the progression of vulnerability (Blaikie & Brookfield, 1987).

9th assumption: *Disasters occur because of a chain of causality in which root causes interact with structural pressures to produce unsafe conditions. Hazards then interact with unsafe*

(Continued)

> **BOX 3.1 ASSUMPTIONS OF V+ THEORY AND THEIR EMPIRICAL SUPPORT (CONTINUED)**
>
> *conditions to trigger a disaster* (Cardona, 2004; Wisner et al., 2004). A lack of adequate capabilities, or resources and capacities, leads to unsafe conditions and vulnerability (Wisner et al., 2012).
>
> **10th assumption:** *Culture, ideology, and shared meaning are of central importance to the progression of disaster vulnerability.* This assumption is derived from early theorizing on disaster (Barton, 1969; Wallace, 1957). The relationship between society and nature is one of the fundamental pillars of any society's ideological system (Oliver-Smith, 2004). A worldview of dominance over nature can lead to wealth creation (Bankoff, 2004), but also to environmental degradation and unsafe conditions (Robbins, 2012; Wisner et al., 2012).
>
> **11th assumption:** *Environmental capabilities and liabilities, and disaster susceptibility, are related in complex ways to produce the level of community vulnerability.* This assumption is based on stress and coping approaches that acknowledge the complex relationships among severe stress, protective factors (capabilities), and susceptibility. In general, capabilities and liabilities are inversely associated with each other, but some evidence suggests that they are orthogonal, or even positively associated (Norris et al., 2008; Putnam, 2000). Capabilities may cause decreases in vulnerability in an additive fashion, or they may act to buffer liabilities or even hazards to reduce vulnerability (Neria, Galea, & Norris, 2009; Norris et al., 2006).
>
> **12th assumption: global scale:** *The environments of communities are growing in complexity and are increasingly global in scale.* Disasters are the outcome of destructive social and economic processes intensifying on a global scale (Oliver-Smith, 2004). Rapid population change and displacement, rapid urbanization, environmental degradation, declining biodiversity, and global climate change are dynamic pressures and structural constraints leading to increases in community vulnerability (Pulwarty et al., 2004; Wisner et al., 2012). Globalism makes causal chains and processes in the progression to vulnerability more complex and difficult to understand (Robbins, 2012; B. L. Turner et al., 2003; Zakour, 2012).
>
> *From Zakour, M. J., & Gillespie, D. F. (2013). Community disaster vulnerability: Theory, research, and practice (pp. 27–34). New York: Springer Press. Copyright 2013 by Springer Science + Business Media. Adapted with permission.*

EMPIRICAL SUPPORT

In this section, empirical support for V+ theory is summarized. Support for each of the assumptions of V+ theory is summarized in Table 3.1. This table displays the substantial support for V+ theory. The first six assumptions have more support than the last six (Zakour & Gillespie, 2013). We anticipate more balanced coverage through additional research on V+ theory, some of which is reported in the present volume on Hurricane Katrina and New Orleans.

Next, we present a model of community disaster vulnerability and resiliency called V+ theory, based on empirical support (see Fig. 3.2). This model is accompanied by a brief list of some of the more important variables in current vulnerability and resiliency research, and in the development of V+ theory. The model proceeds from the most distal variables, root causes, to the proximal

Table 3.1 Empirical Support for V+ Theory Assumptions

1. Vulnerability of social systems is the reduced capacity of a community, society, or culture to adapt to environmental circumstances	Benight, Ironson, and Durham (1999), Gillespie et al. (1993), Gillespie and Murty (1994)
2. Vulnerability is not evenly distributed among people or communities	Chakraborty, Tobin, and Montz (2005), Gillespie et al. (1993), Mitchell, Thomas, and Cutter (1999), Rogge (1996), Rüstemli and Karanci (1999), Wisner et al. (2004)
3. Disaster vulnerability is multidimensional	Borden et al. (2007), Burnside, Miller, and Rivera (2007), Chakraboty, Tobin, and Montz (2005), Cutter et al. (2003)
4. The availability and equitable distribution of resources in a community decreases disaster vulnerability and facilitates resilience	Burnside et al. (2007), Collins (2008b), Gillespie et al. (1993), Renfrew (2009, 2012), Wisner et al. (2004), Zakour and Harrell (2003)
5. Vulnerability is largely the result of environmental capabilities and liabilities	Bonanno, Galea, Bucciarelli, and Vlahov (2007), Chakraborty et al. (2005), Gillespie et al. (1993), Gillespie and Murty (1994), Kapucu, Augustin, and Garagey (2009), Zakour (2008)
6. Social and demographic attributes of people are associated with but do not cause disaster vulnerability	Bolin (2007), Cutter et al. (2003), Burnside et al. (2007), Girard and Peacock (1997), McGuire, Ford, and Okoro (2007), Peacock and Girard (1997)
7. Unsafe conditions in which people live and work are the most proximate and immediate societal causes of disaster	Borden et al. (2007), Wisner et al. (2004)
8. Root causes, the socio-cultural characteristics of a community or society, historically and in the present, are the ultimate causes of disasters	Burnside et al. (2007), Wisner et al. (2004)
9. Disasters occur because of a chain of causality: root causes interact with dynamic structural factors to produce unsafe conditions. Hazards then interact with unsafe conditions to trigger a disaster	Renfrew (2009, 2012), Wisner et al. (2004)
10. Culture, ideology, and shared meaning are of central importance in the progression to disaster vulnerability	Norris et al. (2008), Rüstemli and Karanci (1999), Simonovic and Ahmad (2005), Tedeschi and Calhoun (2004), Wellman and Frank (2001)
11. Environmental capabilities, liabilities, and disaster susceptibility are related in complex ways to produce the level of community vulnerability	Kaniasty and Norris (2009), Simonovic and Ahmad (2005)
12. The environments of communities are growing in complexity and are increasingly global in scale	Girot (2012), Mascarenhas and Wisner (2012), Renfrew (2009, 2012)

From Zakour, M. J., & Gillespie, D. F. (2013). Community disaster vulnerability: Theory, research, and practice (p. 151). New York: Springer Press. Copyright 2013 by Springer Science + Business Media. Reprinted with permission.

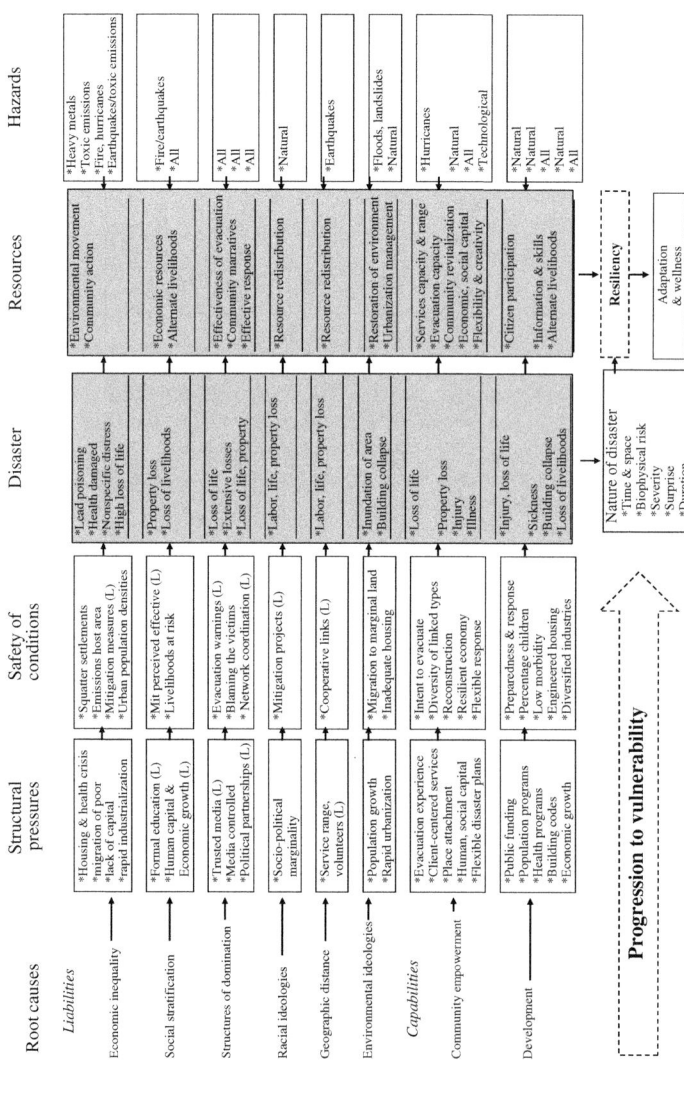

FIGURE 3.2

Theoretical Model of V+ Theory

From Zakour, M. J., & Gillespie, D. F. (2013). *Community disaster vulnerability: Theory, research, and practice* (p. 152). New York: Springer Press. Copyright 2013 by Springer Science + Business Media. Reprinted with permission.

variables describing the safety of locations and fragility of livelihoods (Wisner et al., 2012). Resources relevant to resiliency capabilities are the final set of variables in the causal chain. These resources mediate the severity of disaster exposure, primarily through recovery and wellness. The major variables of V+ theory are listed under the headings as they are presented in the model below. The types of variables are (1) root causes, (2) structural pressures, (3) safety of conditions, (4) resources, (5) hazards, and (6) resiliency.

This V+ model displays causal relationships among variables, which are empirically supported by development, vulnerability, and resilience researchers and theorists (Blaikie & Brookfield, 1987; Blaikie et al., 1994; Norris et al., 2008; Wisner et al., 2004, 2012). Causality flows from left to right and from top to bottom. This diagram of V+ theory follows the causal order of the progression represented by development researchers and their Pressure and Release (PAR) model (Wisner et al., 2004, 2012). Root causes combine with structural pressures and result in conditions with varying degrees of safety. When they occur, hazards interact with the safety of conditions to trigger disasters. Varying degrees of access to resources and networked adaptive capacities occur as a part of the process of disaster. If resources are robust, redundant, and rapid, then resiliency takes place and the system will adapt to new conditions and reestablishes the wellness of individual members (Norris et al., 2008). The characteristics of the disaster and the availability of resources will affect the resiliency of the system.

In the V+ model, horizontal arrows indicate relationships among specific root causes, structural constraints, safety variables, and resilience resources. Causal chains of sets of specific variables are shown along rows in the model. Some root causes, structural constraints, and unsafe conditions are conceptualized as environmental liabilities. The first six root causes at the upper left side of the model are liabilities, as are all of the variables that follow in causal chains along the rows. An (L) following a variable indicates the lack of a variable and represents a liability. Community Empowerment and Development, the two root causes at the bottom of the model, are capabilities, along with the variables they affect in the causal chains. All of the Resource variables are also capabilities.

The model uses the work of Norris et al. (2008) on resources leading to resilient disaster recoveries and a positive trajectory of functioning and adaptation. Community empowerment and development resources are also root causes. Resources are also included in the variables that follow these capability root causes. These resource variables are part of the progression to safety. The resource variables included in the three columns on the left of the model are part of general or inherent resiliency according to resiliency theorists (Tierney, 2014). The resource variables show the overlap between the construct of vulnerability and the construct of general/inherent resilience.

Other resource variables intervene between disasters and their triggering hazards. These resource variables are a part of resilience as process (Norris, 2008), as well as the process of access to resources modeled by vulnerability researchers (Wisner et al., 2004). One kind of resilience is conceptualized as absorptive resilience, which is very similar to reducing the level of vulnerability. The outcome of resilient recovery is adaptation to new conditions, and wellness among community members. Ideally, resilient recovery leads to a transformation of the community or other system, so that not only adaptation and coping take place, but also higher levels of system functioning and population wellness occur.

CAUSAL CHAINS IN V+ THEORY

The sequence of variables in the causal chain of the V+ model follows those of vulnerability theorists (Wisner et al., 2004, 2012), resiliency theorists (Norris et al., 2008), and political ecologists (B. L. Turner et al., 2003; M. D. Turner, 2014). In the progression to vulnerability, root causes are affected by structural constraints to produce the safety of conditions. Hazards interact with unsafe conditions to trigger a disaster. Resilience resources are mediating variables between hazards and disasters. These resource capabilities can help systems resist some aspects of hazards, but more likely they lead to a positive trajectory of resilience and recovery. The predisaster functioning of a system will be restored or improved on, leading to adaptation and wellness outcomes.

ROOT CAUSES

The first stage in the progression to vulnerability, root causes, is deep-seated in a society, temporally distant, and not readily visible. In the V+ model, economic inequality is a major root cause leading to vulnerability. Social stratification, a root cause, refers to large differences in social status for categories of people or populations in a community. Another root cause, structures of domination, refers to long-standing patterns of power of one group or category of people over another. Ideologies are aspects of culture and make up an important category of root causes, and help justify structures of domination.

Both racial and environmental ideologies may be liabilities or capabilities. An environmental ideology of exploitation of nature to produce wealth is a liability, and eventually damages the environment to extract profit. An ideology of exploitation encourages patterns of production that aggressively alter environments to extract short-term value. Racist ideologies and exploitive environmental ideologies help reinforce inequality, stratification, and domination. Geographic distance is a root cause resulting in social inequalities, and reinforcing these divisions. In urban areas with high concentrations of poverty and people of color residing in one geographic segment of the community, the cost of travel for this population to obtain services and other resources can be high.

As root causes, both community empowerment and social development are capabilities. Among resiliency theorists, these two capabilities are conceptualized as aspects of general or inherent resiliency. Empowerment is a root cause capability, encouraging people to gain control over production and access to new resources (Zakour & Gillespie, 2013). Through empowerment and sustainable human development, communities and their populations improve their chances of accessing and networking resources needed for effective disaster recovery and resilience (Norris et al., 2008).

STRUCTURAL CONSTRAINTS

Structural constraints combine with root causes to influence the safety of locations and fragility of livelihoods, and are intermediate between root causes and unsafe conditions. These constraints result from change originating in root causes. Socio-cultural change occurs continuously but the problems resulting from these changes are noticed intermittently and only when a tipping point is reached or research raises awareness of disaster vulnerability. Improved understanding of the relationships among root causes, structural constraints, and safety of conditions will allow vulnerability researchers to better address a community's daily vulnerability and to reduce unsafe conditions (Zakour & Gillespie, 2013).

Examples of structural constraints in the V+ model resulting from the root cause of economic inequality are housing and health crises, migration of the rural poor to urban areas, lack of money and physical capital, and rapid industrialization. These structural constraints are caused by economic inequality, and interact with this root cause to produce unsafe conditions such as squatter settlements and high urban population densities. These unsafe locations and conditions coincide with high levels of environmental pollution and a lack of disaster mitigation projects.

UNSAFE CONDITIONS

Unsafe conditions refer to the dangerous locations and fragile livelihoods of populations or communities. Fragile livelihoods include tourism and natural resource extraction such as mining. Tourism and natural resource/mineral extraction are often disrupted after a disaster, and few alternate livelihoods are available in communities dependent on these industries. This is the situation in many Central Gulf Coast communities in the United States.

V+ theory is useful for reducing the effects of structural constraints on safety of conditions. Persistent economic failings and low levels of human capital, including formal education, are structural constraints that interact with social stratification to produce unsafe conditions. Economic weakness compromises the resources needed to improve the safety of conditions, as well as the resources needed for resilient recoveries after disaster. Dangerous locations and fragile livelihoods, and the perceived ineffectiveness of mitigation measures, result from the root cause of social stratification and interacting structural constraints. Vulnerability is reduced through effective disaster mitigation (Gillespie et al., 1993), and the development of sustainable alternate livelihoods. Communities responding resiliently to disasters are more likely to recover rapidly and well, reaching higher levels of sustainable human development than existed predisaster (Zakour, 2010). Recovering to higher levels of functioning and wellness is conceptualized as transformative resilience.

Important types of resources needed for disaster recovery include availability of robust or alternate livelihoods (Wisner et al., 2004), and the level and diversity of equitably distributed economic resources (Norris et al., 2008). Overall, vulnerability results from unsafe conditions produced by a lack of adaptation to the local physical environment (Oliver-Smith, 2004). The use of V+ theory helps researchers examine variables such as economic and social stratification, sociopolitical marginality, and disaster mitigation projects. Other variables involved in safety of conditions are human, economic, physical, natural, social, and cultural capital (Wisner et al., 2012).

HAZARD TYPES

Disasters result from the interaction of unsafe conditions and hazards. Each type of hazard potentially causes a particular pattern of damage. For example, tropical cyclones (hurricanes, typhoons) cause flooding similar to that occurring during heavy rain events. Yet hurricane storm surges are more likely to impact coastal areas and cause long-term salinization and mineralization of coastal land. Natural versus technological hazards also result in different damage. Unlike natural hazards, technological hazards result in disasters lacking a low point, the period of sharply reduced community functioning. The effects of chemicals and toxic emissions on the very young and future generations can potentially unfold over many decades, leading to long-lasting fear about the effects of toxic contamination (Zakour & Gillespie, 2013).

In the V+ model, under the hazards column (Fig. 3.2), the hazard type corresponds to the particular hazard context in which disaster research was conducted, providing support for the causal chain. Some empirical support comes from research on several types of natural hazards. This is indicated by the Natural category. Other support was derived from research including both natural and technological hazards, as indicated by the All category. For example, in the causal chains originating in economic inequality and social stratification, the hazards studied include heavy metal contamination and toxic emissions, fires, earthquakes, and hurricanes. In the social stratification chain, research was conducted in both natural and technological hazards (Zakour & Gillespie, 2013).

It is important to note that hazards such as climate change result from root causes including exploitive environmental ideologies and economic inequality. This relationship illustrates the very close relationships among society and nature. Climate change has recently had a moderate effect in the frequency and severity of natural hazards (IPCC, 2012). Because climate change occurs at the global level, root causes of a community's disaster vulnerability may originate at a great geographic distance from a community (Zakour & Gillespie, 2013). An example of dynamic pressures and structural constraints that interact with environmental ideologies is rapid industrialization and emission of large amounts of greenhouse gases. Increases in greenhouse gases in the atmosphere cause warming of the oceans, resulting in sea-level rise and unsafe conditions in coastal areas. Hazards can be more intense because higher mean sea levels can lead to higher storm and tidal surges.

DISASTER CHARACTERISTICS

As shown in the Disaster column of the V+ model (Fig. 3.2), disasters triggered by specific types of hazards result in particular losses and impacts. Disaster impacts result from combinations of hazard type and particular unsafe conditions (i.e. dangerous locations or fragile livelihoods. Disaster effects include lead poisoning, damaged health, high loss of life, loss of property and livelihoods, and nonspecific emotional distress. A particular set of resilience resources is required for populations and communities to recover rapidly and well from these disasters (Zakour & Gillespie, 2013).

Disasters vary by their level of biophysical risk for a particular place, in the time of day they strike, and with the size of the geographic area they affect. In Fig. 3.2, below the predominant effects of disasters are characteristics labeled Nature of Disaster. Each of these characteristics affects the level of vulnerability and the likelihood of resiliency for the population and community affected by disaster. Disasters also differ in severity, surprise, and duration. If the characteristics of the disaster do not include extreme severity, extent, and duration, a resilient disaster recovery may result in population and community adaptation and wellness.

Sudden, severe, and long-lasting disasters, on the other hand, make a resilient recovery less likely. For example, Hurricanes Katrina and Rita struck Southeastern Louisiana in 2005, resulting in the Murphy Oil Spill in St. Bernard Parish (county). These natural and technological hazards and disasters were followed by the Gulf Oil Spill in 2010. A large area of the Central Gulf Coast was covered and the oil spill disasters affected the area for years. All of these disasters were severe, and the Gulf Oil Spill affected the same geographic region as the other three disasters. The impacts of these four related disasters included high loss of life and property, injury, and illness.

RESOURCES

Resilience resources are part of causal chains and are intervening adaptive capacities. Although hazards and unsafe conditions result in a disaster, resources can moderate the effects of disaster. Availability and access to these resources after a disaster by all constituent populations moves the community along a trajectory of resilient recovery. Such a trajectory has the outcomes of adaptation and wellness. Because both vulnerability and resiliency theorists and researchers have emphasized access to resources and capabilities as a means of coping with disasters (Norris et al., 2008; Wisner et al., 2004, 2012), a number of resources and resource types have been identified. For example, the causal chain that follows economic inequality ends in the resources of environmental movements and community action. These capabilities may include well-managed volunteer programs, an important community resource (Zakour, 1996; Zakour, Gillespie, Sherraden, & Streeter, 1991).

As discussed in the resiliency perspective section the four basic types of resiliency capabilities are economic, social capital, collective action, and information and communication (Norris et al., 2008). When these adaptive capacities are networked and accessed, and become capacities, they act as intervening variables between severity of disaster exposure and the probability of a resilient recovery. For example, alternate livelihoods are necessary for households to resiliently recover from disaster. Some livelihoods, such as tourism and natural resource extraction (fishing, lumbering, hunting, and mining), will be unavailable due to a disaster's damage to the natural environment. Disasters damage economic resources and, without alternate livelihoods, people will be less able to access physical capital (Zakour & Gillespie, 2013). The tourism industry in Southeast Louisiana, particularly in New Orleans, was badly damaged by Hurricanes Katrina and Rita in 2005. Fishing and hunting livelihoods were severely affected after the Murphy (2005) and Gulf (2010) oil spills.

Employment for community members in the disaster recovery effort is one possible means of developing alternate livelihoods for people. Recovery organizations springing from community action and social movements in the local community can hire community members for wages. Even if the community suffers from high levels of economic inequality and social stratification, economic resources and alternative livelihoods can be generated after a disaster. Employment in disaster recovery can help reduce the social inequalities that originally led people to engage in fragile livelihoods.

SUMMARY AND CONCLUSIONS

V+ theory has tremendous potential as a guide to disaster loss reduction. The theory of vulnerability is now more balanced with an emphasis on both predisaster (vulnerability) and postdisaster (resilience) developments. In V+ theory the classification of phenomena into either liabilities or capabilities has helped build continuity across different fields of research. The concepts of V+ theory have been defined operationally, broadly classified into types, and positioned in time. Subsuming the process of resiliency into vulnerability theory is expected to bring more rigorous specification of the concept because in the past resiliency has often been vaguely defined or used as an inspirational idea (Zakour & Gillespie, 2013).

Over the past 20 years, vulnerability and resiliency researchers have become aware of the correspondence between these two concepts, and they have hinted at the integration achieved above (McEntire, 2002; Norris et al., 2008; Wisner et al., 2004, 2012). Empirical support and other evidence for V+ theory has come from different disciplines and professions, including social development, community psychology, sociology, social work, public health, geography, anthropology, and political science (see Table 3.1). Recent vulnerability research has supported a number of relationships among variables (Fig. 3.1). Causal and other relationships among these variables can now be identified and tested. V+ theory is better able to encompass the multidimensionality of disasters. This new theory can exploit potential leverage points for modifying systems, in order to reduce the level of vulnerability and increase the likelihood of resilient recoveries (Zakour & Gillespie, 2013).

Although the resources needed for safe conditions have been examined by vulnerability researchers, and resiliency resources have been largely independently identified by resilience theorists, each theory is fundamentally made up of the same types of resources and capabilities. Those capabilities, which make up both the safety of conditions and resiliency, are the most obvious commonalities between vulnerability and resiliency theories. Capacity variables may offer the best points for intervention to reduce vulnerability and foster resiliency. These capabilities are capitals such as social, human, cultural, economic, political, physical, and natural capital. It is the presence or absence of these types of capital that determine both the safety of conditions and allow access to networked adaptive capacities.

Capacities can be developed and improved in quality to both reduce the level of vulnerability and build resiliency. Making general or inherent resilience capacities available prior to a disaster reduces the level of vulnerability. Building these capacities will also improve the safety of conditions including locations and livelihoods. This will increase availability of resiliency resources. In order to improve the quality of resources, robustness, rapidity, and redundancy of resources must be improved. This is especially true if resources are to remain accessible during and following disaster. These resources must also be networked, or linked to one another, if community resiliency as process is to be achieved. If resource quantity and quality, in addition to their linkages, are developed, disaster vulnerability can be reduced, and communities will move more rapidly and surely to a resilient recovery, and to adaptation and wellness.

REFERENCES

Bankoff, G. (2004). The historical geography of disaster: "Vulnerability" and "local knowledge in Western discourse." In G. Bankoff, G. Frerks, & D. Hilhorst (Eds.), *Mapping vulnerability: Disasters, development & people* (pp. 25–36). London: Earthscan.

Barton, A. H. (1969). *Communities in disaster. A sociological analysis of collective stress situations*. Garden City, NY: Doubleday.

Barton, A. H. (2005). Disaster and collective stress. In R. W. Perry, & E. L. Quarantelli (Eds.), *What is a disaster? New answers to old questions* (pp. 125–152). Bloomington, IN: Xlibris.

Benight, C. C., Ironson, G., & Durham, R. L. (1999). Psychometric properties of a hurricane coping self-efficacy measure. *Journal of Traumatic Stress, 12*(2), 379–387.

Blaikie, P., & Brookfield, H. (1987). *Land degradation and society*. London: Methuen.

Blaikie, P.M., Cameron, J., & Seddon, J. (1979). *The struggle for basic needs in Nepal*. Paris: Development Centre Studies.

Blaikie, P., Cameron, J., & Seddon, D. (1980). *Nepal in crisis: Growth and stagnation at the periphery*. Oxford: Clarendon.

Blaikie, P., Cannon, T., Davis, I., & Wisner, B. (1994). *At risk: Natural hazards, people's vulnerability, and disasters*. London: Routledge.

Bolin, R. C. (2007). Race, class, ethnicity, and disaster vulnerability. In H. Rodriguez, E. L. Quarantelli, & R. R. Dynes (Eds.), *Handbook of disaster research* (pp. 113–129). New York: Springer.

Bonanno, G. A., Galea, S., Bucciarelli, A., & Vlahov, D. (2007). What predicts psychological resilience after disaster? The role of demographics, resources, and life stress. *Journal of Consulting and Clinical Psychology, 75*(5), 671–682.

Borden, K. A., Schmidtlein, M. C., Emrich, C. T., Piegorsch, W. W., & Cutter, S. L. (2007). Vulnerability of U. S. cities to environmental hazards. *Journal of Homeland Security and Emergency Management, 4*(2), 1–21.

Bottero, W. (2007). Social inequality and interaction. *Sociology Compass, 1/2*, 814–831.

Burnside, R., Miller, D. S., & Rivera, J. D. (2007). The impact of information and risk perception on the hurricane evacuation decision-making of greater New Orleans residents. *Sociological Spectrum, 27*, 727–740.

Cardona, O. D. (2004). The need for rethinking the concepts of vulnerability and risk from a holistic perspective: A necessary review and criticism for effective risk management. In G. Bankoff, G. Frerks, & D. Hilhorst (Eds.), *Mapping vulnerability: Disasters, development & people* (pp. 37–51). London: Earthscan.

Chakraborty, J., Tobin, G. A., & Montz, B. E. (2005). Population evacuation: Assessing spatial variability in geophysical risk and social vulnerability to natural hazards. *Natural Hazards Review, 6*(1), 23–33.

Collins, T. W. (2008a). What influences mitigation? Household decision making about wildfire risks in Arizona's White Mountains. *The Professional Geographer, 60*(4), 508–526.

Collins, T. W. (2008b). The political ecology of hazard vulnerability: Marginalization, facilitation and the production of differential risk to urban wildfires in Arizona's White Mountains. *Journal of Political Ecology, 15*, 21–43.

Cutter, S. L., Boruff, B. J., & Shirley, W. L. (2003). Social vulnerability of environmental hazards. *Social Science Quarterly, 84*(1), 242–261.

Gillespie, D. F. (2008a). In Disasters. *Encyclopedia of social work*, (20th ed. Vol. 2, pp. 60–65). New York: Oxford University Press.

Gillespie, D. F. (2008b). *Theories of vulnerability: Key to reducing losses from disasters. Proceedings of the 21st International Conference of Social Work. Social work and human welfare in a changeable community* (pp. 15–26). Cairo: Helwan University.

Gillespie, D. F. (2010). Vulnerability: The central concept of disaster curriculum. In D. F. Gillespie, & K. Danso (Eds.), *Disaster concepts and issues: A guide for social work education and practice* (pp. 3–14). Alexandria, VA: Council on Social Work Education.

Gillespie, D. F., Colignon, R. A., Banerjee, M. M., Murty, S. A., & Rogge, M. (1993). *Partnerships for community preparedness (Program on Environment and Behavior Monograph No. 54)*. Boulder, CO: University of Colorado, Institute of Behavioral Science.

Gillespie, D. F., & Murty, S. A. (1994). Cracks in a post-disaster service delivery network. *American Journal of Community Psychology, 22*(5), 639–660.

Greene, R. R., & Livingston, N. C. (2002). A social construct. In R. R. Greene (Ed.), *Resiliency: An integrated approach to practice, policy, and research* (pp. 63–93). Washington, DC: NASW Press.

Girard, C., & Peacock, W. G. (1997). Ethnicity and segregation: Post-hurricane relocation. In W. G. Peacock, B. H. Morrow, & H. Gladwin (Eds.), *Hurricane Andrew: Ethnicity, gender and the sociology of disasters* (pp. 191–205). London: Routledge.

IPCC. (2012). *Managing the risks of extreme events and disasters to advance climate change adaptation (A Special Report of Working Groups I and II of the Intergovernmental Panel on Climate Change)*. Cambridge, UK: Cambridge University Press.

Kaniasty, K., & Norris, F. H. (2009). Distinctions that matter: Received social support, perceived social support, and social embeddedness after disasters. In Y. Neria, S. Galea, & F. H. Norris (Eds.), *Mental health and disasters* (pp. 175–200). Cambridge: Cambridge University.

Klinenberg, E. (2002). *Heat wave: A social autopsy of disaster in Chicago*. Chicago, IL: University of Chicago.

Laska, S. (2012). Dimensions of resiliency: Essential resiliency, exceptional recovery and scale. *International Journal of Critical Infrastructures, 8*(1), 47–62.

McEntire, D. A. (2002). Coordinating multi-organizational responses to disaster: Lessons from the March 28, 2000, Fort Worth Tornado. *Disaster Prevention and Management, 11*(5), 369–379.

McEntire, D. A. (2004a). Development, disasters and vulnerability: A discussion of divergent theories and the need for their integration. *Disaster Prevention and Management, 13*(3), 193–198.

McEntire, D.A. (2004b, June 8). *The status of emergency management theory: Issues, barriers, and recommendations for improved scholarship*. Paper submitted at the FEMA Higher Education Conference, Emmitsburg, MD.

McEntire, D. A. (2005). Why vulnerability matters: Exploring the merit of an inclusive disaster reduction concept. *Disaster Prevention and Management, 14*(2), 206–222.

Mascarenhas, A., & Wisner, B. (2012). Politics: Power and disasters. In B. Wisner, J. C. Gaillard, & I. Kelman (Eds.), *The Routledge handbook of hazards and disaster risk reduction* (pp. 48–59). New York: Routledge.

McGuire, L. C., Ford, E. S., & Okoro, C. A. (2007). Natural disasters and older US adults with disabilities: Implications for evacuation. *Disasters, 31*(1), 49–56.

Meichenbaum, D. (1997). *Treating Post-traumatic stress disorder: A handbook and practice manual for therapy*. New York: Wiley.

Mitchell, J. T., Thomas, D. S. K., & Cutter, S. L. (1999). Dumping in Dixie revisited: The evolution of environmental injustices in South Carolina. *Social Sciences Quarterly, 80*(2), 229–243.

Neria, Y., Galea, S., & Norris, F. H. (2009). Disaster mental health research: Current state, gapes in knowledge, and future directions. In Y. Neria, S. Galea, & F. H. Norris (Eds.), *Mental health and disasters* (pp. 594–610). Cambridge: Cambridge University.

Norris, F. H., Galea, S., Friedman, M. J., & Watson, P. J. (Eds.). (2006). *Methods for disaster mental health research*. New York: Guilford.

Norris, F. H., Murphy, A. D., Kaniasty, K., Perilla, J. L., & Ortis, D. C. (2001). Postdisaster social support in the United States and Mexico: Conceptual and contextual considerations. *Hispanic Journal of Behavioral Sciences, 23*(4), 469–497.

Norris, F. H., Stevens, S. P., Pfefferbaum, B., Wyche, K. F., & Pfefferbaum, R. L. (2008). Community resilience as a metaphor, theory, set of capacities, and strategy for disaster readiness. *American Journal of Community Psychology, 41*(1/2), 127–150.

Oliver-Smith, A. (2004). Theorizing vulnerability in a globalized world: A political ecological perspective. In G. Bankoff, G. Frerks, & D. Hilhorst (Eds.), *Mapping vulnerability: Disasters, development and people* (pp. 10–24). London: Earthscan.

Oliver-Smith, A. (2009). Anthropology and the political economy of disasters. In E. C. Jones, & A. D. Murphy (Eds.), *The political economy of hazards and disasters* (pp. 11–28). Lanham, MD: Altimira Press.

Peacock, W. G., & Girard, C. (1997). Ethnic and racial inequalities in hurricane damage and insurance settlements. In W. G. Peacock, B. H. Morrow, & H. Gladwin (Eds.), *Hurricane Andrew: Ethnicity, gender and the sociology of disasters* (pp. 171–190). London: Routledge.

Pulwarty, R. S., Broad, K., & Finan, T. (2004). El Niño events, forecasts and decision-making. In G. Bankoff, G. Frerks, & D. Hilhorst (Eds.), *Mapping vulnerability: Disasters, development & people* (pp. 83–98). London: Earthscan.

Putnam, R. D. (2000). *Bowling alone: The collapse and revival of American community*. New York: Simon & Shuster.

Queiro-Tajalli, I., & Campbell, C. (2002). Resilience and violence at the macro level. In R. R. Greene (Ed.), *Resiliency: An integrated approach to practice, policy, and research* (pp. 217–240). Washington, DC: NASW Press.

Renfrew, D. E. (2009). In the margins of contamination: Lead poisoning and the production of neoliberal nature in Uruguay. *Journal of Political Ecology, 16*, 87–103.

Renfrew, D. E. (2012). New hazards and old disease: Lead contamination and the Uruguayan battery industry. In C. Sellers, & J. Melling (Eds.), *Dangerous trade: Histories of industrial hazard across a globalizing world* (pp. 99–111). Philadelphia, PA: Temple University.

Robbins, P. (2012). *Political ecology: A critical introduction* (2nd ed.). Malden, MA: Wiley-Blackwell.

Rogge, M. E. (1996). Social vulnerability to toxic risk. In C. L. Streeter, & S. A. Murty (Eds.), *Research on social work and disasters* (pp. 109–129). New York: Haworth.

Rosenberg, M. (1968). *The logic of survey analysis*. New York: Basic Books.

Rosenfeld, L. B., Caye, J. C., Ayalon, O., & Lahad, M. (2005). *When their world falls apart: Helping families and children manage the effects of disasters*. Washington, DC: National Association of Social Workers.

Rüstemli, A., & Karanci, A. N. (1999). Correlates of earthquake cognitions and preparedness behavior in a victimized population. *The Journal of Social Psychology, 139*(1), 91–101.

Solnit, R. (2009). *A paradise built in hell: The extraordinary communities that arise in disaster*. New York: Penguin Books.

Simonovic, S. P., & Ahmad, S. (2005). Computer-based model for flood evacuation emergency planning. *Natural Hazards, 34*, 25–51.

Streeter, C. L. (1992). Redundancy in organizational systems. *Social Service Review, 66*(1), 97–111.

Sundet, P. A., & Mermelstein, J. (2000). Sustainability of rural communities: Lessons from natural disaster. In M. J. Zakour (Ed.), *Disaster and traumatic stress research and intervention* (Vols. 21–22, pp. 25–40). New Orleans, LA: Tulane University.

Taylor, M. (2015). *The political ecology of climate change adaptation: Livelihoods, agrarian change and the conflicts of development (Routledge explorations in development studies)*. New York: Routledge.

Tedeschi, R. G., & Calhoun, L. G. (2004). Posttaumatic growth: Conceptual foundations and empirical evidence. *Psychological Inquiry, 15*(1), 1–18.

Thomas, N. D., & Soliman, H. H. (2002). Preventable tragedies: Heat disaster and the elderly. *Journal of Gerontological Social Work, 38*(4), 53–66.

Tierney, K. J. (2014). *The social roots of risk: Producing disaster, promoting resilience (High reliability and crisis management series)*. Stanford, CA: Stanford University Press.

Turner, B. L., et al. (2003). Illustrating the coupled human–environment system for vulnerability analysis: Three case studies. *Proceedings of the National Academy of Sciences, 100*(14), 8080–8085.

Turner, M. D. (2014). Political ecology I: An alliance with resilience? *Progress in Human Geography, 38*(4), 616–623.

Wallace, A. F. C. (1956/2003). Mazeway resynthesis: A biocultural theory of religious inspiration. In R. S. Grumet (Ed.), *Revitalizations & mazeways: Essays on culture change* (Vol. 1). Lincoln, NE: University of Nebraska.

Wallace, A. F. C. (1956). Revitalization movements. *American Anthropologist, 58*(2), 264–281.

Wallace, A. F. C. (1957). Mazeway disintegration: The individual's perception of sociocultural disorganization. *Human Organization, 16*(2), 23–27.

Wisner, B. (2016). *Vulnerability as concept, model, metric, and tool. Oxford Research Encyclopedia of Natural Hazard Science, August*. New York: Oxford University Press. Available from <http://naturalhazardscience.oxfordre.com/view/10.1093/acrefore/9780199389407.001.0001/acrefore-9780199389407-e-25>

Wisner, B., Blaikie, P., Cannon, T., & Davis, I. (2004). *At risk. Natural hazards, people's vulnerability and disasters* (2nd ed.). New York: Routledge.

Wisner, B., Gaillard, J. C., & Kelman, I. (2012). Framing disaster: Theories and stories seeking to understand hazards, vulnerability and risk. In B. Wisner, J. C. Gaillard, & I. Kelman (Eds.), *The Routledge handbook of hazards and disaster risk reduction* (pp. 18–34). New York: Routledge.

Zakour, M. J. (1996). Geographic and social distance during emergencies: A path model of interorganizational links. *Social Work Research, 20*(1), 19–29.

Zakour, M. J. (2008). *Vulnerability to climate change in the Nile Delta: Social policy and community development interventions*. Proceedings of the 21st International Conference of Social Work. Social work and human welfare in a changeable community (pp. 425–451). Cairo: , Helwan University.

Zakour, M. J. (2010). Vulnerability and risk assessment: Building community resilience. In D. F. Gillespie, & K. Danso (Eds.), *Disaster concepts and issues: A guide for social work education and practice* (pp. 15–60). Alexandria, VA: Council on Social Work Education.

Zakour, M. J. (2012). Natural and human-caused disasters. In L. M. Healy, & R. J. Link (Eds.), *Handbook of international social work: Human rights, development, and the global profession* (pp. 226–231). Oxford: Oxford University Press.

Zakour, M. J., & Gillespie, D. F. (1998). Effects of organizational type and localism on volunteerism and resource sharing during disasters. *Nonprofit and Voluntary Sector Quarterly*, 27(1), 49–65.

Zakour, M. J., & Gillespie, D. F. (2013). *Community disaster vulnerability: Theory, research, and practice*. New York: Springer Science.

Zakour, M. J., Gillespie, D. F., Sherraden, M. S., & Streeter, C. L. (1991). Volunteer organizations in disasters. *Journal of Volunteer Administration*, 9(2), 18–28.

Zakour, M. J., & Harrell, E. B. (2003). Access to disaster services: Social work interventions for vulnerable populations. *Journal of Social Service Research*, 30(2), 27–54.

CHAPTER 4

A systems approach to vulnerability and resilience in post-Katrina New Orleans

Nancy B. Mock[1], Melissa Schigoda[2] and Paul Kadetz[3]

[1]*Tulane University, New Orleans, LA, United States*
[2]*City of New Orleans, New Orleans, LA, United States* [3]*Drew University, Madison, NJ, United States*

CHAPTER OUTLINE

Systems Approach to Vulnerability and Resilience 79
    Key Features of Complex Systems in Post-Katrina Recovery 81
Recovery as a Complex Adaptive Social System Problem 85
Signals and Information in Post-Katrina Recovery 86
    Information Blackout 87
    Monitoring Recovery: *The New Orleans Index* 87
    Resiliency, Recovery Planning and Collective Action 92
    Aid, Culture, and Resilience 93
    Resilience and Footprint 93
    In Need of a "New Deal" 93
Conclusion: The Road to Complex Recovery 94
References 95
Further Readings 96

SYSTEMS APPROACH TO VULNERABILITY AND RESILIENCE

Systems thinking provides a framework for understanding the complex linkages between vulnerability and resilience in the response to and recovery from disasters. Social systems cannot be accurately represented as a closed box, in which variables behave in a predictable, unilinear, unidirectional manner. Social systems are dynamic, open systems in which components interact in unpredictable, multidimensional, and multilinear ways. In other words, social systems are complex rather than simple systems. Complex systems thinking facilitates an understanding of open dynamic systems in which independent and dependent variables can interact with one another in unpredictable ways forming new variables and outcomes via a process known as emergence. This complexity is not only evidenced in the extremely dynamic situations of disasters, but also can be witnessed in the interplay between local, state, and federal governance. Thus far, the opportunity to understand the post-Katrina recovery of New Orleans as a Complex Adaptive Social System (CASS) has been largely overlooked. This chapter focuses on disaster recovery as a complex systems issue and argues that although many organizations and subsystems demonstrated resilience during the response and recovery process of Hurricane Katrina, the most important drivers of vulnerability—of the socio-ecological system of New Orleans—were never intentionally and strategically targeted. In this chapter, we examine how a complex systems framework can facilitate an understanding of both the recovery process that unfolded after Hurricane Katrina and why this recovery has been so uneven.

Systems thinking has generated great interest in science and professional studies as a means to integrate the understanding of human and social interaction with the environment in order to produce sustainable outcomes for individuals and communities (Mock et al., 2015; Snyder, 2013). Systems thinking builds on the platforms of complexity "science," chaos theory, cybernetics, and general systems theory. Cybernetics, or the st'udy of feedback in regulating systems, was first defined by Weiner (1948) and focuses on the interaction of system components regulated by feedback. General systems theory, as elaborated by van Bertanaffy—articulated in his landmark 1968 text—identified systems as a set of independent, but interacting, parts that perform a function or pursue a mission. These early efforts draw a distinction between closed systems, that were highly predictable and deterministic, and open systems; where individual components or agents have their own motives that self-organize through their interactions and thereby, can facilitate the emergence of new system states. Key to this emergence is positive and negative feedback. Positive feedback moves the system state toward new goals, while negative feedback keeps system states inert.

Complex Adaptive Systems (CAS) are open evolving nonlinear systems that cannot be reduced to their component parts "because their behaviour is defined to a large extent by local interactions between their components" (Rihani, 2002, p. 7). CAS exhibit several properties: (1) chaos can be created by interfering with whatever order has been established in a CAS; (2) for a CAS to survive and evolve, it must be allowed to change according to changing conditions; (3) CAS often change at their own leisurely pace independent of attempted outside intervention; (4) CAS do not organically evolve toward an optimal endstate (thereby defying teleological interpretations of progress); and (5) in CAS, complexity often increases with time (Rihani, 2002, pp. 7–9). Thus, all of these properties of CAS would seriously hinder any attempts to control them, which according to (1), may engender chaos. The concept of CAS, applied to how social-ecological systems could be managed (Alhadeff-Jones, 2008), emerged as a way to direct social systems toward more resilient states (Auspous & Cabaj, 2014). The intentionality of moving CAS toward particular goals became known as a Complex Adaptive Social System (CASS).

Snowden and Boone (2007) have applied these concepts of systems and intentionality toward leadership decision-making by identifying problems according to their system properties. Their framework termed *Cynafin* provided a simple schematic of different types of problems: simple, complicated, complex, chaotic, and indeterminant as a unifying framework for integrating systems theories with practical management decision-making. These categories can be equated with a certainty level of decision-making, in which the answers to problems are:

1. Known and only require sensing the problem, categorizing, and responding. These are best practices for solving the problem once the problem is clearly identified.
2. Knowable and require sensing, analyzing, and responding.
3. Unknown and require probing, sensing, and responding.
4. Chaotic and require response, sensing, and responding.

Complex systems have a number of characteristics, which are important to understand for developing disaster response and recovery decisions. They are characterized by: different levels and scales of the system components; interactions among these various levels and scales; feedback loops that condition these interactions and create networks; and nonlinearity of change of the larger system based upon these interactions giving rise to tipping points that result in emergence or sudden system change. All of this occurs within a system of systems context, in which there are several systems within the larger system of interest. Each of these subsystems is also subject to the same properties and dynamics of the main system. These subsystems can be described as part of nested hierarchies of systems. We discuss these in Table 4.1, highlighting how each of these features applies to the analysis of Katrina recovery, providing insights to the vulnerability and resilience of New Orleans and its component subsystems.

KEY FEATURES OF COMPLEX SYSTEMS IN POST-KATRINA RECOVERY

We identify several key features of complex systems in general and the complex system of the Hurricane Katrina disaster in particular.

1. We consider the *level and scale of system components*. These levels include: the social level extending from the individual to society; the ecological level from patch to landscape; the jurisdictional level from local to national and beyond; and the temporal level from daily to annually. Specific examples from the Katrina event include: how the post-Katrina response and recovery involved an array of actors at all societal levels, including the involvement of global organizations; the ecological scale of the disaster was vast, but microenvironments were varied according to level of flooding and sociotechnical characteristics; and recovery of various aspects of the city proceeded on different time scales. For example, essential services were established in nonflooded areas within a month, resulting in those areas of the city recovering more quickly. In general, the flooded areas of the city recovered at a pace predominantly linked to the financial and social capital of the neighborhood.

Table 4.1 Indicators Included in *The New Orleans Index* From August 2009

Category	Indicator
Population recovery	
	Total Population Estimates
	Residential Addresses Actively Receiving in New Orleans Metropolitan Statistical Area (MSA) by Parish
	Public School Enrollment Totals in New Orleans MSA by Parish
	Composition of Public School Students in New Orleans MSA by Parish
	Private School Enrollment Totals in New Orleans MSA by Parish
	Composition of Private School Students in New Orleans MSA by Parish
	College Students Enrolled in New Orleans by University
Housing market	
	Number of Single Family Home Sales in New Orleans Metro Area
	Average Sale Price of Single Family Homes in New Orleans Metro Area
	Active Listings of Single Family Homes in New Orleans Metro Area
	Average Days on Market for Single Family Homes in New Orleans Metro Area
	Delinquency and Foreclosure Rates for Sub-prime, Prime and All Loans in Louisiana and the U.S.
	Fair Market Rents in New Orleans MSA by Unit Bedrooms
	Gross Median Rents in New Orleans MSA (in 2007 dollars)
	Affordable Monthly Rent for Select Occupations in New Orleans MSA in 2008
Rebuilding damaged housing stock	
	Residential Building Permits Issued by New Orleans City Hall
	Number of New Residential Housing Units Authorized by Type of Home
	Unoccupied Residential Addresses by Parish
	Status of Louisiana Road Home Applications
	Number of Road Home Closings by Option Selected by Parish
	Number of Active Travel Trailers, Mobile Homes & Park Models in Louisiana
	Total Number of Active Travel Trailers, Mobile Homes & Park Models in Louisiana by Parish
Fiscal and economic conditions	
	Total Sales Tax Collections by Parish
	City of New Orleans Sales Tax Collections by Source
	Labor Force Size
	Unemployment Rates & Total Numbers
	Number of Non-Farm Jobs, in Thousands
	Number of Non-Farm Jobs by Source and Type of Employment, in Thousands
	Number of Non-Farm Jobs in Service-Providing Industries, in Thousands: A–L
	Number of Non-Farm Jobs in Service-Providing Industries, in Thousands: M–Z
	Number of Non-Farm Jobs in Goods-Producing Industries, in Thousands

(*Continued*)

Table 4.1 Indicators Included in *The New Orleans Index* From August 2009 *Continued*

Category	Indicator
	Number of Non-Farm Jobs in Government by Level of Government, in Thousands
	Employment in New Orleans by Industry Sectors
	Average Weekly Wage by Industry Sectors
	Net Change in Total Employers by Parish
	Job Vacancy Rates in New Orleans Regional Labor Market by Occupation
	Number of Unemployment Claims
	Personal Income in New Orleans MSA, Louisiana & the U.S. (in millions of dollars)
	Number of Passengers Arriving and Departing from Louis Armstrong New Orleans International Airport
	Cargo Activity at the Port of New Orleans
Quality and availability of basic public services	
	Open Public Schools in New Orleans Metro Area by Parish
	Open Private Schools in New Orleans Metro Area by Parish
	Open Public Schools in New Orleans by Management Type
	Composition of Public School Students in New Orleans by Management Type and Admissions Policy
	Public School Students Passing High-Stakes LEAP Tests in New Orleans MSA by Parish
	Status of Public Transportation in New Orleans
	Open State-licensed Hospitals by Parish
	Open Child Care Centers in New Orleans Metro Area and in Louisiana
	Open Public Libraries in New Orleans Metro Area by Parish
	Status of New Orleans Police Department Infrastructure
	Status of FEMA Public Assistance Grants for Louisiana by Parish
Recovery of New Orleans by neighborhood	
	Recovery Rate of Residential Addresses Actively Receiving Mail in Orleans Parish by Planning District
	Residential Addresses Actively Receiving Mail in Orleans Parish by Planning District
	Unoccupied Residential Addresses in Orleans Parish by Planning District, March 2009
	New Residential Construction in Orleans Parish by Planning District
	Residential Demolitions in Orleans Parish by Planning District
	Road Home Closings in Orleans Parish by Planning District, June 2009
	Total Employment Located in Orleans Parish by Planning District
	Total Employment Located in Orleans Parish by Planning District, by Industry
	Total Employment Located in Orleans Parish by Target Zone
	Recovery Rate of Residential Addresses Actively Receiving Mail in Orleans Parish by Neighborhood

2. *Cross-level and cross-scale interactions* are identified in the differing agendas at different levels and jurisdictions that led to negative feedback loops and inertia during response and early recovery. Post-Katrina, local faith-based organizations were supported by national and international faith-based organizations. In turn, faith-based organizations provided augmentation of evacuation services to the American Red Cross.
3. *Nested hierarchies* describe how systems are comprised of subsystems that are interconnected and interdependent on one another and the wider environment. Neighborhoods were important subsystems in post-Katrina recovery that shaped and were shaped by city, state, and national interactions.
4. *Feedback and signals* provide information that is critical to a Complex Adaptive Social System (CASS). Signals represent proxy measures for the state of systems, and feedback represents the interactions of system actors based on their use and application of information for action. Learning loops occur within this system of systems. Actors regulate their behavior in response to changes in their environment. Negative feedback can mean actors maintain the status quo, while positive feedback can enable emergence of new system states. In the specific case of Katrina, during the initial stages of response, feedback was hampered by lack of information flow because cellular and other communication systems were not operational for more than a month. Feedback for city-level signals was captured and monitored according to secondary data sources. Neighborhood-level recovery efforts were hampered by the lack of granular data; however, neighborhoods with the resources initiated their own grassroots efforts to gather basic data needed to guide resilient recovery.
5. *Threshold and tipping points* include changes in drivers of a system that cause emergence or changes in system states. In the context of post-Katrina New Orleans, the availability of key services in neighborhoods was a driver of repopulation. For example, emergence of a new K-12 educational system occurred when local governance via charter and private schools became the dominant governance model. However, some neighborhoods still have not achieved tipping points of service availability to attract residents to repopulate those areas.
6. *Social networks* form the groundwork for the key resilience component of social capital. Networks bond, bridge, and link social capital. Bonding consists of networking within communities. Bridging occurs in networking between communities. Linking is vertical networking. These networks lead to the collective action that is required for recovery. In post-Katrina New Orleans, networks were a crucial driver of the recovery process. Bonding social capital was particularly evident in faith-based organizations and through neighborhood associations, which resulted in collective action. Some ethnic groups, such as the Vietnamese community of New Orleans East, were particularly effective in building bonding, bridging, and linking capital through utilizing ethnic identity ties to achieve collective action (see Chapters 11: Resilience among vulnerable populations: The neglected

role of culture and Chapter 13: Collective efficacy, social capital and resilience: An inquiry into the relationship between social infrastructure and resilience after Hurricane Katrina (in this volume)). Latino networks effectively mobilized people of Latin American origin for construction efforts. While some ethnic groups experienced lower levels of bonding social capital, resulting from years of structural violence and horizontal inequality. However, the complexity of the interplay between ethnic identity and resilience is best illustrated in the concept of *intersectionality* (see Chapter 9: Problematizing vulnerability: Unpacking gender, intersectionality, and thenormative disaster paradigm (in this volume) for a full discussion).

7. *Nonlinearity of system change* is related to thresholds and tipping points. Change in complex system states is typically nonlinear. Early post-Katrina recovery in the nonflooded areas occurred almost instantly when key essential services were available. Nonlinear change and increasing vulnerability and unpredictability in ecosystem health are an outcome of climate change.
8. *Self-organization*, emergence of new system states, and unanticipated change are key features of CAS. In the case of post-Katrina New Orleans, self-organization at the grassroots level was prominent among neighborhood associations, faith-based organizations, and nonprofit groups.
9. Disaster recovery systems are *sensitive to the initial conditions* of the pre-event system. This is evidenced in the social and ecological vulnerability of New Orleans before Katrina struck. Environmental and social vulnerabilities interacted to produce an extremely complex recovery problem and the ensuing multidimensionality of vulnerabilities complicated recovery efforts.
10. *Attractors* are a set of values toward which a system evolves. For example, New Orleans has a strong cultural identity. Preserving this identity acted as a strong internal and external attractor.
11. System states are limited to a few drivers; usually no more than five. This is known as the *Rule of Hand (Yorque. et al., 2002)*. In the case of New Orleans, vulnerability is largely a function of the two drivers of ecological fragility and income inequality.

RECOVERY AS A COMPLEX ADAPTIVE SOCIAL SYSTEM PROBLEM

These key features illustrate why postdisaster recovery is best understood as a Complex Adaptive Social System (CASS) issue. Comfort, Bolton, and Stolcis were among the early disaster scholars who framed the study of disaster response and recovery as a CASS problem (Bolton & Stolcis, 2008; Comfort, Cigler, Birkland, & Nance, 2010). They link the intentionality of guiding the system to disaster response and recovery. Therefore, due to the important role of human behavior/leadership in guiding the development of social-ecological systems after a disaster, it is best to consider disaster response and recovery as a CASS problem. Comfort et al. (2010, p. 669) summarize:

> *The tasks of recovery can be viewed as generating a complex system of interacting jurisdictions, public agencies, private and nonprofit organizations, and households that are engaged in a shared effort to rebuild a community following disaster. The process is dynamic, as interactions among actors at any one point may facilitate or hinder possible actions of other actors at the next point of decision.*

They further argue that policy makers must learn to harness complexity to achieve a robust recovery. This requires a combination of building social capital toward collective action and constant situational awareness through good systems to monitor signals of the drivers of systems. These two key features of collective action toward some goals and constant situational awareness are imperative to understand the uneven recovery of New Orleans, as well as the way forward.

SIGNALS AND INFORMATION IN POST-KATRINA RECOVERY

We draw from the literature on Complex Adaptive Systems to describe the role of signals and information in disaster recovery systems. Stakeholders perceive information and signals from their environment and through other actors (Gell-Mann, 1994; Holland, 1995). They then filter or condense the information, incorporate it into their mental model (Holland, 1995) or schema (Gell-Mann, 1994), and act on that mental model/schema. Two types of knowledge are employed in this process. Explicit knowledge is commonly associated with scientific (or authoritative) knowledge, while tacit knowledge is derived from common observations and awareness of one's community. Because actors within a disaster recovery system simultaneously send signals to and receive signals from one another, they produce a large number of signals that must be sifted through (Holland, 2006). As actors process and act on this information, their actions impact other stakeholders, as well as their environment, and produce new signals and convey information to other parts of the system (Gell-Mann, 1994; Holland, 1995).

According to Mitchel (2009), the attributes of information processing in CAS can be summarized as follows. First, information processing involves spatial and temporal sampling, since no one agent can perceive the whole system at once. This has significant implications for how we understand systems and why neighborhoods became one of the most important basic building blocks of the recovery of New Orleans. People living in neighborhoods had a great deal of tacit knowledge about these neighborhoods. However, neighborhood residents had very little explicit knowledge at the neighborhood level on which to rely. Second, information processing is often random and probabilistic, starting with a series of random searches for signals. Residents were innovative with regard to their search for signals. During the initial recovery period, client feedback sessions were sometimes established to determine needs and priorities of residents. Third, because CAS are comprised of a large number of parts providing feedback to one another, information processing allows the system to continuously adapt, based on the information it obtains. In post-Katrina New Orleans the exchange of information across actors was ad hoc and there were few clearly identified knowledge hubs for these exchanges. Finally, CAS involve "a continual interplay of unfocused random explorations and focused actions driven by the systems perceived needs to obtain information and exploit that information to successfully adapt" (Mitchel, 2009, p. 182). In the case of Hurricane Katrina, random explorations were dampened by the lack of a central collection and analysis points for the synthesis of neighborhood level data.

INFORMATION BLACKOUT

One of the most striking features of Hurricane Katrina was that it completely disrupted telecommunications throughout New Orleans. Over 60% of telecommunications networks were inoperable, even nearly a month after the storm (Leitl, 2006). Cell phone service was interrupted for anyone who had a greater New Orleans (504) area code, regardless of where they were relocated. This was an outcome of the destruction of both telecommunications equipment and of backup power. Total failure of communications during response and early recovery resulted in handicapping local systems within the city and city-wide actors, thereby favoring those from the outside who had access to more information during the early recovery period. For the first few months after Katrina, residents had become accustomed to using street posters and notices on communications boards to exchange information. The disruption of information flows resulting from a disaster event significantly slows response and can disadvantage local actor/agents in damaged systems, as again witnessed after Hurricane Maria hit Puerto Rico.

MONITORING RECOVERY: *THE NEW ORLEANS INDEX*

To determine the effectiveness of the recovery effort, it must, as with any intervention, be monitored and evaluated. Appropriate monitoring of recovery also requires a complex systems approach. By late 2005, a recovery monitoring project was created at the systems level. Given the enormity and novelty of the New Orleans recovery, the Brookings Institute initiated a city-level recovery monitoring project in December 2005. It was initially named *The Katrina Index* and was later changed to *The New Orleans Index*. The Greater New Orleans Community Data Center (The Data Center), a local nonprofit with data expertise, began collaborating with Brookings Institution on the *Index* and later took full ownership of the project. This proved to be an important decision because the *Index* became owned and financed by local stakeholders, who were able to adapt the *Index* over time to emerging information needs.

Each edition of the report consisted of about 30—60 recovery indicators presented in data tables. In addition to the data tables, the report included a "Summary of Findings" or analysis of key recovery trends and remaining challenges. The data and analysis in the *Index* were designed "to serve as an independent, fact-based, resource for leaders to monitor and evaluate rebuilding efforts" (Liu, Mabanta, & Fellowes, 2006, p. 3). The recovery indicators that formed the basis of the *Index* were primarily derived from administrative data sources and tracked on a monthly basis. Sources included unemployment claims from the department of labor, home sales from the Louisiana Association of Realtors, and Army Corps demolitions from the US Army Corps of Engineers among others.

According to Schigoda (2016), *The New Orleans Index* was widely used by disaster recovery leaders, such as policy makers and the local public and nonprofit organizations, in several ways related to a complex systems approach to recovery. For instance, it was used to provide situational awareness of post-Katrina recovery at a systems level of analysis and served to provide a common-operating picture of recovery. The *Index* was also employed to identify and prioritize needs; to communicate and collaborate with others; and to make the case for new investments, policies, and programs through both advocacy and grant proposals. High percentages of disaster recovery leaders who used the *Index* reported that it positively impacted emerging conversations (including policy conversations) concerning cross-cutting issues in recovery, as well as in funding and other resources received by New Orleans.

The *Index* consistently illustrated the uneven repopulation of neighborhoods and the slow return of public services and housing. Fig. 4.1 illustrates the dramatic nature of early recovery failure to meet basic needs of lower income and working-class residents as measured by the *Index*. The data identify that poverty traps were being recreated across many sectors of post-Katrina New Orleans.

The *Index*'s indicators, publication schedule, authors, format, accompanying products, and even its name, changed over time, reflecting the dynamics of information need throughout the recovery process (see Table 4.2). The number

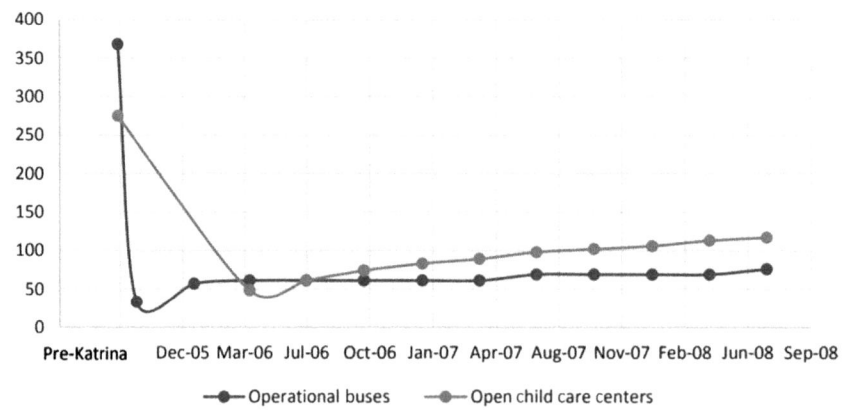

FIGURE 4.1

Operational Buses and Child Care Centers in New Orleans

Source: *From New Orleans Regional Transit Authority, Agenda for Children, and Louisiana Department of Social Services Bureau of Licensing.*

Table 4.2 Timeline of Changes in *The New Orleans Index*

Year	Month	Edition	Notes on Changes
2005	Dec	1	Brookings begins publishing *The Katrina Index*. The full report includes about 40 indicators, accompanied by a 2–3-page executive summary.
2006	Jan	2	
	Feb	3	
	Mar	4	
	Apr	5	
	May	6	
	Jun	7	
	Jul	8	
	Aug	9	At 16 pages, this first anniversary special edition of *The Katrina Index* is much shorter than the others published since December 2005. It included a variety of graphs and was not accompanied by an executive summary or data tables.
	Oct	10	*The Katrina Index* resumed its initial format with about 40 indicators, accompanied by a 2–3-page executive summary.
	Nov	11	
	Dec	12	
2007	Jan	13	The Data Center began formally collaborating with Brookings on *The Katrina Index*.
	Feb	14	
	Mar	15	
	Apr	16	The geographic focus of *The Katrina Index* became the City of New Orleans and the surrounding parishes, with data for Louisiana wherever possible. Mississippi was no longer included within the geographic scope. The recovery indicators were streamlined and organized into new categories including: housing, population, services and infrastructure, economy, and emergency response. Indicator data were mapped to highlight the differences in recovery by geography. The indicator analysis was bolstered by local insights provided by The Data Center.

(*Continued*)

Table 4.2 Timeline of Changes in *The New Orleans Index* Continued

Year	Month	Edition	Notes on Changes
	May	17	
	Jun	18	
	Aug	19	For the second anniversary edition the name was changed from *The Katrina Index* to *The New Orleans Index* with the goal of shifting the emphasis away from hurricane recovery to the rebuilding of an American city. New indicators were added and other indicators were eliminated as the authors learned more about the data sources available and the needs of their readers. A new section tracking programmatic response to rebuilding efforts, including data on the availability of key public services, programs, and infrastructure, was added. Data were provided for the key neighborhoods and commercial corridors targeted for redevelopment by the City of New Orleans' Office of Recovery Management.
	Nov	20	"At a glance" graphical representations were added, highlighting a small number of high-level indicators to direct attention to particularly important trends.
2008	Jan	21	
	Apr	22	
	Aug	23	The third anniversary edition of the *Index* included more than 50 indicators, and presented many key indicators by planning districts. There are 13 planning districts in New Orleans and these subdivisions were used in the development of the Unified New Orleans Plan. This version of the *Index* also included data from the United States Postal Service to estimate population recovery, population distribution, and unoccupied housing. Another addition was data from the Army Corps of Engineers on their completed and projected work to reduce flood risk in New Orleans.
2009	Feb	24	This edition included outcome data from the Road Home program (the federal program that offered residents in flooded areas a grant to rebuild or the opportunity to sell their property to the state and move elsewhere) and residential demolitions by planning district. It also included a map of the last 77 years of wetland loss in South Eastern Louisiana. To accompany this report, The Data Center released its first 10-minute briefing on New Orleans recovery, which presented the findings in a popular online video format.
	Aug	25	The fourth anniversary edition included several new types of data and maps. For example, it included population recovery indicators for all 76 neighborhoods in the City of New Orleans. These data were also made available at an even finer geographic grain via an interactive mapping system on The Data Center's website.

of indicators, grouping of the indicators, and the indicators themselves varied from edition to edition reflecting the dynamic and evolving recovery situation with the emergence of new factors to evaluate. Table 4.1 identifies the indicators and categories used to group the indicators from the August 2009 edition of the *Index*.

From December 2005 to June 2007, the *Index* was published on a monthly basis with few exceptions. The release of the second anniversary edition marked a transition to a quarterly publication and a name change from *The Katrina Index* to *The New Orleans Index*. In August 2008 a semiannual publication schedule was adopted. And in August 2009 the *Index* became an annual publication. The fourth anniversary edition, published in August 2009, included several new types of data and maps, including a map of population recovery indicators for all 76 New Orleans neighborhoods. The fifth anniversary edition, published in August 2010, focused on just 20 indicators organized into four sections: economic growth, inclusion, quality of life, and sustainability. In addition to tracking recovery over the past 5 years, this edition compared the post-Katrina recovery trends to pre-Katrina trends dating to 1980. Different geographic comparisons were employed including: parishes; the 7-parish classification; an average of 57 "weak city" metros considered to be peers of New Orleans; and national-level comparisons. This edition was accompanied by a series of seven essays written by leading local scholars and practitioners that aimed to systematically document major post-Katrina reforms. The sixth anniversary edition of the *Index* was released along with *Resilience and Opportunity: Lessons from the U.S. Gulf Coast after Katrina and Rita*, a book compiled by Brookings Institution Press that was developed from essays released with previous editions of the *Index*. The fifth, sixth, and eighth editions of the *Index* were accompanied by YouTube videos highlighting key trends in the data. Table 4.2 details some of these changes to the *Index* that developed over time.

The *Index* reflected the complex changing needs of information while maintaining consistent monitoring of core measures. Each publication also contained thematic analyses that were particularly germane to the information needs during the process of recovery, reflecting different information needs at a given time. However, the *Index* contained few indicators for federal assistance and other flows of assistance to New Orleans, which was attributed, by one of the *Index*'s creators, to a lack of access to data concerning resource flows (A. Liu, 2016, personal communication). The limited monitoring of recovery resource flows during the early recovery period proved an important limitation. Another limitation of the *Index* is that it provided very limited information on neighborhood-level recovery. Aside from the measurement of re-population of neighborhoods, there was not a systematic effort to measure recovery at the neighborhood level that would reflect the granularity of recovery that was actually being conducted at the neighborhood level.

RESILIENCY, RECOVERY PLANNING AND COLLECTIVE ACTION

Resiliency is an outcome of multidimensional factors and, thereby, also requires a complex systems perspective. While strong and consistent city-level information was available, as discussed, less information was available at the less aggregated level of the neighborhood. This resulted in hidden populations, such as the population of Latino migrant workers hidden in the central city neighborhood (for a full discussion, see Chapter 14 in this volume). This lack of neighborhood-level data could weaken the targeting of interventions to the most vulnerable and ultimately weaken their resilience.

Furthermore, the lack of a strategic recovery plan hampered the effectiveness of the city's initial recovery (Brand & Seidman, 2008). Communications among city-level agencies (planning and recovery) and between City and neighborhood organizations were also problematic. The City's investment resources for revitalizing neighborhoods have not sufficiently targeted the reduction of poverty and inequality.

The first phase of the city's recovery was not guided by a strategic recovery plan that aimed to address the drivers of vulnerability. While numerous planning processes took place, resulting in the Unified New Orleans Plan and later the adoption of the City Master Plan in 2010, it is difficult to say if consensus development of a common vision for the city was ever really achieved. Therefore the *Index* serves as a nagging reminder that, at the city system level, social and ecological vulnerability was being reconstructed. Even basic mitigation measures were not applied consistently.

There are, however, some rays of hope for eventual system transformation For example, K-12 education, entrepreneurialism, and some facets of the city's economy demonstrate progress. Whether city-wide resilience will emerge is still uncertain, but resilience will remain a challenge as long as the combination of socio-environmental vulnerability continues.

On the other hand, different component systems often achieved resilient outcomes. In heavily flooded neighborhoods, such as Lakeview, Broadmore, Mid-City, and the Village de L'Est, a combination of leadership, and in some cases financial resources, resulted in vibrant and resilient neighborhood redevelopment. Many local organizations also became more resilient because they became stronger learning organizations. The New Orleans Disaster Recovery Partnership (NODRP) was an ephemeral coordinating body aimed at improving cooperation and interorganizational learning. Attendance was good because many nonprofit organizations had a strong need for situational analysis, and in fact, most were stressed by the demands of piecing together a common-operating picture. As local organizations began competing for recovery funding, they were required to enhance their evaluation and learning capacities, which many have achieved (see Chapter 14: Dynamics of early recovery in two historically low-income New Orleans' neighborhoods: Tremé and Central City (in this volume), for a discussion of the experience of the RALLY Foundation in developing local monitoring and evaluation capacities). However, neighborhood organizations as well as local nonprofits started at a relatively low baseline of capacity compared with national norms. Thus, while these organizations may be considered stronger, their resilience in the face of significant social and environmental vulnerability remains uncertain.

AID, CULTURE, AND RESILIENCE

Post-Katrina resilience was also impacted by the dimension of aid relief. The unique culture of New Orleans and Southeast Louisiana encouraged response aid from the rest of the United States, North America, and the World. New Orleans had a very appealing culture including a unique array of music, food, architecture, and fêtes. This culture developed over centuries from the colonial past to the more recent history of the city. Dramatic electronic media coverage of the disaster and catastrophe gained the attention and sympathy of world audiences, and this sympathy and identification was reinforced by New Orleans intriguing culture. Tourism had been a significant factor in spreading exposure to New Orleans' unique culture. Perhaps the most important cultural output for the visibility of New Orleans came from music. Starting in the 1920s, New Orleans is considered the birthplace of jazz, arguable the most internationally recognizable indigenous American form of music. Cultural outputs, such as music itself, fosters resilience, as discussed in detail in Chapter 10, Culture and resilience: How music has fostered resilience in post-Katrina New Orleans (in this volume).

RESILIENCE AND FOOTPRINT

Since the 1960s, the built environment of levees and floodwalls encouraged the development of new residential neighborhoods and urban sprawl throughout New Orleans. Urbanization in these areas exposed many more people to the risks of flooding and hurricanes. Paradoxically, levees can create more severe disasters, because: they encourage people to live in unsafe conditions; they can inevitably fail; they cause land to subside; they hold water, which must be pumped out; and they channel water into less protected areas.

IN NEED OF A "NEW DEAL"

The greenspacing (ie., the attempted replacement of lower lying residential areas with nonresidential parks by the City council) of at-risk residential areas in New Orleans failed because there were no alternative options offered for those returning to homes and neighborhoods. People were told that they must defend the continued existence of their neighborhoods to avoid greenspacing. However, no funds were provided to facilitate relocation to safer areas of the city. New Orleans did not have the funds for this, the federal Road Home grants were not adequate for this task, and the federal government was involved in the costly invasion of Iraq at the time.

Although it was possible that a new Progressive Era could have helped people relocate to safer areas, and a "New Deal" type program could have offered aid for doing this, no reconstruction funds were available to promote the shrinking of the New Orleans footprint. Instead, people could either return to their pre-Katrina homes and try to rebuild, or they could sell their nearly worthless lots to developers and hope for some small amount of Road Home funds to relocate to another part of Louisiana or the United States. Clearly, these circumstances exacerbated vulnerability and compromised resilience.

CONCLUSION: THE ROAD TO COMPLEX RECOVERY

This chapter reviewed systems thinking and its implications for a better understanding of, and intervention with, postdisaster recovery. We reviewed systems thinking, building from an appreciation of general systems theory to the Complex Adaptive Social System (CASS) concept and the role of information/feedback. We examined the importance of intentionality and collective action as necessary for harnessing information for postdisaster recovery. We demonstrated how vulnerability and resilience are best understood as complex systems in themselves and are imperative to utilize in order to understand the greater complexity of postdisaster recovery.

While *The New Orleans Index* has proved a novel and reliable source of information for system recovery, there has not been a sufficiently strong coordination and decision structure to effectively utilize this information in order to address the key vulnerabilities of the city; although the Index was a valuable source of information for many organizations working at the level of subsystems of New Orleans. Though recovery was largely neighborhood driven, granular information at the neighborhood level was largely derived from tacit knowledge. Small-scale collective action and organizational learning through use of *The New Orleans Index* and other ad hoc sources did occur; however, these efforts would be better supported by more granular information that is specifically germane to the level of the system aimed to be improved. These oversights of the full complexity of post-Katrina recovery have led to mixed results in rebuilding a more resilient New Orleans. Lastly, unless we capture the key drivers of vulnerability and resilience in the recovery process, we may miss important information. However, one of the main tasks to emphasize is the differentiation of the important signals of resilient recovery from the noise. Although the signals identified for the *Index* are useful, it is not clear if these are the most appropriate for the measurement of resilient recovery. If we don't find signals at the appropriate levels of analysis, then we are probably not fully capturing the signals that are important for resilient recovery. A good measurement strategy requires that we identify the key signals and, for greater rigor, add local knowledge to the interpretation of these signals.

We conclude that the CASS framework is essential for recovery planning and management, as it is capable of stressing the importance of strategy and intentionality, in addition to monitoring key drivers of systems change. The determinants of New Orleans' vulnerability have been clear for decades. The increased resource flows during disaster response and recovery should be directed toward addressing key vulnerabilities. Thus, *The New Orleans Index* project would prove valuable in similar contexts. The management of the *Index* through a nonprofit entity with technical credibility would be recommended for its sustainability. In recognition of the complexity of disaster response and recovery, mechanisms that facilitate the collection and analysis of more granular information need to be prioritized in order to better guide postdisaster communities.

REFERENCES

Alhadeff-Jones, M. (2008). Three generations of complexity theories: nuances and ambiguities. *Educational Philosophy and Theory, 40*(1), 66−82.

Auspous, P., & Cabaj, M. (2014). *Complexity and community change: Managing adaptiability to improve effectiveness*. Washington, DC: The Aspen Institute Roundtable on Community Change, Aspen Institute.

Bolton, M. J., & Stolcis, G. B. (2008). Overcoming failure of imagination in crisis management: The complex adaptive system. *The Innovation Journal: The Public Sector Innovation Journal, 13*, 2−12.

Brand, L., & Seidman, K. (2008). *Assessing post Katrina recovery in New Orleans*. Boston, MA: Community Innovators Lab, Department of Urban Studies and Planning, Massachussets Institute of Technology.

Comfort, L. K., Cigler, B. A., Birkland, T. A., & Nance, E. (2010). Retrospectives and prospectives on Hurricane Katrina: Five years and counting. *Public Administration Review, 70*(5), 669−678.

Gell-Mann, M. (1994). *The Quark and the jaguar: Adventures in the simple and the complex*. New York, NY: W.H. Freeman and Co.

Holland, J. (1995). *Hidden order: How adaptation builds complexity*. New York: Helix Books.

Holland, J. (2006). Studying complex adaptive systems. *Journal of System Science & Complexity, 19*, 1−8. Available from <http://deepblue.lib.umich.edu/bitstream/handle/2027.42/41486/11424_2006_Article_1.pdf?sequence=1>.

Leitl, E. (2006). Information technology issues during and after Katrina and usefulness of the Internet: How we mobilized and utilized digital communications systems. *Critical Care, 10*(1). Available from <http://dx.doi.org/10.1186/cc3945>.

Liu, A., Mabanta, M., & Fellowes, M. (2006). *Katrina Index: Tracking variables of post-Katrina reconstruction*. Washington, DC: The Brookings Institution.

Mitchel, M. (2009). *Complexity: A guided tour*. New York: Oxford University Press.

Mock, N., Bene, C., Constas, M., Frankenberger, T. (2015). *Systems analysis in the context of resilience*. Resilience Measurement Technical Working Group. Technical Series No. 6. Rome: Food Security Information Network. Available from <http://www.fsincop.net/fileadmin/user_upload/fsin/docs/resources/FSIN_TechnicalSeries_6.pdf>.

Rihani, S. (2002). *Complex systems theory and development practice. Understanding non-linear realities.* London: Zed Books.

Schigoda, M. (2016). *The use and impact of disaster recovery indicators from the perspective of complex adaptive systems theory: A case study of the New Orleans Index* (Unpublished Dissertation), Tulane University.

Snowden, D.J., and Boone, M.E. (2007). A Leader's Framework for Decision Making. Available at <http://hbr.org/2007/11/a-leaders-framework-for-decision-making/>.

Snyder, S. (2013). The simple, the complicated and the complex: Educational reform through the lens of complexity theory. OECD Education Working Paper, No 96, OECD Publication, Available from <http://dx.doi.org/10.1787/5k3txnpt1lnr-en>.

Weiner, N. (1948). *Cybernetics: Or control and communications in the animal and the machine.* Paris: John Wiley and Sons.

Yorque, R., Walker, B., Holling, C., Gunderson, L., Folke, C., & Carpenter, S. (2002). Toward integrative synthesis. In L. Gunderson, & C. Holling (Eds.), *Panarchy: Understanding transformation of human and natural systems.* Washington, D.C., USA: Island Press.

FURTHER READINGS

Brown, D. S., Platt, S., & Bevington, J. (2010). *Disaster recovery indicators.* Cambridge: Cambridge University Centre for Risk in the Built Environment (CURBE).

Brown, D.S., Saito, K., Spence, R., Chenvidyakarn, T., Adams, B., Mcmillan, A., & Platt, S. (2008). Indicators for measuring, monitoring and evaluating post-disaster recovery. Paper presented at the 6th International Workshop on Remote Sensing for Disaster Applications. Pavia: University of Pavia.

Comfort, L. K. (1994). Self-organization in complex systems. *Journal of Public Administration Research and Theory, 4*(3), 393−410.

Comfort, L. K. (1999). *Shared risk: Complex systems in seismic response.* Oxford: Pergamon.

Comfort, L. K. (2002). Rethinking security: Organizational fragility in extreme events. *Public Administration Review, 62,* 98−107.

Comfort, L. K. (2003). Governance under fire: Organizational fragility in complex systems. In A. Roberts (Ed.), *Governance and public security* (pp. 113−127). New York: Campbell Public Affairs Institute, Maxwell School of Citizenship and Public Affairs, Syracuse University.

Comfort, L. K. (2006). Cities at risk: Hurricane Katrina and the drowning of New Orleans. *Urban Affairs Review, 41*(4), 501−516.

Comfort, L. K. (2007). Crisis management in hindsight: Cognition, communication, coordination, and control. *Public Administration Review, 67,* 189−197.

Comfort, L. K., Ko, K., & Zagorecki, A. (2003). *Modeling fragility in rapidly evolving disaster response systems.* Berkeley, CA: Institute of Governmental Studies, University of California at Berkeley.

Comfort, L. K., Ko, K., & Zagorecki, A. (2004). Coordination in rapidly evolving disaster response systems: The role of information. *American Behavioral Scientist, 48*(3), 295−313.

Further Readings

Comfort, L. K., Sungu, Y., Johnson, D., & Dunn, M. (2001). Complex systems in crisis: Anticipation and resilience in dynamic environments. *Journal of Contingencies & Crisis Management*, *9*(3), 144.

Comfort, L. K., Waugh, W. L., & Cigler, B. A. (2012). Emergency management research and practice in public administration: Emergence, evolution, expansion, and future directions. *Public Administration Review*, *72*(4), 539–547.

FEMA. (2011). *The national disaster framework*. Available from <http://www.fema.gov/national-disaster-recovery-framework>.

Holland, J. (2000). *Emergence: From chaos to order*. Cambridge: Perseus.

Holland, J. (2012). *Signals and boundaries: Building blocks for complex adaptive systems*. Cambridge, MA: MIT Press.

Holte, J. (Ed.). (1993). *Chaos: The new science*. London: Gustavus Adolphus College, University Press America.

CHAPTER 5

"Built-in" structural violence and vulnerability: A common threat to resilient disaster recovery*

Shirley Laska[1,2], Susan Howell[1] and Alessandra Jerolleman[2]

[1]*University of New Orleans, New Orleans, LA, United States*
[2]*Lowlander Center, Gray, LA, United States*

CHAPTER OUTLINE

Introduction	100
Important Terms and Concepts	100
Agency	100
Resiliency	102
Agency Disrespect as Structural Violence	103
Trauma	104
Delay	105
Examples of Structural Violence	106
"Louisiana Road Home" and New York City's "Build It Back"	106
Shrinking the Footprint or Not?—How *Not* to Have That Discussion	108
The Extreme Structural Violence on the Economically and Politically Powerless	111
Evacuation Nightmare	111
Health Care Scarcity	112
Postdisaster Housing	113
How to Stop the Violence	114
Owner-Occupied Housing Recovery	114
Recovery Planning	117
Successful Evacuation That Supports Return	118
Expediently Provided, Postdisaster Health Care	121

---

*This chapter has been expanded with the assistance of the two other authors, Howell and Jerolleman, from presentations made by the first author in the Hurricane Katrina plenary session of the July, 2015 University of Colorado's Natural Hazards Workshop. It was then also presented to the White House Sub-Committee on Disaster Reduction, November 3, 2015 by invitation of the Sub-Committee. Because of the project's initial purpose—plenary presentation, the data for the examples presented herein of structural violence were drawn from media reports so that the visuals could be shown in the PowerPoint presentations for those events.

Creating Katrina, Rebuilding Resilience. DOI: http://dx.doi.org/10.1016/B978-0-12-809557-7.00005-3
© 2018 Elsevier Inc. All rights reserved.

Survivor Aid and Return of Agency to Economically and Politically
Disenfranchised 121
Concluding Remarks 123
References 125

INTRODUCTION

"The most important actors in disaster recovery are the survivors themselves."

This chapter builds on the above statement, a seeming truism that has *not* inspired the United States to respond to disasters and disaster victims as if the statement were true. The authors describe what takes place when the statement is not respected and what survivors' experiences are under those circumstances. Other researchers' work on this topic of observing other disasters and critical situations will be blended with our own observations.

Then we examine what subgroups of disaster survivors are most harmed by a society's lack of respect for them and their agency, and how their mistreatment is grounded in the broader disrespect for all survivors. Next, we ask and suggest why restoring agency is ignored in most disaster/crisis situations and what can/should be done to rectify the situation. Others have offered suggestions about how to restore agency, which we summarize, although they struggle, like ourselves, to find useful and implementable solutions. The challenges *run deep* in our societal institutional framework; thus, the solutions fail to alter behavior unless the commitment to do so is strong and comprehensive.

Finally, we conclude with implications of ignoring the truism quoted in the introduction, projecting forward to anticipated enhanced extreme weather conditions with accompanying unfathomable costs, human, and material. If survivors are not honored in their quest to participate in their own recovery, their recovery will in fact, under these future climate change conditions, be unlikely to happen.

IMPORTANT TERMS AND CONCEPTS
AGENCY

"Agency" is not commonly used, even in the academic arena. However, there appears to be no better term to describe "the power to act as the actors determine." It implies a control of one's actions, the determination by the actor of the purpose of the action, the control of the direction of the effort, and the degree of effort that will be expended. It also implies that the actor has the power to adjust their own behavior to achieve the particular goals for which they are aiming. And there is an assumption of the competency of the actor to assess the situation and to use their power of self-guided action to better their situation.

This chapter commits to the use of the term agency in the positive tone that has been used in the above paragraph. However, there are those that believe that disaster-affected people do not act in their own best interests, a myth that has been dispelled by most scientific studies for the majority of disaster incidents (Wilson et al., 2007). Instead, people act in what *they believe* to be their best interests, based upon the information that they have, as has been shown regarding the decision to purchase insurance (Kunreuther & Pauly, 2005). Yet, there are disconnects between the immediate needs of affected populations and their longer-term well-being, an area where traditional science may be able to provide disaster survivors with help such as understanding the "lifetime costs" associated with their home. Lifetime costs include construction costs, elevation costs, insurance costs, and damage costs, all tallied to appreciate the benefits of elevation and insurance over the home's life or the homeowner's occupancy.

When a government dismisses the survivor's right to have agency, or ignores how successful agency can be enhanced with the help of government and non-profit agencies, a survivor's agency might not be used to enhance the survivor's condition (Lipsky, 1980). In some cases, the exercise of bureaucratic discretion by local government officials also impacts the survivors' agency by limiting their options and directing them toward particular outcomes that local bureaucrats feel is best. Sadly, this exercise of street level discretion can lead to increased risk, constraining survivors' options while making them feel as if they have been granted agency.

An example of harmful, misguided survivor agency is found in Katrina recovery. Public officials in the city of New Orleans were concerned that those who had fled the city would not return in sufficient numbers to achieve recovery (Russell & Donze, 2006). To encourage evacuated residents to return to the city, the then mayor, Mayor Ray Nagin, decided to "assist" people's return by making it possible for homeowners with official estimates of "substantial damage" to rebuild without conforming to the Federal Emergency Management Agency (FEMA) requirements of elevation to the Base Flood Elevation (BFE). Substantial damage was defined as over 50% damage to a home as determined by the FEMA and estimated by City officials and/or hired contractors. In order to do so, homeowners were required to present City Hall an estimate from a private contractor stating that the home was flooded less than 50%. That assessment permitted the homeowner to repair the home while NOT elevating it. It is doubtful that the original estimates were incorrect (Russell & Donze, 2006). In this case, street-level bureaucrats exercised their agency with the intention of supporting homeowner return and recovery by reducing the burden of onerous regulations. However, as a result, homeowners were not given sufficient information and allowed to choose; they were instead steered toward a particular outcome that the bureaucrats thought might be more beneficial but which increased their risk.

Today as you drive around the city, you can see "slab-on-grade" one-story homes side-by-side with homes that have been elevated. The inconsistency is due to a bad decision made by City officials without sufficient education of homeowners with substantially flooded homes about what that decision would mean to their future risk. It is not an example of assisting survivors in achieving successful agency. Rather it is an example of misleading survivors to make decisions that were against their long-term interest and against the long-term interest of the entire city. The decision increased the number of at-risk homes not elevated after the Katrina flooding that have a higher chance of being flooded in the future.

That guidance has put the entire city at significantly more risk in future inundations. It is harmful to *individual* agency. And, it is harmful to *community* agency, both enabled by a harmful use of bureaucratic agency and discretion. Should the state and federal officials have appreciated the harmfulness of the decision and encouraged that the decision not be implemented? Or was the disaster as a catastrophe too uncertain in its impact and the situation too chaotic to allow effective guidance from higher levels of government, with perhaps more disaster experience, for community leaders to support long-term safer community agency? Would the return of the populace have been severely restricted by enforcing the substantial damage elevation requirement as the city government leaders feared?

RESILIENCY

The outcome of successful *agency* becomes the core of resilience. This is especially so when we expand the concept of agency to reflect the capacity of survivors at the various levels of social organization: individual, household, neighborhood, community, etc. While not often included specifically in discussions of resiliency, we posit that agency is the important element of community resiliency as well as that of individual survivors. All the way up the inverted pyramid of social organization, in terms of numbers and magnitude of societal control (Fig. 5.1), agency is critical. Community participation in risk reduction and sustainability resulting in resilience can only be achieved if societal members

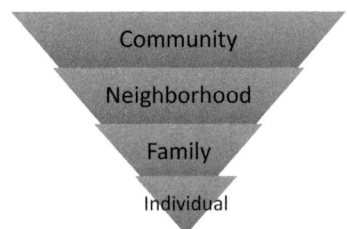

FIGURE 5.1

Levels of Agency.

at all levels of organization, as well as individually, take the actions that will reduce the risks in the future (Kapucu, 2015). Resilience is not achieved by others *doing for* those who want to be resilient.

A dramatic example of the opposite of agency leading to resilience, and one that manifested itself in a horribly dangerous and harmful process, was the removal from community control of water management in Flint, Michigan. Community agency was removed and assigned to an economic "emergency manager" appointed by the state when economic decline caused city bankruptcy. As was assessed and concluded by Dolan and Egan (2016), removing agency from the Flint government to manage the drinking water source and treatment has led to permanent injury of untold numbers of children who have ingested the lead-contaminated water. This method for addressing severe community budgetary challenges by removing *community* agency is a clear example of how agency constraints at higher levels of organization do harm.

AGENCY DISRESPECT AS STRUCTURAL VIOLENCE

> Structural violence *is one way of describing social arrangements that put individuals and populations in harm's way. ... they are embedded in the political and economic organization of our social world; they conspire to constrain* agency. *(Paul Farmer in Burtle, n.d.)*

More refinement of the concept structural violence is presented by Soron (2007): "Normal, unexceptional, anonymous, and often un-scrutinized violence woven into the routine workings of prevailing power structures."

A concept used even less frequently than the concept *agency* is *structural violence*. The first way to approach appreciating its meaning is to consider *institutional racism*. Simply defined, institutional racism is the impact of racist treatment that does not require individual persons who are racist causing the impacts. Rather, institutional rules/laws result in racist impacts. An example would be the different incarceration rates for black and white drug offenders. "Despite similar rates of drug use, Blacks are incarcerated on drug charges at a rate 10 times greater than whites" (American Civil Liberties Union, 2013). Finally, the society—on all points of the political spectrum—has acknowledged this institutional racism and is moving to address it. The harm of this institutional racism will require generations and unfathomable amounts of public resources to rectify, if it ever does. The society is not in such agreement about the changes in voting opportunities being an example of institutional racism as well. Restrictions to voter registration and early voting prevent a traditional pattern of African-American voting, and lowers rates of voting by them. State legislatures have passed the racist laws and courts have overturned them. (The Supreme Court, a Federal District Court, and the Court of Appeals overturned three state voting laws—Wisconsin, Texas, and North Carolina—in late summer of 2016 because they were seen as directly targeting African American voters to prevent them from voting.) It too is institutional racism.

Replace *racism* with *violence* and *institutional* with *structural*, and the similarities should enhance one's understanding of *structural violence*. No actor need have personal inclinations to violently harm a disaster survivor for that survivor to be irrevocably harmed. The laws, policies, and the implementation practices accomplish the violence. Are they written with the intent to harm? That is likely not the case.

But what we do believe is that the laws, policies, and implementation practices are *not* written with the care needed to avoid the violence. And there is definitely a lack of commitment by sufficient numbers and levels of government officials to prevent the violence from happening. In effect, other vested interests "trump" survivors' interests in how aid to survivors is implemented. These interests include corporate greed as discussed by Klein (2007), but also concerns of efficiency, which on the surface promote good governance but can, in addition, perpetuate structural inequalities. The woefully inadequate degree of advocacy for survivor agency creates the violence (i.e., extreme harm to them). How do we know it is *institutional?* The pattern of structural violence against disaster survivors has been repeating itself across different disasters in extreme (extensive) ways. As one example, there has been wide documentation of delays in providing assistance to impacted residents in both Hurricane Katrina and Superstorm Sandy, with applicants voicing the exact same concerns in both cases (Adams, 2013; New York City Comptroller, 2015; Rand Corporation, 2008). And the action's violence has no conscious intention, a quality that is usually attributed to violence (Soron, 2007). More will be said about these observations and then in the last section of the chapter conclusions from the observations will be drawn.

Before turning to examples of structural violence in the form of restrictions of disaster survivor agency, one caveat must be offered. "When harm [such as restrictions on survivor agency] is avoidable, violence is present" (Soron, 2007). "When the actual [restrictions on agency] is unavoidable, then violence is not present" (Galtung, 2009). In the case of the recovery process from a catastrophe, withholding agency support might not have been avoidable in the early phases of the event due to it overwhelming the response capabilities of all the levels of government. For example, take Hurricane Katrina. As the recovery time passed and the issues of restoration of agency still continued, the role of the magnitude of the storm became less of an explanation. There were continuing reports by survivors of inadequate commitment to support their agency even when the recovery process became more organized. And when the delays in assisting survivor support to regain agency were seen to be worse in the Hurricane Sandy aftermath, a disaster and not a catastrophe (Hammer, 2015), the conclusion of avoidability of curbing survivor agency became stronger: It was structural violence in both cases.

TRAUMA

Structural violence traumatizes survivors who are subject to it. One of the paths by which the trauma occurs is by creating a feeling of powerlessness among survivors subjected to it (Norris, Stevens, Pfefferbaum, B., Wyche, & Pfefferbaum, R., 2008). It is a paralysis of not being able to move forward in what seems to be a logical, functional way.

Such a linkage can be seen in all of the examples later, both those that are generic (i.e., not related to lower income or racial subgroups) as well as those that are not. Minorities and those who have lower incomes should be more traumatized because structural support is more important to the recovery of their agency. But as some poor responded: "We experience structural violence daily so it was just another regular day." Such an analysis should reinforce to us the magnitude of the trauma. *Even* in a catastrophe, the trauma for the poor and minorities continues unabated.

For those unaccustomed to structural violence, they are traumatized for the opposite reason: they cannot use their usual agency to achieve their goals. Because they assume they should be able to, it takes them longer to recognize that they are being subject to structural violence, whereas individual violence such as domestic violence or violent robberies, for example, would be clearly evident to them.

It is not the intent of this chapter to explore the concept of trauma, but rather to recognize it as an outcome of structural violence. Marks (2008) draws attention to the importance of recognizing *long-term trauma* as opposed to more fleeting, temporary trauma that would be expected with a catastrophe. He labels the role that the problematic FEMA recovery, insurance company responses, and unscrupulous contractor behaviors played as "chronic assaults" on the survivor. With chronic assault comes chronic trauma.

DELAY

Delay can be an example of structural violence. Reid (2013) uses the term "temporal domination" to describe the ability to make someone wait for resources valuable to them. For incarcerated prisoners the violence of waiting is purposeful punishment. Its use clearly demonstrates the "harm" that waiting imposes. In other situations where government is providing a service to citizens, delay can be used as a punishment for those who are suspected of being undeserving (Auyero, 2010), but it also discourages eligible individuals from receiving assistance (Reid, 2013). (It is important to note that the ways in which contracts are structured between the state and the case management—or grants management—contractor can create perverse incentives that discourage the contractor from assisting those individuals whose cases are more complicated in favor of those whose cases can be closed more readily. As a result, delays are perpetuated for those who can least afford them (Jerolleman, 2013)). Identifying the undeserving may be a justifiable objective; however, a more thoughtful, tested approach to achieve that goal should be developed. When the very poor experience a lengthy, extended catastrophe, any delay in receiving the services for which they are eligible because of a lack of identification, address, extensive waiting lists, etc. leaves them behind as happens to "the poorest and most isolated" of the abject poor in nondisaster situations (Lein, 2015).

Another example of delay that may have had a punishment element is the delay in sending in the National Guard to rescue those who were not able to exit the city before the storm. Gun violence was reported and suggested as a reason for the delay of National Guard and Army, as they did not want to come under fire. The reasoning was, until the violence was stopped, the military would not enter. The counter response to this is that the earlier the arrival, the more likely that order could have been restored and thus fewer residents harmed. Some could say that this delay was institutional violence.

The psychological impacts of waiting are well documented. Reid (2013) found that "waiting has a negative impact on emotional and material well-being of those in need of assistance." Continuing, Reid (2013) found that

> [s]urvivors who were subjected to waiting experienced a complicated loss of agency that resulted in stress, worry, instability and loss of ability to make long-term plans, compiling the basic loss of ability to act for their own recovery.

The frustration of experiencing delay in government response is reflected in a survivor's comments:

> They are not helping us. You know ... We call FEMA. 'Call back in 7 days, call back in 14 days, 15 days, still pending.' What is there to be pending about? There are just two people in the household!

Alfreda's application had been held up for months with no explanation from FEMA (Auyero, 2010). "Time, being invisible, is a handy and effective tool to use [as structural violence] as, behind the language of procedure and bureaucracy, it is easy to forget its corrosive properties" (Kenny, 2015). The delays that became the "signature" of both Katrina and Sandy were perceived to be intentional (see next section).

EXAMPLES OF STRUCTURAL VIOLENCE
"LOUISIANA ROAD HOME" AND NEW YORK CITY'S "BUILD IT BACK"

The Louisiana Road Home program was created to manage federal funds that were awarded by congressional appropriation as a result of the extreme extent of damages. The program was designed by the state of Louisiana and managed by private contractors. Similar extraordinary funding was also given to the state of Mississippi (the eye of Hurricane Katrina passed just south of New Orleans and then directly into the coastal towns of Mississippi). The program was extremely generous with the total federal funds provided exceeding $76 billion. In just the first 2 years of the Road Home, 117,000 homeowners received a total of $7 billion (Hammer, 2015). The tremendously high amount awarded was due to the extent of the damage in Louisiana due to the density of population and the collapse of the flood walls protecting New Orleans, in addition to the damage in the southern coastal communities thought to be less well protected.

The affront to agency that was repeated literally thousands of times was the delay in implementing the program to restore owner-occupied housing. The reason given to survivors as they tried to navigate their way through the process became the "mantra" of frustration: "We are unable to locate your paperwork." This was after survivors had submitted their paperwork repeatedly. The illogic of paperwork loss happening repeatedly to the same survivor created the affront. "How could it happen?" "Why was it happening?" "Why couldn't officials fix their aid program?" "Don't we count?" "We have been through so much; how can the government treat us this way?" (Fig. 5.2).

"The grant-making process was slow, in large part, because it was not designed to be fast," said Rick Eden, lead author of the study and a senior research analyst at RAND, a nonprofit research organization. "It was not designed to ensure that each application would be handled in a timely manner" (Eden & Boren, 2008). However, a key outcome measure for grant-making utilizing federal dollars postdisaster is often speed, which is privileged over risk reduction and over efforts to support survivor agency. In many cases the contractors are paid by the case and have no incentive to assist persons whose cases are more complicated as success is judged by the speed at which money is moved (Jerolleman, 2013).

New Orleans Advocate, July 20, 2015

FIGURE 5.2

The Challenges of Survivor Recovery: "It's more than the money. It's the hoops we had to jump through to do it"

Source: From http://www.theadvocate.com/baton_rouge/news/article_f9763ca5-42ba-5a62-9935-c5f7ca94a7c4.html

Similar mistreatment of Sandy survivors was reported in the audit by the New York City Comptroller (2015) for the single-family program in the recovery after Hurricane Sandy (Stringer, 2015). "Superstorm Sandy victims were shuffled from one staff person to the next, many of whom were not familiar with their cases... nearly 90 percent of the applicants ... went over one month before getting a follow-up call" and "scores of applicants echoed the findings that confusing forms, resubmissions and the lack of returned phone calls contributed to lengthy delays in getting aid" (New York City Comptroller, 2015). The Build It Back program in New York City had over 100 procedural changes between August 2013 and July 2014, a fact that resulted in case management errors and further delays (New York City Comptroller, 2015).

SHRINKING THE FOOTPRINT OR NOT?— HOW *NOT* TO HAVE THAT DISCUSSION

Over 80% of the city of New Orleans had been flooded. The depth of the flooding varied due to the variable elevations of the land onto which the water poured and due to the pathways into the city that the water took before the water "leveled off" across the city. The areas along the river and close to the lake (artificial land dredged to create lakeside neighborhoods in the 1930s and 1940) did not flood. While the City officials were permitting repair of homes that likely had been damaged over 50%, questions were arising as to whether certain parts of the city should not be reinhabited. Experts within the state and city as well as outside of the area quickly expressed their opinions about the best decision to make—to permit recovery of all neighborhoods or only those with certain qualities such as higher elevation and lower percentage of homes destroyed. "The solution devised by Joseph Canizaro," local developer charged with developing the City's recovery plan, "was borrowed from the outside planners in plentiful supply in New Orleans in those first months after Katrina..." (Rivlin, 2015). The perception by these nonresidents of the affected areas that they had the right, the license to propose, to mostly encourage that parts of the city not be permitted to be restored, was evident in the ease with which the assessments were offered (Rivlin, 2015). It appeared that there was very little consideration by public officials and the experts who offered their opinions of the impact that the public outcry would have, not just the substance of the comments but how the conversation about this extremely impacting, "sensitive" decision was (to be) undertaken. Important issues in this conversation included (1) Who should be expressing opinions? (2) How should those opinions be expressed? (3) Did some have more rights to express their opinions than others? (4) What should the decision-making process be?

The "conversation" unfolded in 2006 with input from Mayor Nagin's Bring New Orleans Back Commission identifying areas targeted for commercial reinvestment "and backed by the widely respected Urban Land Institute, drew howls

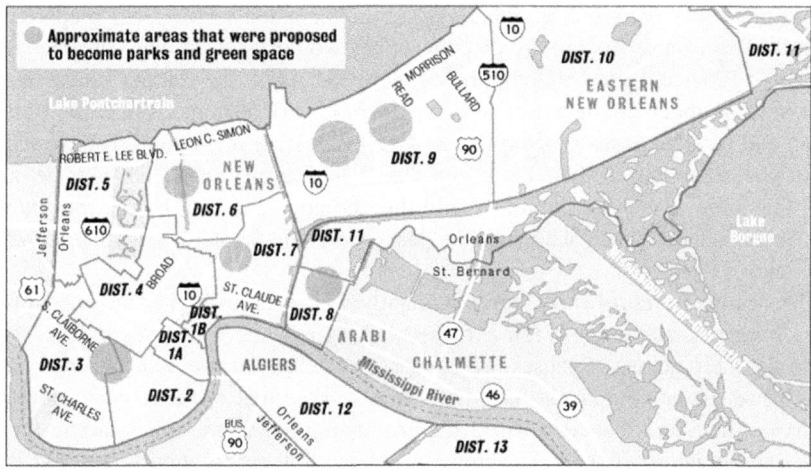

FIGURE 5.3

A Smaller Footprint: The Green Dots

Source: From http://www.nola.com/katrina/index.ssf/2010/08/
many_areas_marked_for_green_space_after_hurricane_katrina_have_rebounded.html

from residents who found their neighborhoods represented on maps by green dots (Figure 5.3) that denoted redevelopment as perpetual green space" (Krupa & Russell, 2007).

When the protest outcry became city-wide the mayor modified the plan and informed each neighborhood that if it did not want to be converted into green space, it would have to demonstrate by its engagement and plan output that it deserved to be redeveloped. The green dots were particularly devastating because of their "graphic" representation of what was proposed to happen.

The use of these two planning approaches—visual identification by a reputable outside planning organization of the areas that would not be redeveloped, and the order by the mayor that residents had to prove their neighborhoods worthy of redevelopment—contain the same qualities of structural violence discussed earlier. One could argue that the outcome of extreme community engagement should be the focus rather than *how* that outcome was accomplished. Using this logic (ends justify means), it does not matter how harmed the residents were by the means, the ends of engagement were accomplished.

This argument does not ring true. The exacerbation of trauma through the engagement process leads to further disengagement and a reduction of agency. Some 80% of the residents who lived in the neighborhoods identified by the green dots were African American (Rivlin, 2015). Due to their economic challenges, most were not able to return to the city to participate in the community meetings. (An effort by a consortium of Louisiana universities to provide outreach to residents who had not returned to support their participation in neighborhood engagement was thwarted by a federal contractor when FEMA was prepared to fund it. The excuse given was that the effort would "slow down" the recovery process. FEMA did not challenge this conclusion.) And the trauma (more in the later section about this term) of learning from afar (Houston, Atlanta, and other urban and rural areas away from the New Orleans area) is obvious. They worried what would become of their homes—be they owned or rented? Of their belongings that were still salvageable? Would they be able to return before their home was demolished? Those who had the least were the least able to defend what they had.

Those able to return early—predominantly white and of higher income—often came back in time spurts to retrieve their belongings and to begin the process of restoring their homes. Spouses were separated; children were in school somewhere else. If the companies for whom they worked had opened again, they had to divide their time between work, repair, commuting to the other family members, and attending unending meetings to "defend" their neighborhoods and thus their homes.

The first author of this chapter has indelibly marked into her brain the image of a salvaged church on lower Franklin Avenue, the postflood now-exposed concrete floor barely cleaned of flood water and mud, with barely any lights lit for an evening meeting in which regional residents were sitting numbly listening to an out-of-town planner hired to seek their input for a vision of their neighborhoods that was lofty, expensive, and has never come to fruition. They sat there like mere shadows of themselves, believing that physically they had to be present, but almost incapable of doing so due to the exhaustion, frustration, fear, and trauma. Their experience was structural violence as well. It could have been prevented if respect for the agency of survivors had been honored and care taken not to harm survivors further as recovery was going on.

If public officials and other professional experts had understood what process of consideration of such a drastic move needed to be undertaken, would an "organic" conversation about shrinking the city have been more likely to have occurred?

It is evident from the example of the owner-occupied rebuilding processes in both New Orleans and New York, and the "green dot" neighborhood recovery in New Orleans, that being of middle income does not protect one from structural violence. That this is the case is a prime indicator that the harm comes from the social/bureaucratic structure (government, nonprofit, professional organizations) rather than being perpetrated by individual organizational officials, by their intent to harm (more below on intent). No one in the society is immune. Those with personal wealth and political resources also experience the violence (lecture by Kai Erikson, October 26, 2015, Environmental Studies Program, Tulane University). The way that the society is structured and the values which underlie this structure result in the society, the recovery process, not truly caring for the fate of the citizens, the disaster survivors. This is what generates the violence.

THE EXTREME STRUCTURAL VIOLENCE ON THE ECONOMICALLY AND POLITICALLY POWERLESS

All of the other examples used in this chapter—evacuation / long-term sheltering, access to health care postdisaster, and recovery planning for lower-income neighborhoods—focus on the particularly virulent impact of structural violence on lower income disaster survivors (White, 2015). The delay in assisting lower income, politically disenfranchised, and female-headed householder disaster survivors to recover agency is by far the most serious (Jones-DeWeever & Hartman, 2006). This resistance is manyfold greater in the harm that it does, than the harm experienced by those with economic and political power. These delays are exacerbated by the systemic issues that are built into contracting processes, as described earlier in the chapter. The latter have personal and group resources to overcome the harm. They could go elsewhere to get assistance, or had the resource time to wait for the assistance without such dire consequences (Auyero, 2010). And it is very likely that the harm is compounded going forward in terms of the condition of the survivor's life and that of their children after the disaster. For example, fully half of the African-American children in New Orleans lived in poor households in 2015, a higher percentage than when Katrina hit (Quigley, 2016a). Is it likely that household poverty has been made worse by the legacy of families being inadequately assisted after Katrina? How much did structural violence against the parent(s) perpetrate structural violence on their children? And it is a major question whether this violence is reversible for those survivors so impacted.

EVACUATION NIGHTMARE

Everyone in touch with American television media the last week of August 2005 retains in their memory the images of Katrina survivors clinging to rooftops for days after the event and crouched about and within the New Orleans Super Dome and the Convention Center. They remained for days with little food, water, and respectful treatment. Troops did not enter the city until 3 days after the levees broke and when they did, it was with guns continually drawn. The survivors were treated like war enemies until the Katrina military coordinator Army Lt. General Russell Honoré entered the city and told the troops: "Put those **** weapons down! I'm not going to tell you again, **** it. Get those **** weapons down" (Quigley, 2016b).

As survivors escaped the flooding they were dispersed around the country in a disorganized fashion. This challenge prevented systematic and thus successful family recoveries. Never had such an experience occurred to Americans within their own country. Respect for those fleeing, and professional recognition of what would be required for survivors to restore their lives, were not present. And again, the question arises: was it the magnitude of the event that caused these challenges, was it the ill preparation for a severe hurricane evacuation response, or was it structural violence? These authors believe it was all three: overwhelming challenges of a catastrophe that the country had never experienced before, lack of anywhere near an appropriate evacuation plan for those New Orleans residents who were without cars, and equally important disrespect for the survivors, predominantly poor African-American residents of New Orleans.

HEALTH CARE SCARCITY

Once the water that had entered through the breached flood walls and levees had been pumped out, public health officials and others charged with assisting in the recovery (i.e., the National Guard and regular Army) began the process of cleaning out the primary health care facility for lower income residents. Until Hurricane Katrina, Louisiana retained a separate health care system for indigent residents. Established in 1736, this system relied on "charity" hospitals, rather than simply paying for health care service at facilities that serve a spectrum of incomes. The hospital that served this function in New Orleans was actually called Charity Hospital, because of its purpose but also because it was operated by the Daughters of Charity, a Catholic religious order.

The ground and basement floors that had been flooded were restored to usability by the military. However, much to the surprise of those who had readied it, the Governor of Louisiana, Kathleen Blanco, refused to permit the building's use (Fig. 5.4). It never opened nor has it since Hurricane Katrina.

The reason for this turn of affairs was without consideration of the immediate and prolonged impact on the patients, lower-income residents. Instead it was due to the "prize" that the private-sector contractors and public-sector boosters wanted; to use the federal recovery funds to construct a new public/private hospital facility (Ott, 2012).

Speculations at the time about the reasons for it not reopening were expressed by General Russel Honoré, commander of Joint Task Force Katrina. Honoré said: "This is about business, man. This is about rich people making more money. This is not about providing health care" (Burdeau, 2009). Former Charity Hospital emergency ward doctor James Moises said it was about getting a new medical complex; it was an orchestrated plan. It was "How can we manipulate the disaster for institutional gains?" (Burdeau, 2009).

Uncertainty increased concerning the continued quality support of health care for lower income residents. Contributing to this uncertainty were the (1) changes in expectations of services to be provided to the poor, (2) the decrease in number of patients expected to be served, (3) lack of knowledge about how the new hospital University Medical Center—New Orleans would be supported by the changes in federal funds for health care, and (4) the extent to which the hospital would serve patients of all incomes (Ott, 2012). That it took a decade to construct and that the health care system that evolved during that decade was not financially sustainable (Reckdahl, 2014) demonstrates how fragile survivor health care agency was during that period. Concern for the health care needs of so many

FIGURE 5.4

Dramatic Delay in Post-Katrina Restoration of Health Care for the Poor.

Source: From https://www.thenation.com/article/why-was-new-orleanss-charity-hospital-allowed-die/

indigent survivors after Katrina was not debated, advocated for, or truly vetted by government officials in the decision-making process. Decisions were made at the same time health care needs were extreme because of the stress, unattended illnesses, and disruption of the regularly limited life-lines of care that they had experienced before the storm (Gratz, 2011; Ott, 2012). This was indeed a case of structural violence.

POSTDISASTER HOUSING

Emphasis on recovery of single family homes was at the expense of renters (Butterbaugh, 2005). In effect there was a bias in disaster recovery toward those "who had something" before the disaster (Landry, 1987; Pattille-McCoy, 1999; Pentarki, 2013). This fact becomes a very large barrier to addressing the improvement in the challenges described in the next paragraph. To the extent the media focused on owner-occupied housing recovery, the recovery challenges for lower income survivors became invisible. In some cases the poor survivors were simply viewed as undesirable and undeserving of recovery (Lafer, 2008).

Housing lower-income survivors after a disaster has appeared to be even more of a challenge to the government than has an efficient, timely assistance program for homeowner survivors. The challenges are threefold: (1) distribution of housing assistance funds to those survivors, especially families who need it; (2) providing temporary housing after a survivor is initially housed in a public shelter; and finally (3) creating, and restoring inhabitable housing in a timely manner.

The challenges that were faced in terms of housing assistance centered around the one-household-head declaration. If a couple had separated, there became competition about who would be the recipient. Often the male partner was designated the recipient when the mother and her children should have been. Sorting this challenge out caused much of the delay in awarding the funds. Delays were compounded by the challenges of survivors in establishing an address and an "identity" with government personnel, who were charged with controlling fraud by not awarding funding to those who did not "deserve" it.

Creating housing opportunities for lower income survivors is the next key issue. Assumptions were made that survivors would have sufficient funds for the deposit for an apartment while they were applying for government housing assistance. No logic existed in that assumption. In addition, the two means of creating/expanding the housing market for lower income residents, the repair of existing storm-damaged housing and creating more housing, were both implemented at a minimal level. In the wake of Katrina the rental housing repair program in Louisiana floundered (Hammer, 2015). Some 145 thousand travel trailers were purchased by FEMA for the Katrina/Rita impact areas. Not all were used and the survivors left them as quickly as possible once the discussion of chemical contamination of them began as early as 2006. The alternative housing option of Katrina cottages aspired to construct 3500 small houses in coastal Gulf design, principally in Mississippi but some in Louisiana. Fewer than 100 have been built (Alter, 2015).

HOW TO STOP THE VIOLENCE

We will examine changes that would/could improve the structural violence toward agency recovery in the five examples described earlier: (1) owner-occupied housing recovery, (2) planning for the community's recovery, (3) evacuation and long-term evacuee housing, (4) health care immediately after the disaster, and (5) survivor aid and return of agency to the economically and politically disenfranchised.

The recommendations encompass a wide variety of social organizational and behavioral changes. We use each topic to emphasize those changes relevant to it.

OWNER-OCCUPIED HOUSING RECOVERY

Government should improve management and supervision of recovery contractors by tightly controlling deliverables/payments. Speedy recovery of survivor agency must be paramount. Contractors paid to manage and staff housing recovery have inflicted much pain on survivors. Lessons on how to manage such a responsibility were not handed down well from Hurricane Katrina to Hurricane Sandy. The same complaints were heard. Accountability for service delivery was as absent after Sandy as it was after Katrina. Contractors were paid an exorbitant amount of the recovery funds ($17 million) before even 500 Sandy survivors were provided with benefits to begin repairing their homes (Fig. 5.5). Similarly, almost 6 years after Katrina, only 86% of eligible homeowners had received any funding despite the fact that contractors had received substantial payments. Organizationally, contracts between the government and private contractors need to reflect the very best of business practices and oversight including the use of performance measures and financial penalties. Contracts need to be structured based upon policy goals such as equitable recovery. Contracts need to avoid outcome measures that only include the number of interactions or hours spent with residents.

The lack of an adequate number of government officials knowledgeable about *how* to achieve successful recovery of agency is also a problem. Research from Hurricane Katrina from a Rockefeller Foundation funded project suggested that such experts should be trained before a disaster by being embedded within the disaster recovery organizations of each state. Agency recovery experts should be prepared to deploy to the disaster, especially the large disaster/catastrophe

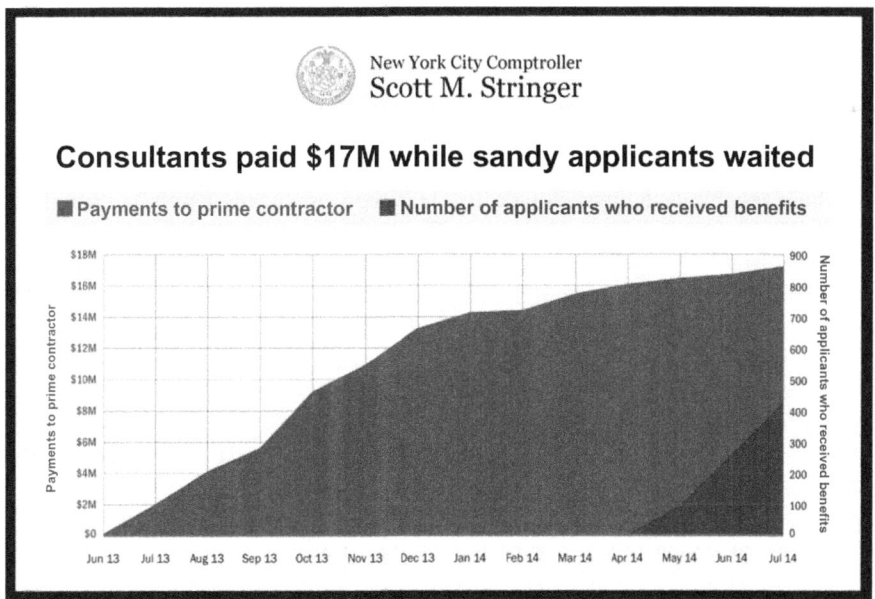

FIGURE 5.5

Consultants Paid $17M While Sandy Applicants Waited.

(Montjoy, Farris, & Devalcourt, 2010). It is acknowledged that the speed of ramp-up that is expected is difficult to accomplish without such a cadre of pretrained government officials. It is not possible to leave such a large team idle until a disaster, and thus the recommendation that they be deployed to develop disaster recovery management skills in conjunction with state officials in the interim.

Draw from best practices of recruiting, training, and supervising case workers for effective implementation of the housing recovery. Many agencies not associated with disaster recovery have as their responsibility to assist those in need of a social, economic, or psychological service. The method to provide these services that are used is "case management." Each intended recipient of the service is provided with a case manager and data are collected for their file so that the services can be provided effectively, efficiently, and timely. There is no need to reinvent a bureaucratic method that has been applied and honed for decades.

However, it is important to emphasize that case management practices for managing disaster survivors have unique qualities. Because the ramp-up of staff must be done quickly and efficiently, those recruited for the staff positions should not be employed at the time of the disaster, and thus should be able to accept employment immediately. They may never have served as a case manager in previous work. Thus their training cannot extend over a significant time span, and must include more than simply processing paperwork. Their responsibilities often include offering services and resources blended from *different* agencies. Another unique quality of the staff is that they too have been traumatized by their own disaster experience, yet must of course listen continuously to survivor after survivor's description of their disaster experiences.

Despite these unique qualities of disaster response staff, recommendations for how to overcome the challenges so often expressed by disaster survivors, with regard to how their cases are managed, are definitely available from the regularized practices of case management and the studies conducted of them.

These recommendations include to (1) screen applicants carefully including for their capacity to be empathetic, a quality that creates trust by the survivor that they will be assisted; (2) appreciate the importance of having competent case managers and provide the training and environment for them to serve the disaster survivors in a manner that enables efficient restoration of survivors' agency; (3) train and provide refresher training; (4) supervise well and create an implementation process that is doable by the staff and achieves the respect for survivor's agency recovery that this chapter describes.

In addition, case manager qualities that will benefit the success of an agency recovery program include: likeability, acceptance, encouragement, respectfulness, helpfulness (Conte, Ratto, Clutz, & Karasu, 1995; Rhudy, 1999), caring, empathy, clear communication (Cleary & McNeil, 1988), responding appropriately to the client's recovery process (Draucker & Petrovic, 1997), and having an "available" and humane demeanor (Sixma, Spreeuwenberg, & Van der Pasch, 1998). Clients need to feel welcome and validated, and believe that the case workers are empathetic. It is not only important that the worker gives the message "I value you," but that the client received and internalized the message "I feel valued." Finally, disaster survivors, like any clients of an organizations, need to have a sense of control over the process of disaster recovery assistance to have a sense of trust and connection (Wright, 2011). It is with regret that many of these qualities were not observed for the case managers of both the Katrina catastrophe and the Sandy disaster.

RECOVERY PLANNING

Honor and support true survivor engagement in community planning and recovery not based on additional threats to them. In a review of New Orleans residents' civic engagement activities post-Katrina, Campanella (2009) applauds the city-wide engagement that ensued. He sees the outcome as producing extensive ideas for bettering the city (117 emerged just in the rebuilding projects proposed by the City's "recovery czar" Dr. Edward Blakely). Campanella states: "Crises often spawn eager participation in civic affairs; few things motivate like a threat" (Campanella, 2009, p. 41).

There is a difference, however, between the natural fear of a city not recovering resiliently, and a government/professional organization-induced fear engendered out of ignorance by government/organizational officials of the psychological and physical harm that such threats would produce. Threats to shut down neighborhood recovery, and green dots to indicate whose neighborhoods would be first on the list, were not thought through responsibly before they were publicized. They were structural violence. The trauma and physical and psychological harm such a threat posed was a heartless way to encourage residents to engage in planning. It is very clear that survivors actually died from the stress of having to balance family recovery with civic "engagement" to save their neighborhoods. Others who observed the dynamics from afar because they could not return also likely suffered life-threatening trauma by virtue of feeling helpless to do anything about the threat.

Unfortunately, lessons do not seem to have been learned adequately from the Katrina experience. The Urban Land Institute continues to be involved in planning post-Katrina "redevelopment" of New Orleans. For its 2015 ULI/Hines competition for planning graduate students nationwide, it utilized the site near the new hospital described earlier as the target for proposals for redevelopment. The call for submissions recognized the challenge: "Due to its unique location, the neighborhood faces the challenge of retaining the existing communities..." (Kreuger, 2015). The competition referred the proposal submitters to the ULI's "Ten Principles for Building Healthy Places" as a resource for developing proposals (Eitler et al., 2013). Unfortunately, this document does not address the challenges of development that often sees the redevelopment area as a "blank slate" devoid of residents and commercial activities for original residents.

Part of the area targeted for redevelopment near the hospital is an historic African-American commercial area that was initially badly harmed by the construction of Interstate-10 some 50 years ago. This led to the decline of the area

(Jones, n.d.). Concern that another "green dot" process would occur now was expressed by Tremé resident, Muskoghee Alibaamuu during a community meeting, "You're gonna come down here with your degrees, do what you want to do, then go back where you came from. How do we know this is no different?" (Jones, n.d.). Structural violence, unless it is addressed continually, will be the tendency of the social/political/economic system of this society (the winning submission, "The Crossing," prepared by a University of Maryland graduate student team does include some consideration of the local residents. http://uli.org/general-posts/uli-hines-competition-2015-finalist-crossing-university-maryland/).

Engagement in recovery planning benefits from an informed survivor who has true agency participating in the planning process, not "lip service" as was often the case in the Katrina recovery (Campanella, 2009). Citizens and government officials / nonprofit professionals / government contractors should create collaborative learning experiences to achieve the goal of more knowledgeable participants all around in the recovery process. Citizens must be honored as rightful participants, not tokens.

Controlling recovery either by nonprofits, as was the case after the Haitian earthquake of 2010 (Edmonds, 2012) and somewhat after Katrina, or by governmental processes harms the survivors. Instead, they should be encouraged to take part in the changes (Benight, 2004; Fullilove & Saul, 2006; Landau & Saul, 2004; Norris et al., 2008; Perez-Sales, Cervellon, Vazquez, Vidales, & Gaborit, 2005; Van den Eynde & Veno, 1999). In addition to the contribution that survivors can make, their involvement in recovery programs reduces their profound feeling of powerlessness, and thereby also reduces their trauma (Norriset al., 2008).

SUCCESSFUL EVACUATION THAT SUPPORTS RETURN

The horrific nature of the Katrina evacuation, and the scattering of lower income survivors around the country into unfamiliar cultures, left a legacy of harm. Survivors were separated from the support networks that were their base for survival in their financially limited lives. Many families have never returned. The African-American population within New Orleans area is 90,000 fewer than before the storm. Only 9000 fewer whites are in the region than before Katrina (The Data Center, 2015).

While more always needs to be done, the improvements since Katrina of the evacuation from the greater New Orleans area, especially the city of New Orleans itself, are a major means of empowering agency. A brief summary follows of some of the key elements.

Measures that are empowering agency. For those individuals and households with cars the improved management of the highways in the New Orleans area as well as Baton Rouge and Mississippi, and the contraflow (reversing incoming lanes) of the interstates in all directions and into Mississippi, have significantly improved evacuation time. For those who need assistance to evacuate, the plan

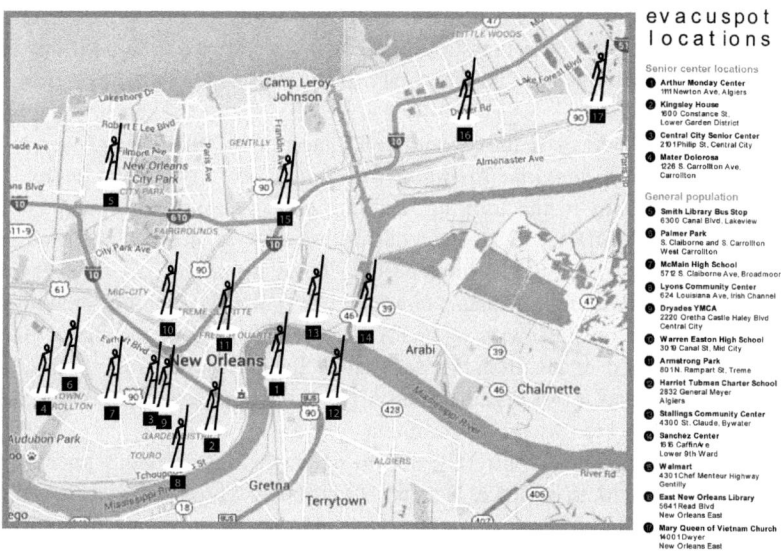

FIGURE 5.6

Transportation-Needing Citizens Collection Sites for City-Assisted Evacuation.

Source: http://www.nola.gov/ready/evacuspots/map/

has two main parts: those able to congregate at one of the 17 "Evacuspots" in the City (see Fig. 5.6) will be assisted by a cadre of volunteers called "Evacuteers" to be transported to shelters out of the coastal area. The estimated number including tourists in the City who will need evacuation is 40,000, about twice as many as for Hurricane Gustav in 2008 (personal communication with Dev Jani, Chief, Planning & Preparedness, New Orleans Homeland Security and Emergency Preparedness Office, August 10, 2016). There is an enhanced outreach program to determine who will need special transportation, and their contact information is being refined. Approximately 2000 are expected to need this special assistance. The challenges of successfully accomplishing this portion of the evacuation are extensive, beginning with identifying private transport companies and contracting them for the services in a manner that will have them able to achieve their commitment when an event occurs.

Evacuation sites inland for both those participating in city-assisted evacuation as well as those with cars but limited resources are being configured by the state. In the case of an extreme storm mandating full evacuation, some residents will be flown out of the state.

Research and anecdotal reports of the Katrina experiences demonstrate the extreme hardships of evacuation for lower-income residents who do not have money for fuel and motels, especially at the end of the month (Jenkins, 2015). One-half of African-American New Orleans residents live in poverty. This figure is higher than before Katrina. And 55% of the population of New Orleans are African American (The Data Center, 2015). The pressure on these families is extreme and thus all of the measures currently in place are necessary to try to "salvage" their agency during a hurricane.

Two area of responses require continued, determined attention. *First, government officials should continue to retain the "edge" on evacuation plans.* This encouragement applies to the training, drilling, and practice of government officials and staff, and citizens. A storm requiring evacuation has not occurred since 2008, over 8 years ago. Such a long period "takes the edge" off of everyone's response. For government officials, continued preparation includes maintaining the myriad of relationships with agencies who serve the groups who will need government-managed evacuation. These include local nonprofits as well as government agencies and those organizations who will organize the receiving facilities for those transported by the government.

A second concern that is directly tied to agency is less obvious and has had limited attention paid to it since Katrina. *If long-term housing away from New Orleans is required as it was after Katrina, guidance to successfully provide it with agency in mind needs to be developed for the receiving communities.* Some of the recommendations are to pay careful attention to retaining the social network of family and friends and other safety net elements that the poor had before the disaster, and to support their efforts to add to this fabric after the disaster, especially if they are sheltered afar. "In the months and years that followed Katrina, the force of those winds and the rush of those waters were matched by a powerful need to sustain old social networks and to create new ones—to draw individuals back into a center of kin, family, and familiarity. Unable to return to the *place* that had once been home, they tried to reconnect with the *persons* whom they identified with it" (Litt, 2012).

We are not privy to how working class citizens and disaster survivors manage their worlds. This is because these social networks providing safety are not immediately evident, other than to specialists on how the lower-income manage their reality. Excellent research was conducted post-Katrina to create a much more robust body of knowledge about these dynamics. Included in this are Pardee (2014), and the contributors to the edited volume by Weber and Peek (2012). "Low-income women throughout the world exchange goods and assistance within their networks of family and friends as a way of coping with scarce resources" (Fussell, 2012). Evacuation plans that dislocate survivors far away from their home communities and isolate people from one another, add extreme challenges to the survivor being able to gain agency. Government agencies that manage disasters have not invested adequately in learning about the challenges of the poor, and applying what they would learn to creating disaster recovery programs that serve their needs. Given the paucity of financial resources provided to survivor evacuees, knowing about these ways in which lower income citizens survive is critical for their recovery. "Tending to the culture of the survivors, preserves agency and resiliency" (Browne, 2016). Restoring the safety networks as well as supplementing them correctly *is critical* for restoring the lower income survivor's agency.

EXPEDIENTLY PROVIDED, POSTDISASTER HEALTH CARE

Require much more in-depth review of the use of recovery funds for public assistance, as to whether the decisions harm large numbers of survivors when use of the funds is being delayed for the "best" economic development outcome. The purpose of recovery funds must be reaffirmed. The long delay in replacing health care for the poor came about as described above because the decision was made to seek massive recovery support for an economic development initiative that would *in addition* provide medical care for the poor. The order was not the reverse: to restore medical care as quickly as possible—to restore survivor agency for health care—and in addition reap the supplementary benefit of economic development. Concern about economic development being viewed as more important than survivor recovery has a long history (Howitt, 1991; Lawrence, 2007). To avoid this structural violence, more needs to be done to review the process of recovering basic services that have been destroyed or badly damaged. A decadal delay in restoring the ability of the lower income to seek quality medical care is not acceptable, regardless of how "wonderful" the eventual replacement facility might be (Fig. 5.7). Why was not a more careful review of how health care would be provided in the interim being required before the decadal delay was approved? "Seventy-one billion dollars was received by the State of Louisiana for Katrina repairs, rehabilitation and rebuilding. One look at this ['pain'] index and you see who did NOT get the money" (Quigley, 2016a).

SURVIVOR AID AND RETURN OF AGENCY TO ECONOMICALLY AND POLITICALLY DISENFRANCHISED

Consider the economically challenged FIRST in delivering services and restoring agency after a disaster. Because they lack adequate "back up" resources, the poor, often female-headed households (Pardee, 2014) need the assistance and the recovery of their agency *first*. This has rarely happened. Homeowner recovery has been invariably emphasized. Middle and upper income citizens have more political power than the poor. The entire FEMA recovery framework has centered around homeowners. After homeowner aid, assistance to recover the public infrastructure is emphasized, one of the reasons being how lucrative the construction contracts can be. Both of course are important but no more than addressing the recovery of lower income and female-headed households who are so much at risk (Ross, 2015). Case management of lower income people poses challenges no different than the challenges the poor experience themselves. As with the case management of the owners of their own homes, the same recommendations for successful case management should be used for others.

FIGURE 5.7

Foyer of University Medical Center of New Orleans.

Source: *From http://www.nola.com/health/index.ssf/2015/06/look_inside_the_new_1_billion.html*

The examples used to describe the resistance to placing priority on restoring agency to disaster survivors are clearly described by Soron (2007):

Taken together, such scenarios provide us with a distilled image of contemporary structural violence, in which poor and oppressed people are tossed headlong into calamities that have been produced by a social order over which they have little influence, while being systematically denuded of the right and resources needed to protect and sustain themselves.

Enabling survivors to protect and sustain themselves is what agency is about. No more important an area of functional need to achieve agency is housing (Goldberg, 2015). There are so many dimensions to this topic that the issue of agency could have been covered using housing as the exclusive topic. One important aspect of securing agency for survivors is by making *permanent* affordable housing available to survivors. When the building stock of an area is devastated by a major disaster, much of that housing will not be repaired and reintroduced back into usable housing. Different dynamics occur when this damage happens.

One is abandonment of the housing resulting in blight with the prospect but not likelihood of it being returned to the available housing. Demolition of the structures without a new house or apartment being placed on the lot is another. These dynamics lead to greater demand for the remaining stock, which in turn raises the rent (in 2012 over half (54.1%) of the renters in the New Orleans-Metairie metro area are cost burdened (spending more than 30% of their income) with 32.9% in the "severe cost burdened" category (over ½ of their income) (Chase, 2014)) or ownership costs, putting the different social classes into competition with one another for the remaining housing. This prevents lower income residents from having access to affordable housing within acceptable transportation distances from available employment.

New Orleans is experiencing this challenge in a very extensive manner now over a decade following Katrina. Housing for lower income residents is scarce with ability to pay for it hindered by the low salaries of lower income residents. Those units that are available to the poor are often in very poor condition and the upkeep is not adequate with one example from the 2011 Census being that 49,000 rental housing units were in need of major repairs in the previous 12 months (LaRose, 2016). A rental registry is being developed by the City of New Orleans.

Neighborhoods that used to be the homes of those with modest-incomes are now being competed for by more affluent renters and owners, many of whom have come from out of the metro area since the storm. (The New Orleans Advocate perceived this to be such a challenge that it devoted its own editorial column to address it (Advocate Editorials, 2016)). The dynamic is understood as *postdisaster gentrification*. No systematic plan for adding new rental housing was conceived from the recovery context. Public housing continued to be destroyed after Katrina to compound the 5000 units that had been demolished before Katrina. And the Hope Six replacement housing for the demolished units serve only a small number of lower income given the requirement of a distribution of 1/3, 1/3, 1/3 for no rent, subsidized rent, and market rent charges.

Means of addressing these challenges have been put forth. A housing coalition has been created, a report has emanated from the group proposing methods of addressing the inadequate housing. Voluntary inspection of rental housing has been established by the City with its initiation expected to begin in mid-2017 (City of New Orleans and Housing Authority of New Orleans, 2016). On the contrary, gentrification has been welcomed regardless of its effect on lower-income residences.

CONCLUDING REMARKS

The presentation upon which this chapter is based was shared with the White House Subcommittee on Disaster Risk Reduction in November 2015. The first author explained to the assembled group of federal agency representatives that they were going to hear a very harsh presentation, but that it had to be said. We conclude this paper with the same comment. This is not a "naval gazing" exercise to expose the structural violence of government and related nonprofits when agency is taken away from survivors. Too often in disasters there has been a failure to concentrate on restoration of agency after it is snatched from disaster survivors because of embedded policies. Instead, we see a lack of care and respect for the survivors and their ability to have a major role in their own recovery. It is a tremendously important challenge that has not been seen as such by those who could fix it.

If the ideas of harm are important enough to present to professional disaster risk reduction experts including the readers of this volume, then *how* structural violence occurs and *how to mitigate it* must be exposed forcefully. This is no different than the current need to revisit institutional racism in as serious a way possible with the recent public disclosure of its existence, such as in the mistreatment of African-Americans by some significant number of police officers in communities around the country.

New Orleans has made significant strides to provide a successful, agency-affirming evacuation. But the reason is not what one wants to be the motivator: over 1600 identified dead and photos shown around the world of American citizens stranded in life-threatening situations with the military-pointing guns at them "as if they were the enemy" to quote General Honoré. Even these images are fading and years have passed without a significant hurricane to remind one and to practice the agency-affirming activities. And the structural violence of the Louisiana Road Home Program did not act as a deterrent model to the New York State officials who almost replicated the violence in their Build It Back program.

One role model to examine that has flipped the challenges we describe and analyze here is the Coast Guard's response to Katrina. They saved the lives of 35,000 survivors. The form of their organization consists of constantly trained personnel managed by a *centralized* command but *decentralized* execution, giving agency to the personnel once trained, interestingly giving the Coast Guard personnel "perfect agency" in the implementation (Price, n.d.). Perhaps in order to accomplish agency for the survivor, the government, private contractor, and nonprofit personnel should be trained and tasked in such a way that they recognize acts that threaten agency. This recognition can help prevent harm to survivor agency, and fosters pro-active agency-affirming decisions for their survivor clients.

The manner in which the Coast Guard personnel treated each individual rescued exuded respect and a recognition that they were returning agency to the survivors (Price, n.d.). A documentary produced about the Coast Guard's Katrina rescue experiences, "Paratus 14:50" (Video: Paratus 14:50 | Watch Alabama Public Television Presents Online | APT—Alabama Public Television Video) shows clearly this respect of survivors. In addition to fully resourcing the disaster response staffs in all those domains that depend on staff, perhaps inculcating well-trained agency into them will help them remain committed to supporting survivors to recover theirs, in other words helping them resist being the *instruments* of structural violence. It certainly has worked with the Coast Guard.

For those who believe that such an organizational structure and commitment to survivors is not possible because the ramp up of the recovery organization must be so rapid, we ask this question: Why does it seems so easy to create an implementation plan that provides large profits to private contractors before success is achieved by them to aid survivors; but our society is not capable of creating an organizational implementation structure that treats survivors with respect, compassion, and strong, efficient support to restore their agency? The authors believe that it is the will to do so, rather than developing the way. This is what we mean by structural violence. (While this research did not consider the question of which are able to do a better job—government employees or employees of private companies contracted with the government to provide effective recovery services, this question is asked and answered in research conducted by a coauthor of this work (Jerolleman, 2013).

It is expected that more extreme weather events will occur with the evolution of climate change. Disasters, even major catastrophes, will likely become more frequent. If this dire prediction occurs, the various levels of government will not be able to provide the levels of recovery assistance that is provided now and that is wasted now. Learning how to recover agency efficiently and successfully is a capability that will be increasingly needed for community resilience to occur. Getting survivors "back on their feet" rapidly will be a very valuable skill. An example of such agency lauded during the initial Katrina rescue was repeated again in Louisiana during the widespread August 2016 floods: the "Cajun Navy." Private citizens with recreational fishing boats initiated searches for survivors stranded in flood waters (Visser et al., 2016). The culture has again thanked them and celebrated them for the large rescue contribution which they made. The issue of whether their behavior should be regulated has receded as an issue, as recognition of the importance of their agency has occurred. Like the Coast Guard's rescue action, these clear examples of agency and honoring agency must be institutionalized broadly across disaster response.

REFERENCES

Adams, V. (2013). *Markets of sorrow, labors of faith*. Durham, NC: Duke University Press.

Advocate Editorials. Our views: A long-term dilemma for housing after the floods. *The Advocate*. (2016). <http://www.theadvocate.com/baton_rouge/opinion/our_views/article_4f621d48-6ba4-11e6-8822-6fd43c324de8.html> Accessed 28.08.16.

Alter, L. So whatever happened to Katrina cottages? *Treehugger*. (2015). <http://www.treehugger.com/tiny-houses/so-what-ever-happened-katrina-cottages.html> Accessed 12.08.16.

American Civil Liberties Union. *Drug sentencing and penalties*. (2013). Available from https://www.aclu.org/criminal-law-reform/drug-sentencing-and-penalties; U.S. Department of Health & Human Services, Substance Abuse and Mental Health Services Administration.

Auyero, J. (2010). Chuck and Pierre at the welfare office. *Sociological Forum, 25*, 851–860.

Benight, C. (2004). Collective efficacy following a series of natural disasters. *Anxiety, Stress, and Coping, 17*, 401–420.

Browne, K. Presentation as panelist on "Plenary: The Relevance of Culture in Disaster: Expanding the Boundaries of Research and Practice," University of Colorado Natural Hazards Center Workshop, July 13, 2016.

Burdeau, C. (2009). *Honoré: Ex-La. Governor halted hospital reopening*. New York: The Associated Press.

Burtle, A. *Structural violence*. (n.d.). <http://www.structuralviolence.org/> Accessed 10.06.15.

Butterbaugh, L. (2005). Why did hurricane Katrina hit women so hard? *Off Our Backs, 35*, 17–19.

Campanella, R. (2009). Bring your own chairs: Civic engagement in postdiluvial New Orleans. In A. Koritz, & G. J. Sanchez (Eds.), *Civic engagement in the wake of Katrina*. Ann Arbor, MI: University of Michigan Press/Michigan Publishing. <www.press.umich.edu/923684/civic_engagement_in_the_wake_of_katrina> Accessed 27.07.16.

Chase, S. Over half of renters in the metro area are 'cost burdened.' *Curbed NewOrleans 2*. (2014). <http://nola.curbed.com/2014/7/10/10077572/over-half-of-renters-in-the-metro-area-are-cost-burdened> Accessed 26.09.16.

City of New Orleans and Housing Authority of New Orleans *Assessment of fair housing tool*. (2016). Available from https://www.google.com/search?Source = hp&q = city + of + new + orleans + and + housing + authority + of + new + orleans + %282016%29 + assessment + of + fair + housing + tool.&oq = City + of + New&gs_l = psy-ab.1.0.35i39k1 j0j0i20k1l2.602.2545.0.6380.12.11.0.0.0.0.149.1187.0j10.10.0....0...1.1.64.psy-ab..2.10.1186.0..0i131k1.7MMfxGSzzLE[SL1]

Cleary, P. D., & McNeil, B. J. (1988). Patient satisfaction as an indicator of quality of care. *Inquiry: A Journal of Medical Care Organization, Provision, and Financing, 25*, 25–36.

Conte, H. R., Ratto, R., Clutz, K., & Karasu, T. B. (1995). Determinants of outpatients' satisfaction with therapists: Relation to outcome. *Journal of Psychotherapy Practice and Research, 4*, 43–51.

Dolan, M. & Egan P. *U.S. Rep. to Snyder: emergency manager law brought Flint 'to its knees.'* (2016). <http://www.freep.com/story/news/local/michigan/flint-water-crisis/2016/03/17/snyder-defends-emergency-manager-flint/81905614/> Accessed 07.08.16.

Draucker, C. B., & Petrovic, K. (1997). Therapy with male survivors of sexual abuse: The client perspective. *Issues in Mental Health Nursing, 18*, 139–155.

Eden, R. & Boren, P. *Timely assistance: Evaluating the speed of road home grant making*. Rand Corp. (2008). Available from http://www.rand.org/pubs/documented_briefings/DB557.html.

Edmonds, K. (2012). Beyond good intentions: The structural limitations of NGOs in Haiti. *Critical Sociology, 39*, 439–452.

Eitler, T. W., McMahon, E. T., & Thoerig, T. C. (2013). *Ten Principles for building healthy places.*. Washington, D.C: Urban Land Institute.

Fullilove, M., & Saul, J. (2006). Rebuilding communities post-disaster in New York. In Y. Neria, R. Gross, R. Marshall, & E. Susser (Eds.), *9/11: Mental health in the wake of terrorist attacks* (pp. 164–177). New York: Cambridge.

Fussell, E. (2012). Help from family, friends, and strangers during hurricane Katrina: Finding the limits of social networks. In L. Weber, & L. Peek (Eds.), *Displaced: Life in the Katrina diaspora* (pp. 150−166). Austin, TX: University of Texas Press.

Goldberg, E. Psychologist's 'crazy' idea is now ending homelessness. *Huffington Post*. (2015). <http://www.huffingtonpost.com/2015/05/07/housing-first-_n_7226218.html> Accessed 29.09.16.

Gratz, R.B. *Why was New Orleans's Charity Hospital allowed to die?* (2011). <http://www.thenation.com/> Accessed 15.06.15.

Galtung, J. (2009). Violence, peace, and peace research, *Journal of Peace Research*, 6(3), 167−191.

Hammer, D. Examining post-Katrina road home program: "It's more than the money. It's the hoops we had to jump through to do it." *New Orleans Advocate*. (2015, August 23). <http://www.theadvocate.com/content/tncms/live/> Accessed 07.08.16.

Howitt, R. (1991). Aborigines and restructuring in the mining sector: Vested and representative interests. *Australian Geographer*, 22, 117−119.

Jenkins, P. Preparing for storms in Louisiana: manual for nonreaders. *Chart Publications, Paper 46*. (2015). Available from http://www.scholarworks.uno.edu/chart_pubs/46.

Jerolleman, A. (2013). The privatization of hazard mitigation: a case study of the creation and implementation of a federal program, University of New Orleans Theses and Dissertations, Paper 1692.

Jones, L. Many residents leery of 'Claiborne corridor' study. *The New Orleans Tribune*. (n.d.). <http://www.theneworleanstribune.com/main/many-residents-leery-of-claiborne-corridor-study/> Accessed 31.07.16.

Jones-DeWeever, A., & Hartmann, H. (2006). Abandoned before the storms: The glaring disaster of gender, race and class disparities in the gulf. In C. Hartman, & G. D. Squires (Eds.), *There's no such thing as a natural disaster: Race, class and Katrina* (pp. 85−102). New York, London: Routledge.

Kapucu, N. (2015). Emergency Management: Whole Community Approach. In *Encyclopedia of public administration and public policy* (3rd ed.). CRC Press, Taylor and Francis.

Kenny, A. (2015). Time and punishment. *West Space Journal*, 3(Autumn). <www.westspacejournal.org.au/issue/autumn/2015> Accessed 28.01.16.

Klein, N. (2007). Disaster capitalism. *Harper's Magazine*, 315, 47−58.

Kreuger, R. (2015). *New Orleans site selected as study area for 2015 Hines competition*. Urbanland: The Magazine of the Urban Land Institute. <https://americas.uli.org/press-release/new-orleans-2015-hines-competition/> Accessed 27.07.16.

Krupa, M., & Russell, G. (2007). N.O.-K blueprint unveiled: Plan puts most cash in east, lower ninth. *Times Picayune*, A1.

Kunreuther, H., & Pauly, M. (2005). Insurance decision-making and market behaviour. *Foundations and Trends in Microeconomics*, 1(2), 63−127.

Lafer, G. (2008). From Baghdad to the Bayou: Neoliberalism and the shrinking of democracy. *New Labor Forum (Routledge)*, 17(1), 108−121.

Landau, J., & Saul, J. (2004). Facilitating family and community resilience in response to major disaster. In F. Walsh, & M. McGoldrick (Eds.), *Living beyond loss: Death in the family* (pp. 285−309). New York: Norton.

Landry, B. (1987). *The new Black middle class*. Oakland: CA: Univ. of California Press.

LaRose, G. (2016). New Orleans rental registry would root out slumlords. *The Times Picayune*. <http://www.nola.com/politics/index.ssf/2016/05/new_orleans_rental_registry_wo.html> Accessed 26.09.16.

Lawrence, R. (2007). Corporate social responsibility, supply chains and saami claims: Tracing the political in the Finnish forestry industry. *Geographical Research, 45*, 167–176.

Lein, L. (2015). Still waiting for help. *The Conversation*. <http://theconversation.com/still-waiting-for-help-the-lessons-of-hurricane-katrina-on-poverty-46666> Accessed 29.12.16.

Lipsky, M. (1980). *Street-level bureaucracy: Dilemmas of the individual in public services.*. New York: Russell Sage Foundation.

Litt, J. (2012). Section introduction to "social networks.". In L. Weber, & L. Peek (Eds.), *Displaced: life in the Katrina diaspora* (pp. 145–149). Austin, TX: University of Texas Press.

Marks, R. E. (2008). Canoeing home: A personal and professional journey through Murky Waters. *Traumatology, 14*, 14–20.

Montjoy, R., Farris, M., & Devalcourt, J. Achieving successful long-term recovery and safety from a catastrophe: recommendations for public assistance. *CHART Publications. Paper 4*. (2010). Available from http://scholarworks.uno.edu/chart_pubs/4.

New York City Comptroller. *Audit report on the administration of the New York City Build It Back single family program by the Mayor's Office of Housing Recovery Operations* (2015).

Norris, F., Stevens, S., Pfefferbaum, B., Wyche, K., & Pfefferbaum, R. (2008). Community resilience as a metaphor, theory, set of capacities, and strategy for disaster readiness. *American Journal of Community Psychology, 41*, 127–150.

Ott, K.B. (2012). The closure of New Orleans' Charity Hospital after hurricane Katrina: A case of disaster capitalism, University of New Orleans Theses and Dissertations, Paper 1472.

Pattillo-McCoy, M. (1999). *Black picket fences: Privilege and peril among the black middle class.*. Chicago: University of Chicago Press.

Pardee, J. W. (2014). Surviving Katrina: The experiences of low-income African American women. Boulder & London: First Forum Press.

Pentarki, M. (2013). The class impact of post disaster restorations policies: The example of Ilia, Greece and the need for a politics of disaster. *International Social Work, 56*, 761–774.

Perez-Sales, P., Cervellon, P., Vazquez, C., Vidales, D., & Gaborit, M. (2005). Post-traumatic factors and resilience: The role of shelter management and survivors' attitudes after the earthquakes in El Salvador (2001). *Journal of Community and Applied Social Ps/ychology, 15*, 358–382.

Price, S., Sr. *U.S. coast guard and hurricane Katrina*. (n.d.). <https://www.uscg.mil/history/katrina/docs/DarkestDay.pdf> Accessed 11.08.16.

Quigley, B. New Orleans Katrina pain index at 10: Who was left behind. *Huffington Post*. (2016a). <http://www.huffingtonpost.com/bill-quigley/new-orleans-katrina-pain_b_7831870.html> First accessed 15.10.15; re-accessed 11.08.16.

Quigley (2016b, July). Baton Rouge: "Put Those Damn Weapons Down!" Huffington Post. <http://www.huffingtonpost.com/entry/baton-rouge-put-those-damn-weapons-down_-us_57856aeee4b0e7c8734f0639>. Accessed 22.08.16.

Rand Corporation. (2008). *Long, Unpredictable Delays Found in Louisiana 'The Road Home' Grants to Homeowners*. <http://www.rand.org/news/press/2008/05/27.html>. Accessed 16.06.15.

Reckdahl, K. (2014). Ice chests on sidewalks: New Orleans' Groundbreaking Post-Katrina Healthcare System. *Next City*. <https://nextcity.org/daily/author/katy-reckdahl> Accessed 28.08.16.

Reid, M. (2013). Social policy, "deservingness," and socio-temporal marginalization: Katrina survivors and FEMA. *Sociological Forum, 28*, 742−763.

Rhudy, D. (1999). Satisfaction with brief or time-limited therapy in our Texas community mental health centers (brief therapy). *Dissertation Abstracts International Section A: Humanities and Social Sciences, 60*(4-A), 1029.

Rivlin, G. *Why the plan to shrink New Orleans failed.* (2015). <http://fivethirtyeight.com/features/why-the-plan-to-shrink-new-orleans-after-katrina-failed/> Accessed 27.07.16.

Ross, T. (2015). A disaster in the making: Addressing the vulnerability of low-income communities to extreme weather. *Center for American Progress*. <https://www.americanprogress.org/issues/poverty/report/2013/08/19/72445/a-disaster-in-the-making/> Accessed 31.07.16..

Russell, G. & Donze, F. Officials tiptoe around footprint issue. *BGR > In the News*. (2006). <http://www.bgr.org/news/archives/officials-tiptoe-around-footprint-issue/> Accessed 02.08.16.

Sixma, H. J., Spreeuwenberg, P. M., & Van der Pasch, M. A. A. (1998). Patient satisfaction with the general practitioner: A two-level analysis. *Medical Care, 36*, 212−229.

Soron, D. (2007). Cruel weather: Natural disasters and structural violence. *Transformations: Journal of Media and Culture*. Available from http://www.transformationsjournal.org/issues/14/PDF/Soron_Transformations14.pdf.

Stringer, S. M. *New York City Comptroller Press Release*. (2015). <http://comptroller.nyc.gov/newsroom/comptroller-stringer-audit-of-build-it-back-reveals-millions-paid-out-for-incomplete-work-double-billing-undocumented-travel-costs/> Accessed 12.01.16.

The Data Center *Facts for features: Hurricane Katrina recovery*. (2015). <http://www.datacenterresearch.org/data-resources/katrina/facts-for-features-katrina-recovery/http://www.datacenterresearch.org/data-resources/katrina/facts-for-features-katrina-recovery/> Accessed 11.08.16.

Van den Eynde, J., & Veno, A. (1999). Coping with disastrous events: An empowerment model of community healing. In R. Gist, & B. Lubin (Eds.), *Response to disaster: Psychosocial, community, and ecological approaches* (pp. 167−192). Philadelphia, PA: Brunner/Mazel.

Visser, S., Jackson, A., Yan, H., & Flores, R. Louisiana flooding: 'Cajun Navy' answers call for volunteers. CNN. (2016). <http://www.cnn.com/2016/08/16/us/louisiana-flooding> Accessed 29.09.16.

Weber, L., & Peek, L. (2012). *Displaced: Life in the Katrina Diaspora*. Austin: Univ. of Texas.

White, G. A long road home. *The Atlantic*. (2015). <http://www.theatlantic.com/business/archive/2015/08/hurricane-katrina-sandy-disaster-recovery-/400244/> Accessed 31.07.16.

Wilson, S., Temple, B., Milliron, M., Vazquez, C., Packard, M., & Rudy, B. (2007). The lack of disaster preparedness by the public and it's effect on Communities. *The Internet Journal of Rescue and Disaster Medicine, 7*. Available from http://ispub.com/IJRDM/7/2/11721.

Wright, M. (2011). Client satisfaction and the helping/healing dance. *Qualitative Social Work., 11*, 644−660.

PART II

Disaster Vulnerability

CHAPTER 6

Setting the Stage for the Katrina Catastrophe: Environmental Degradation, Engineering Miscalculation, Ignoring Science, and Human Mismanagement

Ivor L. van Heerden

Agulhas Ventures, Inc., Reedville, VA, United States

CHAPTER OUTLINE

Introduction	134
Origin and Function of Louisiana's Coastal Wetlands	134
The "Natural" Cycle of Wetland Development and Maintenance	134
Subsidence and Relative Sea-Level Rise	137
Surge and Storm Wind Reduction	138
Management of the Coastal Wetlands and the Resultant Dilemma	139
Control of the Lower Mississippi River	139
Oil and Gas Extraction and the Disruption of Natural Hydrology; Enhanced Subsidence	141
The Mississippi River Gulf Outlet Navigation Channel	141
The MRGO "Funnel"	144
New Studies—Wetland Storm Reduction Value	145
Wetland Role in Surge Reduction	145
Computer Modeling to Reconstruct Surge and Waves Along the MRGO	145
Effects of MRGO Reach 2 on Waves	146
Hurricane Betsy and the 1965 Flood Control Act	148
Forensic Investigations—Levee Failures	149
Polder Levee Failures	149
Accounting for Subsidence and Sea-Level Rise	150
Recommendations for Sustainability	151
Conclusions: Habitat Sustainability Is Needed to Support Human Resilience	154
References	155
Further Reading	158

INTRODUCTION

The coastal wetlands and estuaries of Louisiana are one of the world's great ecosystems (Fig. 6.1). For millennia, the Mississippi River has supplied the coast with sediment, freshwater, and nutrients to build a vast expanse of marsh and swamp. These wetlands have been altered by natural erosional processes. The dynamic interplay of land and water—where new wetlands are continuously built and old lands remolded and lost—has produced an environment rich in natural habitats and biological productivity. Millions of people rely directly or indirectly on the coastal wetlands for their livelihood and, most importantly for this chapter, for protection against hurricanes and winter storms. The Delta Coast also produces some 26% of the US domestic oil production and almost 20% of its natural gas and, because the combination of ports that line the lower 300 km of the Mississippi River rank as the greatest port in the world, the coast is critically important to US export industries (van Heerden, 2007).

In the last several decades, humans have impacted this ecosystem in several ways. As a result the natural surge buffeting offered by the wetlands has been severely reduced. Coastal Louisiana is, therefore, very vulnerable to hurricanes and major hurricane surge flooding; as 70% of the population lives on 36% of land at or below sea level (van Heerden, 1994).

Hurricane Katrina made landfall in southeast Louisiana as a fast-moving Category 3 hurricane at 6:10 a.m. on August 29, 2005 (Fig. 6.1). Of the populated areas that constitute greater New Orleans (GNO), 80% of Orleans Parish, 99% of St. Bernard Parish, and approximately 40% of Jefferson Parish were flooded, in some cases for weeks (van Heerden et al., 2007). This flooding costs the lives of more than 1600 residents (van Heerden et al., 2007). More than 100,000 families were rendered homeless, the great majority of whom had heeded evacuation orders. Overall, Hurricane Katrina was the costliest and one of the deadliest disasters in US history. But, what factors precipitated this catastrophe and compromised human survival and resilience in the wake of this storm? Was it self-inflicted? Was it natures' "fault"? And most importantly, was the magnitude of destruction preventable? To better understand the factors that impacted the destructive capacities of Hurricane Katrina, we need to review both the natural geological evolution of this region and the impact of human intervention and the changing built environment.

ORIGIN AND FUNCTION OF LOUISIANA'S COASTAL WETLANDS

THE "NATURAL" CYCLE OF WETLAND DEVELOPMENT AND MAINTENANCE

Over the past 7000 years, the Mississippi River has created seven deltas (Fig. 6.2) by a process of delta switching, in which the river abandoned its main channel

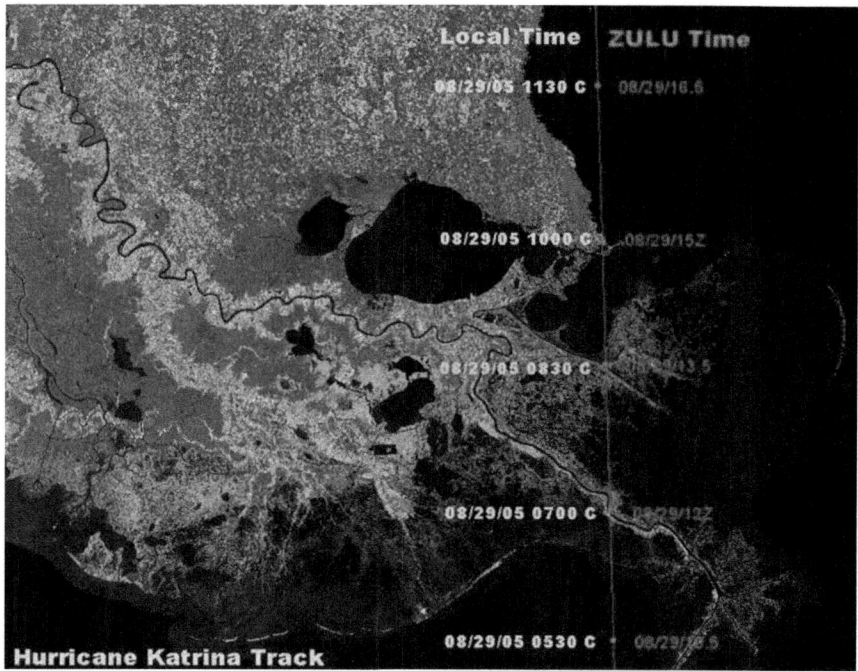

FIGURE 6.1

Satellite image of Coastal Louisiana showing track of Hurricane Katrina.
Source: *Modified from van Heerden, Kemp, Mashriqui, Sharma, Prochaska, Capozzoli, et al. (2007). The failure of the New Orleans Levee System during Hurricane Katrina. A report prepared by TEAM LOUISIANA for Secretary Johnny Bradberry, Louisiana Department of Transport and Development, Baton Rouge, Louisiana.*

approximately every 1000 years in favor of a shorter route to the Gulf of Mexico (Kolb & Van Lopik, 1958). The abandoned delta then entered an erosional phase, while the locus of sediment deposition slowly shifted to the new active watercourse. Prior to the early 1900s, the Mississippi River overtopped its banks for approximately 90 days a year. This annual river flood introduced sediment and freshwater into the adjoining wetlands, which aided in their maintenance. In addition, distributaries (the coastal branches of rivers) delivered sediment and freshwater to more distant wetlands. Switching from an inefficient older watercourse to a newer one can take up to 500 years, and even thereafter, the abandoned channel may act as a distributary for many hundreds of years (van Heerden, 1994); diverting sediment and freshwater to the wetlands—initially created when the distributary was the main channel of the river (Fig. 6.3). Two such former courses of the Mississippi—Bayou Lafourche and Bayou La Loutre—were distributaries of the Mississippi River in the late 1600s (van Heerden, 1994).

CHAPTER 6 Setting the Stage for the Katrina Catastrophe

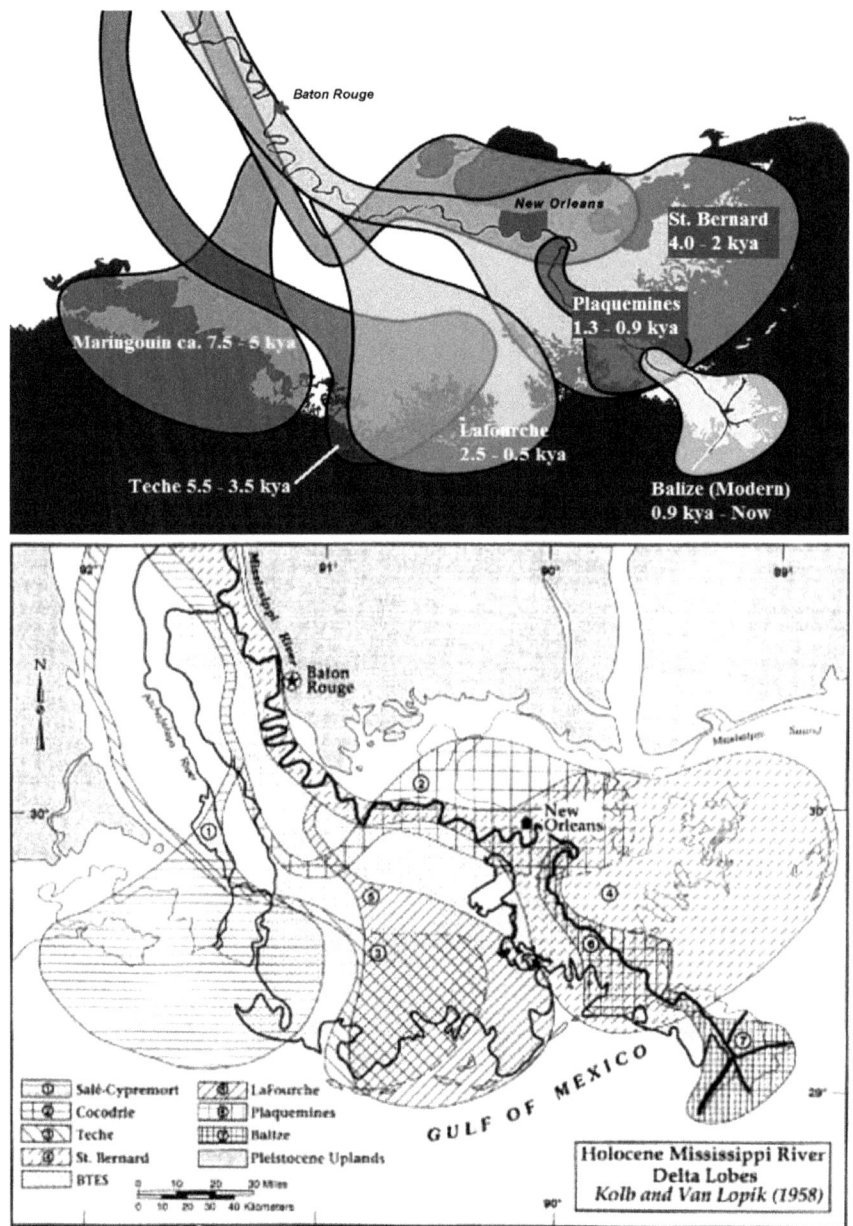

FIGURE 6.2

Sequence of Holocene delta development (Kolb & van Lopik, 1958).

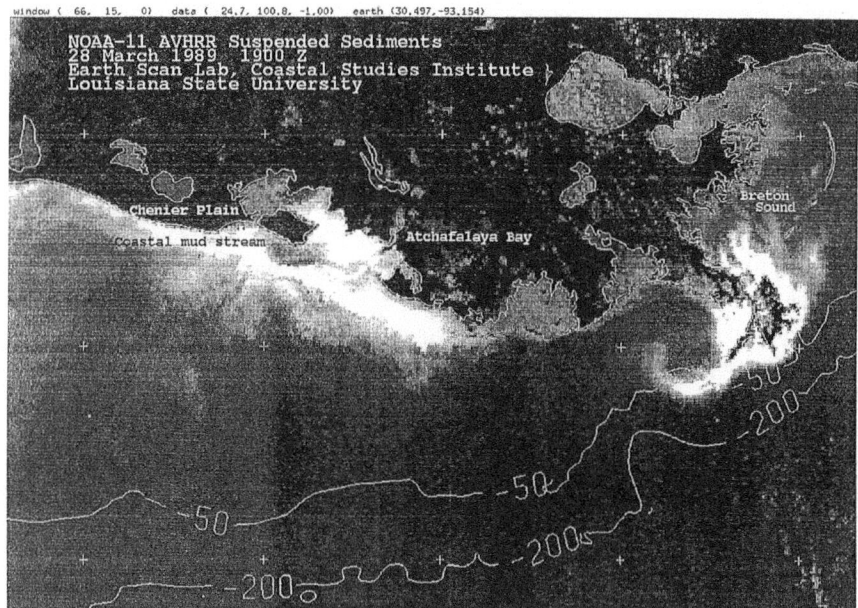

FIGURE 6.3

Suspended sediment distribution iii Louisiana Coastal waters as revealed by an enhanced NOAA-11 AVHRR image.

Source: *Courtesy: Dr. Nan Walker, Earth Scan Lab, Coastal Studies Institute, L.S.U.*

Thus, large-scale switching of the loci of deposition fashioned the distributary ridges and extensive interdistributary (i.e., land between coastal branches) wetlands that comprise the vast deltaic plain of South Louisiana. Until human intervention reversed the trend about 90 years ago, the delta switching process, coupled with the shorter-period spring flood sedimentation, accounted for a net significant gain of land, of the magnitude of 3 km^2/yr (Templet & Meyer-Arendt, 1988).

SUBSIDENCE AND RELATIVE SEA-LEVEL RISE

An important part of shaping the delta cycle is surface subsidence (or the lowering of land surface elevation, in this instance relative to sea level) and global sea-level rise (i.e., the lowering of land elevation with the rising of the sea level). Estimates exist for relative sea-level rise (RSLR) for Louisiana. The most comprehensive study of RSLR for Louisiana (Ramsey & Penland, 1989) utilized data from two tide-gage networks in Louisiana and the northern Gulf of Mexico to determine local and regional trends in RSLR. Two tide-gage data are a network of instruments that measure the tidal amplitude signal. Tide-gage data are used to measure the impacts of global warming. The US Army Corps of Engineers (USACE) maintains a network of 83 tide-gage stations throughout coastal Louisiana. The National Ocean Survey (NOS) maintains nine tide-gage stations throughout the northern Gulf of Mexico in Texas, Louisiana, Mississippi, Alabama, and Florida. The data gathered from these tide-gage stations illustrate how coastal Louisiana is sinking more rapidly compared with neighboring coastlines.

Based on the USACE data set, the rates of RSLR in Louisiana (Ramsey & Penland, 1989) range from 0.34 to 1.77 cm/year (or from 1.1 to 5.8 feet/century). The central portion of the coast is "sinking" the fastest; reflecting the fact that it sits on top of the thickest sequence of younger sediments. Representative water-level histories from outside of the central coast indicate that the regional rates of RSLR decrease to the east and the west. In the surrounding coastal states the rates of RSLR, based on the NOS data set, decrease from 0.62 cm/year at Galveston, TX, to 0.15 cm/year at Biloxi, MS. Hence, mean RSLR in Louisiana is more than five times the average for the Gulf of Mexico and is rising 10 times faster in the central coast than in the rest of the world.

The rapid rate of RSLR observed in Louisiana can be attributed to the compactional subsidence, (in which the earth's surface is lowering due to the compacting of sediment) of the Mississippi River delta plain, and the signal from the rise in sea level globally due to global warming. However, the major component of the rise is due to the fact that Louisiana directly overlies the entrenched Pleistocene valley of the Mississippi River, which is filled with younger (Holocene) deltaic sediments that are more than 150 m thick. These sediments compact and dewater (i.e., the sinking or subsidence of land due to the loss of water, and thereby the loss in bulk of the land) with time, and along with a series of growth faults, and the activities of subsurface oil and gas extraction, the whole system of coastal Louisiana is subsiding (sinking) at a very rapid rate (van Heerden & Bryan, 2006) and will continue to do so into the future. As will be discussed shortly, the human activities of oil and gas extraction are markedly enhancing this subsidence.

SURGE AND STORM WIND REDUCTION

Since the middle of the 20th century, there has been an understanding that wetlands buffer storm-surge and hurricane winds. The USACE (1965) published a graph illustrating that, "on average, land lying between the coast and human inhabitants, reduced storm surge by 6.9 cm/km (1 foot / 2.75 miles). This regression is based on hurricanes hitting the Louisiana coast from 1909 to 1957" (Shaffer, et al., 2009, p. 209). However, prior to Hurricane Katrina, little research or modeling had been undertaken to determine the actual storm reduction that wetlands offer in coastal Louisiana. van Heerden et al. (2007) determined that the 1992 Hurricane Andrew winds lost about 50% of their energy as they crossed the 35 km of healthy wetlands separating Morgan City from the coastline.

Forested wetlands are more effective at surge and wave reduction than low-lying marshes. The sheltering effect of forested wetlands affects the fetch (i.e., the distance the wind blows over the water) over which wave development takes place. Shallow water depths attenuate waves via bottom friction and breaking, while vegetation provides additional frictional drag and wave attenuation, and also limits static wave set-up, which limits the height of waves. Thus, destroying the wetlands apron and/or impacting its ability to maintain itself significantly enhances the risk of hurricane impacts on the built environment.

MANAGEMENT OF THE COASTAL WETLANDS AND THE RESULTANT DILEMMA

CONTROL OF THE LOWER MISSISSIPPI RIVER

When the French settled the region in 1699, the Mississippi River was not confined to the present narrow bird-foot delta. Rather, the river flowed "sweet and wild" into the Gulf of Mexico through a series of distributaries, which extended for more than 260 km along the coast (Condrey, 1993). Superimposed on the distributary network were two major channels, the right fork being the present Mississippi River and the left fork, Bayou Lafourche. As discussed earlier, at the larger system scale, the land was built up as a consequence of delta switching; however, frequent river flooding events effected some wetland creation and certainly were responsible for the vertical growth of wetlands (i.e., wetland aggradation) and maintenance.

The political and economic control of the Mississippi was a driving force for early European explorers. Flood control became an issue after the first settlements were established. Europeans had developed the technology of constructing levees along the Po, Danube, Rhine, Rhone, and Volga rivers, prior to the establishment of New Orleans. Thus, Louisiana's new immigrants had the technical skills to start immediately building levees, and in 1717 the Mississippi River levee system was born (van Heerden, 1994). While flooding was of concern to Louisianans' along the river, national interest in the first half of the 19th century was focused on enhancing river navigation.

In 1849, as a consequence of a series of severe floods, the flood control problem also became a national issue with the enactment of the Swamp Land Act, and by 1851 a continuous 200 km levee existed that was centered in New Orleans. In the 1850s, the state established levee districts and boards, and by 1858 more than 3218 km of levees lined the Mississippi River, centered on New Orleans (Davis, 1993). These levees failed catastrophically during the flood of 1927; the product of abnormally high rainfall in the Mississippi's drainage basin. Approximately 800,000 individuals were driven from their homes and in excess of 59,570 km^2 were inundated (Simpich, 1927). The severity of the 1927 flood resulted in passage of the 1928 Flood Control Act and the construction of the USACE's present

Mississippi River guide levees; which have, to date, effectively eliminated overland flooding, but also sealed most of the Mississippi River distributaries south of Baton Rouge (Fig. 6.2).

Natural overbank flooding and distributary-initiated wetland accretion (i.e., the lateral growth of the wetlands) and maintenance was thus terminated by these very effective artificial river levees. Sediment originally destined for shallow wetland locations now flows down an effective conduit, the leveed Mississippi River, where it is deposited in deep Gulf of Mexico water and, thereby, completely lost from the system. In this manner, human intervention has terminated the system's natural ability to maintain itself through areal spreading of sediments in order to facilitate the expansion of "our" world; the built environment. These activities and their deleterious consequences are a reflection of the ongoing folly between humans and the environment.

There is, however, one bright spot; The Atchafalaya River, which first captured Mississippi discharge in the mid-1500s, offers a much shorter course to the Gulf. Because of a very real concern that the Mississippi River might abandoned its course through New Orleans, a control structure was erected at the point of diversion (van Heerden, 1983; Fig. 6.1). Consequently the Atchafalaya discharge is now held at 30% of the Mississippi discharge; thereby restricting the Atchafalaya's ability to capture the flow of the Mississippi. Between 1550 and 1950 the Atchafalaya Basin, once a large lake, was filled with river sediment and converted to a riverine forest and swamp (Roberts, Adams, & Cunningham, 1980). Channelization (i.e., man-made navigation channels dredged in the basin) within the basin in the 1950s, forced Atchafalaya sediment, originally destined for the basin, to be deposited as a subaqueous delta in Atchafalaya Bay; thereby setting the stage for the development of new land (van Heerden, 1983). In 1973, the Atchafalaya delta became a subaerial feature; now growing above the water with vegetation (Roberts et al., 1980). Wax Lake Outlet also connects the river to Atchafalaya Bay, so that two deltas are now developing in the Bay (van Heerden, 1983). Their combined subaerial expression exceeds 17,300 acres and represents the only real area of natural wetland growth in Louisiana. Hence, this is the only coastal area of Louisiana that is actually *gaining* wetlands (Roberts & van Heerden, 1992a; van Heerden & Bryan, 2006).

In addition, Atchafalaya River sediment is dispersing over a much greater area than the two deltas in Atchafalaya Bay. Sediment that bypasses the two deltas is being deposited on the shelf seaward of the bay, or pushed westward by longshore currents. The current system transports so much sediment the feature is now known as the "Coastal Mud Stream" (Fig. 6.3). Some of this sediment is deposited along the Chenier Coast, where mud flats are forming. As these flats become vegetated, the Chenier Coast annually gains about 150 acres of new wetland (Roberts & van Heerden, 1992b). Sediment is also migrating out of Atchafalaya Bay into the adjacent marshes. Increased freshwater inputs from Atchafalaya Bay are forcing a plant succession from brackish to intermediate marsh types.

Engineering the Mississippi River resulted in the shutdown of the maintenance and expansion of natural wetlands by redirecting the sediment load into the Gulf of Mexico. Thus, humans have very effectively begun to "starve" the wetlands; initiating their demise. The Atchafalaya River delta distributary system remains the only small ray of hope; offering an example for future coastal management concerning major river diversions.

OIL AND GAS EXTRACTION AND THE DISRUPTION OF NATURAL HYDROLOGY; ENHANCED SUBSIDENCE

The 20th century exploration and mining of oil and gas sealed the fate of Louisiana's wetlands. Thousands of kilometers of access and pipeline canals and navigation channels were dredged allowing saltwater intrusion into sensitive freshwater wetlands with very destructive direct consequences, as well as significant disruption of the natural wetland hydrology. Within the delta, about 15,000 km of canals led to pervasive alterations of hydrology (Day et al., 2000; Turner, Costanza, & Schaife, 1982). Canal spoil banks interrupt sheet flow, impound water, and cause deterioration of wetlands. Long, deep navigation channels lessen freshwater retention time and allow greater inland penetration of saltwater. This concept is discussed further in regard to the impacts of larger navigation channels.

While many researchers (see, Gagliano, Kemp, Wicker, Wiltenmuth, & Sabate, 2003) believe that most of the subsidence in Louisiana is due to natural processes, Morton, Buster, and Krohn (2002) and Morton, Tiling, and Ferina (2003) conclude that the historical subsidence and wetland loss were primarily induced by fluid withdrawal and, therefore, the future impacts are qualitatively predictable. Results from a further study by Morton, Bernier, Barras, and Ferina (2005) confirmed that the most likely explanation for historical wetland losses in south-central Louisiana is regional subsidence and local fault reactivation induced by hydrocarbon production. These impacts are an important lesson for any other coastal state looking to expand oil and gas extraction—for which there will be a cost.

THE MISSISSIPPI RIVER GULF OUTLET NAVIGATION CHANNEL

By the mid-20th century, artificial channels further cut up the "starving" coastal wetlands and impacted wetland hydrology. Tide and storm-driven flux was enhanced in the early 1960s, when the USACE built the Mississippi River Gulf Outlet (MRGO) project, approved by the US Congress under the Rivers and Harbor Act of 1956, that added a 122 km long deep-draft ship channel to this already modified landscape (Fig. 6.4). This canal directly connected the Gulf of Mexico to the heart of New Orleans. The MRGO project was expected to spur economic development by providing a shorter, slack water alternative to the Mississippi River for vessels calling at the port. For a variety of reasons the MRGO failed to attract shipping, but cost $500 million to maintain while causing

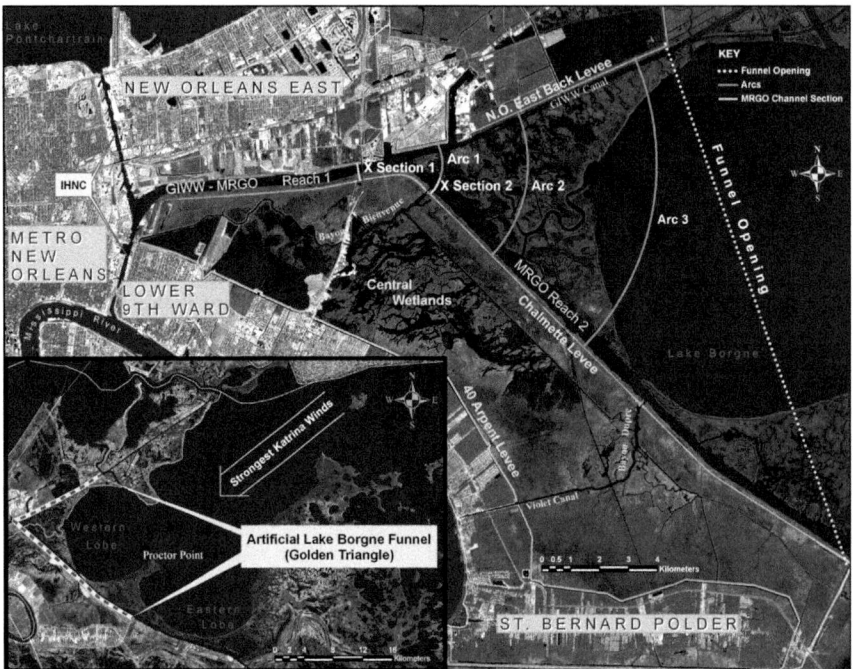

FIGURE 6.4

The MRGO funnel showing channels, levees, polders, and funnel cross-sections used for analysis of surge conveyance.

catastrophic wetland loss and degradation. It was decommissioned in 2008 (Shaffer et al., 2009).

The MRGO was authorized with a 36 foot controlling depth, 500 feet wide at the bottom and 650 feet wide at the top, with a somewhat larger cross-section across the shallow waters of Breton Sound (USACE, 1999). By 2005, the channel had opened to up to 3000 feet in some locations and had been dredged annually (Fig. 6.5). Saltwater intrusion associated with the construction of the channel had killed thousands of acres of surge reducing freshwater marshes and cypress swamp (Shaffer et al., 2009).

The direct impacts of the MRGO included channel excavation, dredged material disposal (i.e., the material that is dug and relocated filling both wetlands and shallow water bodies), and wave erosion. Indirect effects that caused shifts in habitat type and increased wetland loss include salinity intrusion and hydrological changes caused by the MRGO. Numerous studies (see, Coast 2050, 1998; USACE, 1999; van Heerden et al., 2007; Wicker et al., 1982) detail the land loss caused by the MRGO during different periods using various defined areas. All of the studies detail tens of thousands of hectares of land loss. Even the USACE (1999) reported

FIGURE 6.5

Five decades of marsh loss looking east from Bayou Bienvenue at MRGO Reach 2 and the Chalmette Levee with Lake Borgne to left and Central Wetlands (CW) to right (Fitzgerald et al., 2008).

that 6680 hectare (ha) of marsh were filled or excavated during MRGO construction. In addition, 2308 ha of water were filled or deepened. In 1956, 4130 ha of bald cypress forest and 2308 ha of fresh and intermediate marsh existed. Both the bald cypress—water tupelo swamp and the fresh and intermediate marsh have since been replaced with open water and brackish or saline marsh.

THE MRGO "FUNNEL"

The geography of the Mississippi River Gulf Outlet (MRGO) funnel includes the preexisting Gulf Intracoastal Waterway (GIWW) along the northern margin, the enlarged portion of the GIWW referred to as MRGO Reach 1 that serves as an outlet to the Inner Harbor Navigation Canal (IHNC); the MRGO Reach 2 channel along the south margin, and levee embankments paralleling all of the artificial channels on the inland side (Fig. 6.4). The funnel also contains natural features including the southern half of Lake Borgne and thousands of acres of wetlands both on the inboard and outboard sides of the hurricane protection structures. Lake Borgne is a very large shallow bay elongated along a southwest to northeast axis, providing more than 40 miles of open water in this direction—a very long "fetch" over which hurricane winds can build surge and waves that can be funneled toward the city (Fig. 6.4). The Funnel then acts to vertically enhance the storm-surge elevation, as the wind blows the water toward its narrow neck, leading to more extensive flooding due to levee overtopping or failure. The rationale for these developments was to enhance navigation. However, the surge amplification resulting from these changes was ignored. Three highly populated polders (i.e., area of city surrounded by a levee) surround the Funnel (Fig. 6.4). All of these developed polders received floodwaters from the MRGO channel and funnel.

The wetlands within the funnel were dominated by 2.4–3.0 miles of freshwater bald cypress (*Taxodium distichum*)—water tupelo (*Nyssa aquatica*) swamp that have become intermediate to salt marshes and open water ponds since the construction of the MRGO (Fitzgerald et al., 2008). However, their ability to reduce storm surge was severely curtailed by their degradation, which was due to the impacts of the MRGO. St. Bernard has some of the highest land on the East Bank of New Orleans, following, as it does, the natural levee of the Mississippi River and some of its abandoned distributaries (Fig. 6.4). Despite being relatively high by local standards, the St. Bernard polder experienced the most violent, spatially expansive, and deepest flooding in the entire metro area during the Katrina event. Except for a limited contribution from rainfall, all flooding of the St. Bernard polder was caused by water that passed through or across one or more reaches of the MRGO. This water entered the developed area as a result of catastrophic floodwall failures along the IHNC on the western margin, by overtopping of levees on MRGO Reach 1, and by flow through breaches in the federally built levees along the MRGO Reach 2. Thus this extensive flooding was due to the surge height amplification from what is, ultimately, a human-made funnel.

If the control of the Mississippi River and consequential lack of sediment meant starving the coastal wetlands "to death," then navigation and mining activities guaranteed their obliteration. The MRGO offers a textbook example of the consequences of disrupting natural hydrology that enhance flood risk.

NEW STUDIES—WETLAND STORM REDUCTION VALUE

The Katrina catastrophe led to much new research that moves beyond identification of the reasons the levees failed. This research is grouped into the following three categories.

WETLAND ROLE IN SURGE REDUCTION

Emergent canopies, such as provided by forested wetlands, greatly diminish wind penetration, thereby reducing the wind stress available to generate surface waves and storm surge. Mangroves have been shown to reduce wave heights by 20% over distances of only 100 meters (m) (Mazda, Magi, Ikeda, Kurokawa & Asano, 2006; Mazda, Magi, Kogo, & Hong, 1997) and 150 m of Rhizophora-dominated forest can dissipate wave energy by 50% (Brinkman, Massel, Ridd, & Furakawa, 1997). More recently, Krauss et al. (2008) measured storm-surge reduction through a mixture of mangrove and (mostly) marsh and found storm surge was decreased by 9.45 cm/km. Perhaps the best estimate to date of storm-surge reduction by marsh comes from US Geological Survey water-level data taken during Hurricane Rita. On average the wetlands reduced storm surge by 13.5 cm/km (Kemp, 2008). Prior to the construction of the MRGO, there was an average of about 10 km of wetlands between Orleans and St. Bernard parishes and Lake Borgne. These wetlands had the capacity to reduce storm surge by about 1.35 m (4.5 feet). Because a considerable portion of these wetlands was bald cypress—water tupelo swamp, the reduction of storm surge would likely have been even much greater.

COMPUTER MODELING TO RECONSTRUCT SURGE AND WAVES ALONG THE MRGO

The effect of even low-lying marsh vegetation in retarding and attenuating storm surge has been known for as long as the Mississippi River Gulf Outlet (MRGO) has existed and was, in fact, one criterion for the original MRGO Reach 2 levee design (USACE, 1963, 1967). One of the major adverse impacts of the MRGO project was to change marshes to open water and change swamps to marshes (Shaffer et al., 2009). To fully understand the impacts of human-induced environmental degradation, it became very important to incorporate the effects of these MRGO-induced changes on surge and wave dynamics experienced during

Hurricane Katrina, and then to assess whether the presence or absence of swamp and marsh affected the timing and severity of flooding of populated areas. Scientifically, it became critically important to understand the contribution of the MRGO to the flooding. In order to do this, reliance on a suite of computer models was essential (van Heerden et al., 2009). First, an understanding was needed of the surge conveyance potential (i.e., the potential for the surge to be moved down the channel) of the MRGO; and then a determination if the unmanaged state of its width contributed to wave erosion of the MRGO Reach 2 levees (Fig. 6.4).

These modeling efforts are discussed in detail by van Heerden et al. (2009), who modeled two different scenarios. The first scenario, a "MRGO as-is," replicated the actual Katrina storm conditions as accurately as possible, along with a hypothetical "no MRGO" setting with pre-MRGO-construction (c. 1958) wetlands intact. The second scenario can be considered a "Neutral" MRGO setting (van Heerden et al., 2009). The "Neutral MRGO" analyses are based on the rationale that the USACE had a Congressionally directed responsibility to manage the MRGO navigation project in a manner that caused no added, unmitigated impact on the ability of the Hurricane Protection Project to also fulfill its mission—also Congressionally mandated—in order to protect the City of New Orleans and St. Bernard Parish from hurricane-induced flooding. In other words, the second scenario is a navigation channel with minimal environmental impacts.

In the "Neutral MRGO" condition, the surge elevation was reduced slightly and the rise was delayed almost everywhere around the margins of the funnel, but more so toward the west. Even slight reductions in peak surge elevation or delays in peak onset could combine to significantly reduce overtopping of the levee flood protection structures throughout the funnel. When both reaches of the MRGO were removed, overtopping was reduced by about 80% for all of the three developed polders that experienced catastrophic flood damage on August 29, 2005 (van Heerden et al., 2009). Thus, adopting this latter condition would have resulted in minimal flooding from Hurricane Katrina.

EFFECTS OF MRGO REACH 2 ON WAVES

Wind speeds over the Lake Borgne-Chandeleur Sound complex ranged up to a maximum of about 90 knots. On the morning of August 29, 2005, the prevailing wind direction over Lake Borgne was from the northeast (Fig. 6.4) and almost perfectly aligned with the long axis of the Lake from 04:00 to 09:00, resulting in the water being blown directly from Lake Borgne into the funnel (Gautier, Kok, & Vrijling, 2008). The Scenario 1 (as-is scenario) data show that significant wave heights build in the channel up to a maximum of about 9 feet along the whole of Reach 2 in front of the levees; and are thereby capable of destroying the levees, which is what happened (Fig. 6.6). What is very evident is that the very wide Mississippi River Gulf Outlet (MRGO) channel at the time Katrina hit, amplified the waves from a maximum of 5.5 feet with no channel (i.e., no MRGO scenario) to 9.2 feet with the channel. This is what made the difference to these fragile structures of

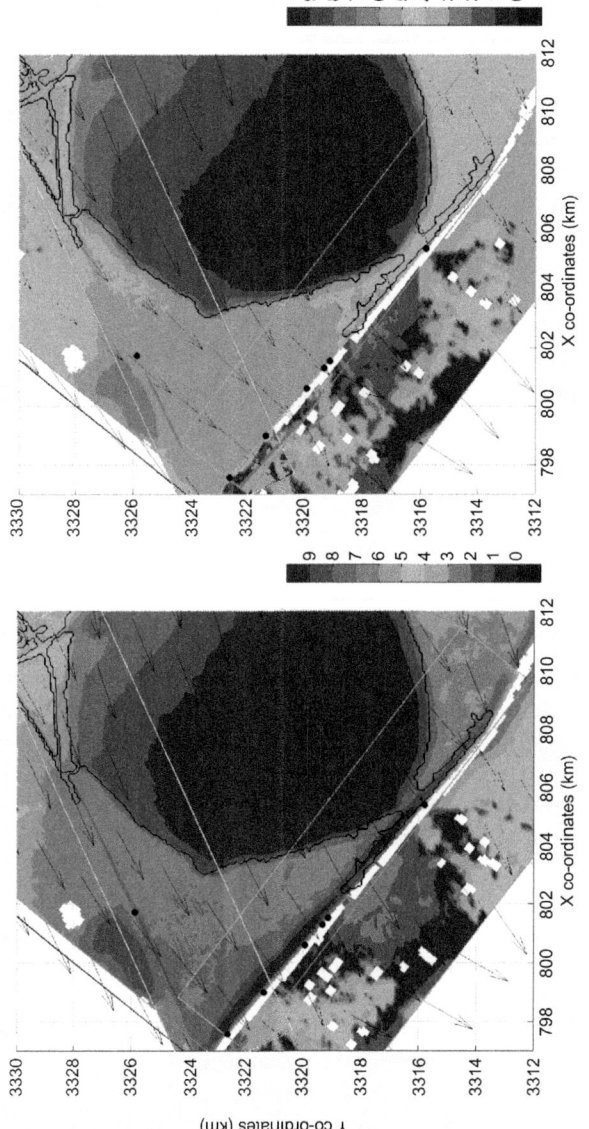

FIGURE 6.6

Scenario 1 and Scenario 2 C SWAN results Hs (feet) and wave direction, 8:00 a.m. LT. Note how the waves in the MRGO channel are almost twice as high in Scenario 1 as compared to Scenario 2 C.

the levees, as well as for the early onset of the flooding in St. Bernard (van Heerden et al., 2009). Doubling the wave height, as a result of the loss of wetlands apron, resulting in turn from mismanagement of the wetlands, meant that the extensive reaches of the MRGO levees were destroyed—and in some places were completely missing—following Hurricane Katrina. This characteristic of the MRGO channel of wave amplification was not considered in the design of the Reach 2 levees. In fact, the original design did not account for the presence of the MRGO at all.

The sheltering effect of forested wetlands also affects the fetch over which wave development takes place. Shallow water depths attenuate waves via bottom friction and breaking, while vegetation provides additional frictional drag and wave attenuation and also limits static wave set-up. In a few circumstances there were patches of trees in front of the levees and even though the levees were still overtopped, they experienced very little damage. Thus, human deforestation clearly rendered the levees more vulnerable. Indeed, overtopped levees flanked by trees received little structural damage from Hurricane Katrina (IPET, 2006; van Heerden et al., 2007).

These new studies have demonstrated just how important the wetland apron fronting the built environment can be in reducing hurricane winds, surge, and waves. They, therefore, need to be considered a fundamental part of any risk reduction planning.

HURRICANE BETSY AND THE 1965 FLOOD CONTROL ACT

The evolution of the built environment can be regulated by appropriate legislation that is enforced. Hurricane Betsy made landfall as a Category 4 storm near Grand Isle (40 km west of Katrina's landfall) on September 9, 1965, and passed south of New Orleans, following the west bank of the Mississippi River. A large section of Greater New Orleans (GNO) was flooded and approximately 75 persons drowned and more than 160,000 homes were flooded.

The Lake Pontchartrain & Vicinity Hurricane Protection Project (HPS) was authorized for construction by the Flood Control Act of 1965, a month after President Johnson viewed the destruction caused by Hurricane Betsy. Unfortunately, the design and construction were still incomplete when Katrina struck (van Heerden & Bryan, 2006). However, with this Act, Congress ordered the USACE to protect New Orleans from "the most severe meteorological conditions considered reasonably characteristic for that region."

As the 20th century progressed, the Louisiana coast lost its coastal wetland at such an ever increasing rate that it reached about 100 km^2 per year in the 1970s. Recently the rate has slowed slightly, but losses of approximately 50 ha/day still occur. Total wetland loss since the turn of the century has been over 500,000 hectare (ha) and predictions are that in less than 30 years only tiny remnants of the wetland apron will remain. Increasingly, life and property are threatened as the populated, low-elevation natural levee lands become more exposed to the Gulf of Mexico. Thus, by the time Katrina struck, wetland loss was exacerbating potential surge impacts and various flood protection projects were accelerating land loss; setting the stage for the catastrophic impact of Katrina. The fact that the coastal zone was a system, was not being incorporated into human planning.

Many observers have noted similarities between the patterns of surge-induced flooding that occurred during Katrina and the previous storm of record, Hurricane Betsy (van Heerden, 2007; van Heerden et al., 2007). The latter precipitated the HPS, which was, in fact, intended to prevent a repeat of the Betsy disaster; however, this system failed during Katrina with a 20-fold increase in loss of life. A further understanding of the reasons for the magnitude of Katrina's damage and destruction can be gleaned through subsequent forensic investigations.

FORENSIC INVESTIGATIONS—LEVEE FAILURES

Numerous investigations into the levee failures have been conducted including: the state of Louisiana's "Team Louisiana" (van Heerden et al., 2007); the National Science Foundation-funded Independent Levee Investigation Team (ILIT, 2006); the self-study Interagency Performance Evaluation Taskforce undertaken by USACE and other federal agencies (IPET, 2006); and a scientific and engineering group partially assembled by the author in support of legal actions in federal court.

POLDER LEVEE FAILURES

van Heerden et al. (2007) demonstrated that the Hurricane Protection System (HPS) was not properly conceived to accomplish the 1965 Congressional mandate to protect against the "most severe combination of meteorological conditions reasonably expected." Incomplete and outdated hurricane science was relied upon to expedite the design process and possibly reduce costs. The initial meteorological and oceanographic analysis based on the 1959 US Weather Bureau (now the National Weather Service, NWS) 1-in-100 year Standard Project Hurricane (SPH)—used as the design template—was known, by 1972, to be obsolete; just as construction of initial parts of the GNO HPS was getting underway. The primary deficiency of the 1959 SPH was in the specification of maximum sustained wind speed, which the NWS had increased by 20%, from 107 to 129 mph. In 1979, the NWS raised the maximum sustained winds to 140 mph; classified as a Category 4. The surge analysis provided a design basis for setting the minimum heights above mean sea level for levee and floodwall crowns to resist overtopping by combined SPH waves and surge. Its sensitivity is reflected in the fact that a 20% underestimate of maximum winds can lead to a 40% reduction in the predicted surge elevation. Furthermore, the USACE missed opportunities to revise the original SPH-based analysis after the NWS revised the SPH in 1972 and 1979, and when the SLOSH storm-surge model came into use in 1979 (Jelesnianski, Chen, & Shaffer, 1992). "The Sea, Lake and Overland Surges from Hurricanes (SLOSH) model is a computerized numerical model developed by the Federal Emergency Management Agency (FEMA), United States Army Corps of Engineers (USACE), and the NWS, to estimate storm-surge heights resulting from historical, hypothetical, or predicted hurricanes by taking into account the atmospheric pressure, size, forward speed, and track data" (National Hurricane Center, 2017). van Heerden et al. (2007) also identified that the floodwall and levee crown elevations were built 30–60 cm (1–2 feet) lower than specified in designs for HPS elements, because of an erroneous assumption by the USACE that an elevation of zero referenced to the National Geodetic Vertical Datum of 1929 was equal to, and interchangeable with, local mean sea level.

The USACE neither followed existing engineering practice, nor even their own guidance, for construction of levees and floodwalls. Weak soil strengths or potential for underseepage, including the presence of thick porous sand layers—which would undermine the stability of the levee system—were evident in strata tested by and subsequently ignored by the USACE during the early 1980s under New Orleans Metro drainage canal floodwall levees that failed (van Heerden & Kemp, 2006). The potential consequences of these layers on levee stability were well known to practicing engineers at the time, yet were overlooked or ignored because of inappropriate averaging of soil strengths on long levee reaches and across layers (van Heerden & Kemp, 2006). Standard engineering practice or USACE guidance were not followed when evaluating whether to protect earthen sea dikes and interior levees from erosion caused by waves or overtopping.

ACCOUNTING FOR SUBSIDENCE AND SEA-LEVEL RISE

Furthermore, van Heerden et al. (2007) revealed that the Greater New Orleans (GNO) HPS was not maintained and operated to assure the required level of protection over time. Local sea level has risen 0.1 m since the 1960s and much of New Orleans has sunk over 0.5 m in the same period for a combined change of nearly 0.6 m relative to sea level, but as IPET (2006) noted: "It was not clear how projected subsidence rates were applied in structural elevation design, if at all. Subsidence was apparently not factored into the design freeboard allowance." "Thus, the USACE simply ignored the science. Prudent engineers operating in coastal Louisiana have made allowances for, subsidence for a century. The USACE was one of the first agencies to directly map coastal wetland loss in Louisiana (May & Britsch, 1987), but this ever continuing diminishment of surge protection was never incorporated into the design philosophies of Sea, Lake, and Overland Surges from Hurricanes (SLOSH) possibly due to a failure to recognize the seriousness of the situation (van Heerden et al., 2007). Crown elevation deficiencies ranging up to 1.5 m (5 feet) at the time Katrina struck, resulted in prolonged overtopping of floodwalls and levees along the Inner Harbor Navigation Canal (IHNC) and to the east in the Lake Borgne funnel, that otherwise would have been overtopped only briefly (van Heerden et al., 2007). Prolonged overtopping reinforced the catastrophic breaches into the Lower 9th Ward on the east and into Orleans Metro on the west, and contributed to the early failures of levees along the GIWW and MRGO."

Researchers in Tulane University's Department of Earth and Environmental Sciences (Jankowski, Törnqvist, & Fernandes, 2017a, 2017b) recently identified that the relative rate of sea-level rise that includes subsidence and the global warming footprint in the region, over the past 6–10 years, amounts to 1.3 cm/yr (half an inch per year); on average four times faster than the global rate. These phenomena signify an ever increasing urgency to solve the coastal Louisiana crises, now; or it will definitely be too late.

The failure of over 50% of the New Orleans levees was the blackest day in the civil engineering history of the United States. One of the United States' prime engineering agencies had been grossly negligent in not following well-established engineering principals with catastrophic results. Only time will tell if the USACE can heal itself to the point where the public may trust its engineering again.

RECOMMENDATIONS FOR SUSTAINABILITY

Coastal land loss, which is historically always part of the landscape—even if only restricted to abandoned delta lobes (i.e., that part of the coast developing through sedimentation)—has accelerated and became a coast-wide phenomenon since the 1930s. Thus, the underlying framework of surge defense, or the lack thereof, whether referring to the wetland apron or the engineered levees, reflect historical mismanagement; mostly by the USACE. This lack of utilization of the available science—ignoring some simple management ideals, such as engaging with the coast as a system—resulted in the catastrophic consequences of Hurricane Katrina. This tragedy could have been avoided, or at least greatly diminished, if the federal government had listened to coastal scientists in Louisiana and elsewhere; who sounded alarms from before the occurrence of Hurricane Betsy, over 40 years before Hurricane Katrina.

The Katrina catastrophe sets off a significant research effort by various groups and agencies. The results of these studies, the lessons learned, and the knowledge gleaned from the Hurricane Katrina catastrophe should prompt the development of a new hurricane protection system that now fully incorporates coastal wetland restoration and will prevent future hurricane flooding. While this chapter has focused on Louisiana, there are many other locations in the United States, and elsewhere in the world, where the threat of a similar catastrophe is looming, especially given the accelerated rise in sea level as a consequence of global warming, due to the human desire to utilize "cheap" fossil energy. For many, this development is a ticking "time bomb."

The following future recommendations were compiled from van Heerden and Bryan (2006), van Heerden et al. (2007), van Heerden (2007), and van Heerden et al. (2009).

Restore the Coastal Wetland Apron

There is a real need to develop a living and modifiable/flexible conceptual plan for long-term protection that includes wetland and barrier island restoration; Dutch-style floodgates and pile-supported barrier levee structures at key locations (van Heerden & Bryan, 2006). The concept must be that the barrier levees and gates protect the built environment; the wetlands protect the levees; and the barrier islands protect the wetlands. In order to protect the levee system, and as a hedge against a rising sea level and climate change, we must concentrate on building and restoring wetlands seaward of the levees, specifically by using the sediment resources of the Atchafalaya and Mississippi Rivers (van Heerden & Bryan, 2006). Even with a major levee system, these wetlands are still the outer defense and must be maintained for this purpose, in addition to their commercial, recreational, and environmental uses. The best procedure is to build the hard structures (levees, floodgates) where they can be protected by an existing "platform" of wetlands, no matter how fragile, and where there is a reasonable supply of good soil building-material for levees. The wetland platforms seaward of the barrier levee and floodgates should be the target sites for future wetlands creation projects, with barrier islands seaward of the expanding wetlands base. Levees—wetlands—barrier islands: This progression assures the survival of the estuarine bays, necessary to keep the commercial and recreation fishing industries alive, as well as to supply the breadth of natural habitats that make up this unique ecosystem (van Heerden & Bryan, 2006).

Activities since Katrina

Since the early 1990s the state of Louisiana has focused attention on restoring its coast (for an example of the first comprehensive planning document, see van Heerden, 1994). Shortly thereafter, a very modest restoration effort began with limited federal funding. Thanks to the fines associated with the BP Horizon oil spill in 2010, Louisiana has finally acquired sufficient funds to undertake some large-scale projects (Schleifstein, 2017a, 2017b, 2017c). These include two very large diversions into the coastal wetlands, downriver from New Orleans, one to the west (Barateria Bay), and one to the east (Breton Sound). In addition, as also advocated by van Heerden (1994), offshore shoal sands (i.e., shallow sand bodies) are being mined to restore the State's barrier islands (Sneath, 2017).

Oversight and use of a science and risk-based engineered levee design process

Katrina was a man-made catastrophe triggered by a relatively fast Category 3 hurricane that missed New Orleans. Due to activities of the USACE, the citizens of the state of Louisiana were basically denied the level of protection mandated by Congress in the 1965 Flood Control Act. Independent Review Panels must be formed to assess the integrity and safety of existing levee systems in Louisiana, as well as the repairs presently being carried out. An independent review panel also needs to be constituted to review all future USACE projects that could possibly have an impact on human life, property, or the natural function of ecosystems. The panel needs to be funded separately from the USACE.

Risk-based management planning

A risk-based approach toward defining the design criteria for hurricane conditions is needed. The SPH needs to be redefined using principles and practices similar to those used in establishing design criteria for other infrequent, but potentially catastrophic, natural disasters, such as earthquakes and floods. Storm-surge numerical models must be used to fully investigate possible surge conditions under a whole range of storm sizes, forward speeds, angles of approach, and so on.

Settling of levees and floodwalls is a recognized factor in construction of these features. The fact that settlement occurs, means that construction is never over—a project is never complete, because even if it is constructed a couple of feet above the required crown elevation, in a decade or two it will be below that "safe" elevation (van Heerden et al., 2007). Hence, construction would have to restart to again lift the levee or floodwalls above the required elevation. Because of settlement and subsidence, it is essentially impossible for the construction of gravity supported levees, or floodwalls that depend on levees for support (e.g., I-walls), to ever be considered "completed" in the GNO area. Actual crown elevations for these structures describe a trajectory over time between enlargements, such that the actual level of protection rarely corresponds to the design level. In the interest of completing projects on a reasonable schedule, pile-supported protective structures should be utilized to a much greater degree than in the past, even if the first construction cost is much higher. Construction lifts—that add new material to raise the levee—could conceivably be limited with deep pile-supported floodwalls and barriers.

This whole physical system—barrier levee with navigation/sediment distribution gates—must be accompanied by legislation that stops development in the wetlands and anywhere in the newly protected areas. Many communities are going to be outside this new levee system. Some retreat of the population will need to take place. It should also be required that all new construction and development within the new levees be elevated a given number of feet, as determined by the design specifics of the whole system. This would provide an extra margin of protection in the event of some levee overtopping, or to reduce flood damage should there be a rainfall flood that exceeds the capacity of the pumps. However, what is really needed is enforceable legislation beyond the existing laws that restricts the new development of communities within and adjacent to these fragile areas.

Coastal wetlands in the United States are estimated to currently provide USD 23.2 billion/yr in storm protection services (Costanza et al., 2008) and function as valuable, self-maintaining "horizontal levees" for storm protection, in addition to providing a host of other ecosystem services that vertical levees do not. Their restoration and preservation is an extremely cost-effective and risk-reducing strategy for society.

CONCLUSIONS: HABITAT SUSTAINABILITY IS NEEDED TO SUPPORT HUMAN RESILIENCE

The infrastructure and engineering failures in the New Orleans area resulted in substantial loss of life, property damage, and destruction. These failures were human-caused and could have been prevented. A combination of engineering errors, ignorance of the value of the wetland apron, and nonscience-based political economic decisions resulted in a hurricane protection system of lesser quality than promised to the citizens of Louisiana in the Design Memos—which serve as a contract between the USACE and the people of Louisiana; represented by the various levee boards (van Heerden et al., 2007). Congress and society need to pay attention to science and accept the design failures of the past in order to ensure that Louisiana is compensated for the federal government's failures and that the full extent of the 1965 Flood Control Act will truly be implemented. The Gulf Coast needs to be restored posthaste—this is the only long-term solution. The implementation of aggressive coastal restoration strategies and the building of a barrier storm protection system will require significant resources. However, it will ensure the continued existence of coastal Louisiana, its inhabitants, and a unique culture. In addition, it presents many new opportunities to expand the regions' job base and, hence, improve its economy. The effort to "save" coastal Louisiana will generate a number of new technologies and stimulate the science of habitat restoration. Lessons learned will be able to be applied all over the world beyond Louisiana.

In the federal government's attempts to enhance the economy of coastal Louisiana through the development of an extensive navigation infrastructure and the development of swamp lands to accommodate new residents, the natural process that shaped the landscape was completely ignored and thereby these natural processes were disrupted, setting the stage for the catastrophe we now know as Katrina. Human resilience is very much impacted by the vulnerabilities of the built environment. The potential for human resilience was thereby undermined through the changes to the built environment identified in this chapter. Resilience and vulnerability are even greater concerns in the immediate and distant future, given the potential impacts of climate change on built environments and of human interventions that compromise sustainability.

REFERENCES

Brinkman, R.M., Massel, S.R., Ridd, P.V., & Furakawa, K. (1997). Surface wave attenuation in mangrove forests. In: *Proceedings of the 13th Australasian Coastal and Ocean Engineering Conference*, Vol. 2 (pp. 941–979).

Condrey, R.E. (1993). *The early explorers' views of the Barataria-Terrebonne System: The fork in the river and its Island*. Baton Rouge: Center for Coastal,Energy, and Environmental Resources, Louisiana State University.

Costanza, R., Pérez-Maqueo, O., Martinez, M. L., Sutton, P., Anderson, S. J., & Mulder, K. (2008). The value of coastal wetlands for hurricane protection. *Ambio, 37*(4), 241–248, 2008 Jun.

Davis, D. W. (1993). Crevasses on the lower course of the Mississippi River. *Coastal Zone '93, 1,* 360–378, July 19–23, 1993, New Orleans, Louisiana.

Day, J. W., Jr, Shaffer, G. P., Britsch, L., Reed, D., Hawes, S., & Cahoon, D. (2000). Pattern and process of land loss in the Mississippi delta: A spatial and temporal analysis of wetland habitat change. *Estuaries, 23,* 425–438.

Fitzgerald, D., et al. (2008). Impact of the Mississippi River Gulf Outlet (MR-GO): Geology & Geomorphology. In S. Penland, A. Milanes, M. Minor, & K. Westphal (Eds.). *Re Katrina Canal Breaches Litigation, Civil Action No. 05-4182, US District Court for the Eastern District of Louisiana, Section "k"(2), Pertaining to Robinson (06-2268)' Judge Duval*. Available from http://katrinadocs.com/report.cfm?r=45.http://katrinadocs.com/report.cfm?r=59.

Gagliano, S. M., Kemp, E. B., Wicker, K. M., Wiltenmuth, K., & Sabate., R. W. (2003). Neo-tectonic framework of southeast Louisiana and applications to coastal restoration: Transactions. *Gulf Coast Association of Geological Societies, 53,* 262–272.

Gautier, C., Kok, M., & Vrijling, J.K. (2008). Wave modeling New Orleans—Mississippi River Gulf Outlet, Hurricane Katrina August 2005. Expert report. In *Re Katrina Canal Breaches Litigation, Civil Action No. 05-4182, US District Court for the Eastern District of Louisiana, Section "k"(2), Pertaining to Robinson (06-2268)' Judge Duval*. Available from http://katrinadocs.com/report.cfm?r=50.

Independent Levee Investigation Team (ILIT). (2006). *Investigation of the performance of the New Orleans Flood Protection Systems in Hurricane Katrina on August 29, 2005*. Draft Final Report, Version 1.2, June 1, 2006, University of California, Berkeley, CA, 2006.

Interagency Performance Evaluation Task Force (IPET). (2006). *Performance evaluation of the New Orleans and Southeast Louisiana Hurricane Protection System*. Draft final report of the Interagency Performance Evaluation Task Force, June 1, 2006. U.S. Army Corps of Engineers, Vicksburg, MS.

Jankowski, K. L., Törnqvist, T. E., & Fernandes, A. M. (2017a). Vulnerability of Louisiana's coastal wetlands to present-day rates of relative sea-level rise. *Nature Communications, 8,* 14792.

K.L. Jankowski, T.E. Törnqvist, & A.M. Fernandes. (2017b). *Vulnerability of Louisiana's coastal wetlands to present-day rates of relative sea-level rise*. Available from https://www.nature.com/articles/ncomms14792.

Jelesnianski, C. P., Chen, J., & Shaffer, W. A. (1992). SLOSH: Sea, lake, and overland surges from Hurricanes, *NOAA Technical Report NWS 48*. Silver Spring, MD: National Weather Service, 71 p.

Kemp, G. P. (2008). Mississippi River Gulf Outlet Effects of Storm Surge, Waves, and Flooding during Hurricane Katrina. *Expert Report 4* (p. 228) New Orleans, LA: Office of Bruno & Bruno.

Kolb, C. R., & Van Lopik, J. R. (1958). *Geology of the Mississippi river deltaic plain, southeastern Louisiana, v. 1*. Vicksburg, MS: United States Army Engineer Waterways Experiment Station.

Krauss, K. W., Doyle, T. J., Swarzenski, C. M., From, A. S., Day, R. H., & Conner, W. H. (2008). Water levels, mangroves, and EW hurricane storm surge. *Wetlands, 29*, 142−149.

May, J.R., & Britsch, L.D. (1987). *Geological investigation of the Mississippi River Deltaic Plain—land loss and land accretion*. Water Ways Experimental Station Vicksburg, MS 39180.

Mazda, Y., Magi, M., Ikeda, Y., Kurokawa, T., & Asano, T. (2006). Wave reduction in a mangrove forest dominated by *Sonneratia* sp. *Wetlands Ecology and Management, 14*, 365−378.

Mazda, Y., Magi, M., Kogo, M., & Hong, P. N. (1997). Mangroves as a coastal protection from waves in the Tong King Delta, Vietnam. *Mangroves and Salt Marshes, 1*, 127−135.

Morton, R. A., Bernier, J. C., Barras, J. A., & Ferina, N. F. (2005). Historical subsidence and wetland loss in the Mississippi Delta Plain. *Gulf Coast Association of Geological Societies Transactions, 55*(2005), 555−571.

Morton, R. A., Buster, N. A., & Krohn, M. D. (2002). Subsurface controls on historical subsidence rates and associated wetland loss in southcentral Louisiana: Transactions. *Gulf Coast Association of Geological Societies, 52*, 767−778.

Morton, R. A., Tiling, G., & Ferina, N. F. (2003). Causes of hotspot wetland loss in the Mississippi delta plain. *Environmental Geosciences, 10*, 71−80.

National Hurricane Center. (2017). *Sea, lake, and overland surges from Hurricanes (SLOSH)*. <http://www.nhc.noaa.gov/surge/slosh.php> Accessed 15.06.17.

Ramsey, K. E., & Penland, S. (1989). Sea-level rise and subsidence in Louisiana and the Gulf of Mexico. *Gulf Coast Association of Geological Societies Transactions, 39*, 491−500.

Roberts, H. H., Adams, R. D., & Cunningham, R. H. W. (1980). Evolution of the sand-dominant subaerial phase, Atchafalaya delta, Louisiana. *American Association of Petroleum Geologists Bulletin, 64*(2), 64−279.

Roberts, H.H., and van Heerden, I.L. (1992a). The Atchafalaya Delta: An analog for thin deltas and subdeltas in the subsurface. Basin Research Institute Bulletin (March 1992), Vol. 2, no. 1, pp. 31−42, Louisiana State University, Baton Rouge, LA.

Roberts, H.H., and van Heerden, I.L. (1992b). Atchafalaya-Wax Lake Delta Complex: The new Mississippi River Delta Lobe. *1st Annual Coastal Studies Institute-Industrial Associates Research Program*. Research Report (CSI - ARP Rept. # 1).

Schleifstein. (2017a). Available from http://www.nola.com/environment/index.ssf/2017/03/gov_edwards_to_trump_speed_rev.html.

Schleifstein. (2017b). Available from http://www.nola.com/environment/index.ssf/2017/04/mid-baratarsia_sediment_diversi_1.html.

Schleifstein. (2017c). Available from http://www.nola.com/environment/index.ssf/2017/01/mid-baratara_sediment_diversio.html.

Shaffer, G. P., Day, J. W., Mack, S., Kemp, G. P., van Heerden, I., Poirrier, M. A., ... Penland, P. S. (2009). The MRGO navigation project: A massive human-induced environmental, economic, and storm disaster. *Journal Coastal Research Special Issue, 54,* 206–224.

Simpich, F. (1927). The great Mississippi flood of 1927. *The National Geographic Magazine, 52*(no.3).

Sneath, S. (2017). Available from http://www.nola.com/environment/index.ssf/2017/03/state_touts_largest_restoratio.html.

Templet, P. H., & Meyer-Arendt, K. J. (1988). Louisiana wetland loss: A regional water management approach to the problem. *Environmental Management, 2*(2), 181–192.

Turner, R. E., Costanza, R., & Schaife, W. (1982). Canals and wetland erosion rates in coastal Louisiana. *Proceedings of the conference on Coastal Erosion and Wetland Modification in Louisiana: Causes, Consequences, and Options* (pp. 73–84). Slidell, LA: Office of Biological Services, U.S. Fish and Wildlife Service, FWS/OBS-82/59.

US Army Corps of Engineers. (1963). *Technical Report No. 2-636; effects on Lake Pontchartrain, LA of Hurricane Surge Control Structures and MRGO Channel—hydraulic model investigation.*

US Army Corps of Engineers. (1965). Available from http://biotech.law.lsu.edu/katrina/hpdc/docs/19650706_SecretaryArmyLetterReport.pdf.

US Army Corps of Engineers. (1967). *Lake Pontchartrain, LA and vicinity; Design Memorandum No.1, Hydrology and hydraulic analysis; Part II—Barrier.*

US Corps of Engineers, New Orleans District. (1999). *Habitat impacts of the construction of the MRGO*, prepared for the Environmental Subcommittee of the Technical Committee convened by EPA in Response to St. Bernard Parish Council Resolution 12-98.

van Heerden, I. L. (1983). *Deltaic sedimentation in eastern Atchafalaya Bay, Louisiana. Special Grant Publication.* Baton Rouge, LA: Center for Wetland Resources, Louisiana State University.

van Heerden, I.L. (1994). *A long-term, comprehensive management plan for coastal Louisiana to ensure sustainable biological productivity, economic growth, and the continued existence of its unique culture and heritage.* NSMEP, Center for Coastal, Energy, and Environmental Resources, Louisiana State University, Baton Rouge, LA, 45 pp. Available from http://www.worldcat.org/title/long-term-comprehensive-management-plan-for-coastal-louisiana-to-ensure-sustainable-biological-productivity-economic-growth-and-the-continued-existence-of-its-unique-culture-and-heritage/oclc/30617836.

van Heerden, I.L. (2007). *The failure of the New Orleans Levee System following Hurricane Katrina and the pathway forward.* Public Administration Review, Supp. to Vol. 67, Dec 2007.

van Heerden, I. L., & Bryan, M. (2006). *The storm—what went wrong and why during Hurricane Katrina—the Inside Story from One Louisiana Scientist* (p. 308) New York: Publ. Penguin/Viking.

van Heerden, I. L., & Kemp, G. P. (2006). The failure of the New Orleans Levee System during Hurricane Katrina. *Loyola L. Rev., 52,* 1225–1245.

van Heerden, I. L., Kemp, G. P., Bea, R., Shaffer, G., Day, J., Morris, C., ... Milanes, A. (2009). How a navigation channel contributed to most of the flooding of New Orleans during Hurricane Katrina. *Public Organ Review, 9,* 291. Available from http://dx.doi.org/10.1007/s11115-009-0093-8.

van Heerden, I.L., Kemp, G.P., Mashriqui, H., Sharma, R., Prochaska, B., Capozzoli, L., ..., & Boyd, E. (2007). *The failure of the New Orleans Levee System during Hurricane Katrina*. A report prepared by TEAM LOUISIANA for Secretary Johnny Bradberry, Louisiana Department of Transport and Development, Baton Rouge, Louisiana.

FURTHER READING

Ramsey, K. (1989). *Sea level rise in Louisiana and Gulf of Mexico, Shea Penland*. Available from http://www.searchanddiscovery.com/abstracts/html/1989/gcags/abstracts/1191.htm.

Westerink, J. J., Luettich, R. A., Jr., & Muccino, J. (1994). Modeling tides in the Western North Atlantic using unstructured graded grids. *Tellus, 46a*(2), 178−199.

CHAPTER 7

Three centuries in the making: Hurricane Katrina from an historical perspective

Michael J. Zakour and Kayla Grogg
West Virginia University, Morgantown, WV, United States

CHAPTER OUTLINE

Significance of an Historical Perspective 160
Conceptual and Theoretical Framework 162
    Vulnerability and Resilience 162
    Political Ecology of New Orleans 163
Historical Events Creating Katrina 165
    The Progression to Vulnerability 165
    Political and Economic Marginalization of People of Color 165
    Oil, Canals, and Environmental Degradation 169
    Migration and Population Displacement Since 1960 171
    Results of the Neglect of New Orleans 172
Vulnerability and Resilience During Katrina 173
    Katrina's Natural, Technological, and Organizational Failure Hazards 173
    The Nature of the Hurricane Katrina Disaster 175
    The Pattern of Disaster Damage and Loss 176
    Katrina as Catastrophe 177
Resilience Resources 178
    Environmental Justice: Recovery by Race, Income, Gender, and Age 178
    Economic Resources 178
    Information and Communication 180
    Social Capital 181
    Collective Action 182
Summary and Conclusions 183
    The Political Ecology of Katrina and Southeast Louisiana 183
    Implications for Vulnerability-plus Theory 184
    Substantive Implications of Katrina 188
    Policy Implications 189
References 191

Creating Katrina, Rebuilding Resilience. DOI: http://dx.doi.org/10.1016/B978-0-12-809557-7.00007-7
© 2018 Elsevier Inc. All rights reserved.

The waves of the Katrina disaster initially unfolded over 3 days in late August 2005. Hurricane Katrina, a natural hazard, made landfall in Louisiana on Monday, August 29, 2005. Its 100 mph winds were felt across the New Orleans metropolitan area that day, causing damage consistent with a Category 2 (96–110 mph) hurricane. By Tuesday, August 30, 2005, technological failures were the major hazard in the disaster. In the Lower Ninth Ward, levee failures allowed a wall of water 20 feet high to surge into that neighborhood, and the failure of other major levees and floodwalls throughout the city resulted in the beginning of widespread flooding. On Wednesday, August 31, 2005, Governor Kathleen Blanco ordered all residents remaining in New Orleans to evacuate, but there were no available buses to carry out this order. Wednesday was the beginning of the failure of agencies at all levels of government in the disaster response, leading to a humanitarian catastrophe (Solnit, 2009).

The Katrina disaster was centuries in the making, and recovery continues more than a decade after Katrina. According to Oliver-Smith (2009)

> *The disaster agent itself functions as a metonym that encompasses the forces that produce the vulnerability to the systemic agent, the event, and its aftermath, including the process of reconstruction. Thus, when we speak of ... Hurricane Katrina, we encompass with those labels a set of social, environmental, and political-economic processes that extend deep into the past and far into the future. (p. 14)*

This chapter uses a historical and longitudinal approach to identify the root societal causes and structural constraints that created a very high level of disaster vulnerability. The Katrina disaster was triggered by natural, technological, and organizational failure hazards. Underlying social structures and root causes had been building for generations and even centuries to make New Orleans highly vulnerable to disaster. This historical process left New Orleans with inadequate resources for a resilient recovery when Katrina struck in 2005 (Oliver-Smith, 2009).

SIGNIFICANCE OF AN HISTORICAL PERSPECTIVE

According to Oliver-Smith, Alcántara-Ayala, Burton, and Lavell (2016, p. 32), "Longitudinal analysis is based on the fact that disasters involve far more than one-off, spatially delimited, temporally demarcated, physical triggering events. Rather, disasters are systemic processes that unfold over time. Their causes are deeply embedded in societal history, structure and organization, including human-environmental relations." To move beyond description of loss and damage, the societal causes of disasters need to be understood. They include the root causes and structural constraints that lead to unsafe conditions for a population, community, or locale. The most important structural constraints are

- population growth and distribution,
- urban and rural land use patterns and processes,
- environmental degradation and ecosystem service depletion, and
- poverty and income inequality (Oliver-Smith et al., 2016).

It is necessary to use an historical approach to identify root societal causes and structural constraints, because these variables may have begun to generate vulnerability many decades or generations ago. Historical processes play out within the particular evolution of each society. They are manifested by culturally specific results. Examples of these results are

- environmental degradation and pollution,
- settlement patterns in hazard-prone areas,
- poor health conditions undermining resilience,
- lack of institutional capacity,
- corruption,
- patterns of social domination, and
- radically skewed wealth distributions (Oliver-Smith et al., 2016).

Beginning with the founding of New Orleans the historical processes and their cultural results have contributed to high levels of disaster vulnerability for the city and region. Earlier adaptations to hazards in the region were either abandoned, or were available only to more affluent populations. Poor and Black populations were often forced to live in unsafe conditions, which grew worse over time.

Environmental degradation and damage from use of the natural environment has been particularly important in the progression to disaster vulnerability in New Orleans. Environmental degradation has shifted disaster risk to the most vulnerable populations and communities in Southeast Louisiana. New Orleans grew on a coastal area near sea level, surrounded by a large lake and the Mississippi River. Although the story of New Orleans and Southeast Louisiana is one of humans attempting to control and exploit nature, the outcome of domination of nature has been the massive destruction of human enterprises and lived environments (see Robbins, 2012).

Socio-cultural processes acting over three centuries contributed to the vulnerability of New Orleans and the populations residing there. An important part of this political–ecological process was New Orleans's role as a major port for imports and exports to a European market, especially for cotton and lumber. In the 20th century the global demand for oil and other resources extracted from coastal Louisiana grew rapidly. These industries dug canals through the marshes to transport resources, including lumber and petroleum, to the Port of New Orleans (Marks, 2010a, 2010b). By the late 20th century, neoliberal pressures intensified the progression to disaster vulnerability.

CONCEPTUAL AND THEORETICAL FRAMEWORK
VULNERABILITY AND RESILIENCE

Katrina is commonly used both as the name of a natural hazard, and the disaster resulting from the combination of hazards and unsafe conditions. A hazard is a force that damages socio-ecological systems. Hazards can be natural forces, such as a hurricane or tropical storm, or they can be human-caused, such as the failure of levees or the collapse of the response system. A disaster is the disruption that occurs in a community or region, triggered by one or more hazards but due equally to social causes (Zakour & Gillespie, 2013).

Vulnerability and resiliency

This chapter uses the concepts of vulnerability and resiliency, and associated concepts, as defined in Zakour and Swager (this volume, Chapter 3: Vulnerability-plus theory: The integration of community disaster vulnerability and resiliency theories). According to Oliver-Smith (2009), vulnerability is a combination of general political—economic forces and particular environmental forces. Poverty and racism produce susceptibilities to specific environmental hazards. Resiliency is the networking of resources that act as adaptive capacities (Norris, Stevens, Pfefferbaum, Wyche & Pfefferbaum, 2008). As with vulnerability, political—ecological forces in a community are important in resiliency, and determine which populations are able to access and mobilize these resources.

Community resilience

Disaster resilience resources are conceptualized as accessible networked capacities (Zakour & Gillespie, 2013). These resources allow communities and their populations to respond and recover in an adaptive fashion, and to restore predisaster levels of wellness to the same or higher levels after disaster. Adaptation and wellness are outcomes of the process of resilience. If resilience is at a very low level, adaptation and wellness will deteriorate postdisaster. If only one or a few populations in a community do not experience a resilient recovery, the entire community lacks disaster resilience (Norris et al., 2008).

Causal chain in Vulnerability-plus theory

Variables increasing the level of disaster vulnerability are root causes, structural constraints, and unsafe conditions (Wisner, Blaikie, Cannon, & Davis, 2004; Wisner, Gaillard, & Kelman, 2012). Root societal causes are the most distal, and originate decades, generations, and even centuries in the past. Over a time scale of generations and decades, structural constraints combine with root causes to produce unsafe conditions. Unsafe conditions are the most proximate causes of disaster vulnerability, and occur within decades or years before a disaster. The hazard of Hurricane Katrina interacted with unsafe conditions to produce the eponymous disaster.

Root causes

Root societal causes of disaster are distal, meaning that they originated far into the past, and they are difficult for people to observe. The root societal causes of disaster consist of both ideologies about access to power and economic resources, and the structures of dominance and wealth distribution in a society. Root causes are the exogenous variables at the beginning of causal chains that lead to disaster (Zakour & Gillespie, 2013).

Structural constraints

Structural constraints, also called dynamic pressures (Wisner et al., 2012) or risk drivers (Oliver-Smith et al., 2016), interact with root societal causes and amplify their effects. Structural constraints are somewhat more recent historically than root causes. Compared to root causes, structural constraints are more apparent, and easier for researchers to observe and measure.

Unsafe conditions

Unsafe conditions are the most proximate and most visible of the causes of the progression to disaster vulnerability. These conditions exist up to the time of impact of a hazard, and they combine with the hazard to produce a disaster. Unsafe conditions include both unsafe and dangerous locations, and fragile livelihoods that are not sustainable (Wisner et al., 2012).

Access to resources

Access to resources is critical for resilient recoveries from disaster (Zakour & Gillespie, 2013). The level of access to resources of populations in the region determined if New Orleans and Southeast Louisiana would progress to a resilient recovery and wellness, or to dysfunction and a higher level of vulnerability. When a system is modified so that its functioning and the wellness of members of the system are at a higher level than predisaster, then the process of transformative resilience has occurred. This can lower the system's vulnerability for future disasters. According to Oliver-Smith et al. (2016, p. 48) "Disaster risk management (DRM) policies and strategies that do not contest current systemic practices may promote or exacerbate vulnerability. Root cause analysis is a virtual necessity if development informed by DRM is to have any transformation potential."

POLITICAL ECOLOGY OF NEW ORLEANS

The metabolism of the New Orleans urban area and surrounding parishes (counties) consists of the flow and distribution of commodities, physical capital, and other types of capital. This includes social, human, built, natural, and cultural capital. The nexus of power and social actors deploying or mobilizing power relations decides who will have access to social and environmental resources, and who will be excluded from access (Heynen, Kiaka, & Swygedouw, 2006).

Because exclusion from access to resources is a social injustice, the environmental justice approach is important for understanding vulnerability and resilience in Hurricane Katrina. The principle of environmental justice guarantees

- protection from environmental degradation;
- prevention of adverse effects from environmental deterioration before harm occurs, rather than after;
- mechanisms for assigning culpability to the agents of environmental degradation, and not to the residents affected; and
- redressing the impacts of environmental damage through remedial action and needed resources (Cutter, 1995).

As conceptualized by urban political ecologists, environmental justice movements challenge control over the flow of social "goods" and "bads" that result in uneven exposure, risk, and opportunity (Robbins, 2012). Through environmental justice, new possibilities emerge, such as "just sustainabilities." New ecological possibilities may be developed in cities where risk and injustice have prevailed.

For the people of New Orleans, development projects such as canals, levees, and seawalls have created and intensified environmental and distributive injustice. A small number of political and economic elites benefitted from these projects, but created serious environmental harm. This harm increased the intensity of Hurricane Katrina, a so-called "natural" hazard, and damaged both natural and social systems. The consequences of Katrina have been the most severe for the many people who did not benefit from development projects, especially for Blacks and the poor (Robbins, 2012). It is the innocent bystanders, and not the politically connected few, who have been punished the most harshly (Freudenburg, Gramling, Laska, & Erikson, 2009).

Freudenburg et al. (2009) have called the impetus for water development projects the *growth machine*. Supported in part by federal tax dollars, the economic development projects of the growth machine in Southeast Louisiana moved relentlessly ahead until they resulted in their own destruction (see Robbins, 2012). Their actions continued unabated until the entire community experienced Katrina, a major disaster caused in part by mistakes and contradictions in development. The water development projects of the growth machine, including canals and levees, initially produced modest gains for a few politically connected elites, and protected other people in an illusory manner (Freudenburg et al., 2009). The key agents of the growth machine are political elites, developers, speculators, and engineers.

The agents of the growth machine adopted an ideology emphasizing the desirability of reshaping the natural world in any way that would produce a profit. In this worldview, environmental harm is an inevitable by-product of economic development. Related to this assumption is the idea that environmentally damaging projects are good for the economy. This assumption is unsupported by evidence, and most environmental harm in the United States is highly disproportionate to the economic benefits of development (Freudenburg et al., 2009).

HISTORICAL EVENTS CREATING KATRINA
THE PROGRESSION TO VULNERABILITY

The root societal causes of the progression to vulnerability and making of Katrina stretch back to the founding of New Orleans. Human modification of the landscape led to Katrina and its severe damage to both nature and the human community. This left the community more vulnerable and exposed. The most consequential actions were in the name of economic growth, supported by the neoliberal policies of the past four decades (Freudenburg et al., 2009; Johnson, 2011).

French colonials founded New Orleans in 1718 as a port city on the Natural Levee of the Mississippi River. Early French settlements already existed at Mobile Bay, and ships from Mobile could sail behind the barrier islands off the coasts of present-day Alabama and Mississippi, and through one of the passes to Lake Pontchartrain. Boats could then use Bayou St. John, connected to the lake, and portage across a short stretch of land to reach the Natural Levee of the Mississippi River (Campanella, 2006).

POLITICAL AND ECONOMIC MARGINALIZATION OF PEOPLE OF COLOR

Racial subjugation through slavery profoundly influenced New Orleans' social and urban geography (Campanella, 2006). After the city's founding in 1718, White colonizers occupied the elevated land in the French Quarter, while the land behind what is now Rampart Street was swampy and used primarily by slaves from Africa. This marginal land, called the "back-of-town," was a recreational area for slaves, allowed by the French and Spanish colonizers. In 1719 the first large group of African slaves arrived in New Orleans. This began over 140 years of slavery in the region, followed by an additional 100 years of Jim Crow laws. The Southern cotton and sugar plantation economies entrenched slavery in the region, and enriched New Orleans into the mid-20th century.

Since its founding, highly inequitable access to political power and economic wealth in New Orleans(root causes of vulnerability) characterized the relationships between the White colonials and Anglos with people of color. Slavery was justified by the ideology of the racial inferiority of Blacks, and a belief that Blacks were less than human. The city's Anglo residents viewed Free People of Color with suspicion, anger, and rejection. Until the Civil War, there was a three-tiered racial system, with Whites at the top,

Black slaves at the bottom, and Free Creoles of Color as an intermediate tier (Campanella, 2006).

Early in the city's history, canal building in New Orleans brought undesirable environmental conditions to established Black neighborhoods. One of the first human modifications of the environment was a canal completed in 1794. The new canal, called either the Carondelet Canal or the Old Basin Canal, was the earliest example of building canals through existing Black neighborhoods. It was the extension of Bayou St. John to the Treme' neighborhood. Faubourg Treme' is the oldest Black neighborhood in New Orleans, located just outside of the French Quarter, and first settled by Creoles of Color (Freudenburg et al., 2009). This early canal was an undesirable industrial development located in an established residential neighborhood.

Before the Civil War and occupation of New Orleans by Union forces, Black and White populations were integrated. Anglo-Americans lived above (upriver of) Canal St. in what was Faubourg St. Marie. Creoles and Creoles of Color resided below (downriver from) Canal Street in elevated land and Faubourgs on the Natural Levee of the Mississippi. In most of New Orleans, enslaved people lived adjacent to their owners in small dwellings, both so that they could be more easily managed, and so that they were available to act as servants and craftsmen for their owners' households (Campanella, 2006).

In the Antebellum period, slaves and free people of color were sometimes relegated to swampy areas beyond the French Quarter. This geographic separation began a trend of locating people of color on marginal land, on the periphery of central New Orleans. Particularly after the Louisiana Purchase, as the city grew in population, affluent Whites settled upriver on the higher land of the Natural Levee of the Mississippi (Freudenburg et al., 2009). Freed slaves, slaves hired out by their owners for projects, and sometimes Creoles of Color lived in shantytowns outside of the natural levee (Campanella, 2006).

After the Civil War the integration of Whites and Blacks grew. With the end of the Civil War, emancipated slaves were forced to live in swampy (back-of-town) parts of the city above Canal Street. Large numbers of emancipated slaves from Louisiana and Mississippi also moved into the city above Canal Street and were drawn to cheap land and rents in these back-swamp areas. Free men of color and Creoles of Color began to move north toward Lake Pontchartrain after these swampy areas were drained. Some Blacks were drawn to the cheaper land and rents caused by nearby nuisances such as fowl-smelling processing plants (Campanella, 2006).

In the postreconstruction era, New Orleans adopted the "classic Southern" pattern of racial geography. Social "bads" (the opposite of social goods) inherent in back-of-town areas were inequitably distributed to people of color and the poor. Whites chose areas lacking in amenities for Black neighborhoods, and built poor-quality houses with low rents specifically for Blacks. Nuisances in black neighborhoods included

- swamps,
- railroad tracks,

- noxious factories,
- flooding,
- mosquitoes,
- unpaved streets,
- open sewers and garbage dumps,
- a lack of city services,
- distance from employment,
- inadequate urban infrastructure, and
- a lack of clean water (Campanella, 2006).

After Reconstruction, Whites instituted harsh laws and practices against the entire Black population. The three-tiered system of Whites, Creoles of Color, and Blacks broke down as all people of color were grouped into a disadvantaged class. Based on an ideology of racial inferiority, Jim Crow laws denied access to political participation and economic resources for the Black population of New Orleans and Southeast Louisiana (Campanella, 2006).

By the late 19th century few Black neighborhoods remained on the Natural Levee of the Mississippi. Intensified de jure' segregation, as well as the high costs of residence, pushed them out of the French Quarter and the Faubourgs nearest the Mississippi River. Whites displaced Creoles of Color from the elevated land near the Mississippi. Because hostility by the White population above Canal Street to the notion of Free People of Color was intense, these Creoles chose to establish settlements below Canal Street with other people with French Creole roots (Campanella, 2006).

In the first half of the 20th century the population of New Orleans grew rapidly. Blacks increasingly moved to former marshlands near Lake Pontchartrain north and east of the downtown. These marshlands had been drained and rapidly subsided. The Black population of New Orleans has been concentrated to the east of a line extending from southwest of the city in Jefferson Parish to the northeast near the Inner Harbor Navigation Canal, known also as the Industrial Canal (Campanella, this volume, Chapter 2: Settlement shifts in the wake of catastrophe). Poor and Black populations, and amenities serving them (i.e., Charity Hospital), were located outside of the French Quarter and the Central Business District adjacent and upriver to it (Fig. 7.1).

In the 1920s and 1930s a project on Lake Pontchartrain created an elevated stretch of land meant to protect the metro area from flooding from the lake. Housing segregation was still widely practiced at that time, preventing Blacks from settling the Lakefront (Freudenburg et al., 2009). The Black population had been moving east (downriver) along the Mississippi River into the Lower Ninth Ward, located in part on the Natural Levee of the Mississippi. With the establishment of the Lakefront area as a White neighborhood, many Creoles relocated to the neighborhoods of the northeastern segment of New Orleans (Campanella, 2006).

After Hurricane Betsy flooded low-lying parts of the city near Lake Pontchartrain in 1965, levees were constructed to protect the New Orleans metropolitan area from surges in the lake (van Heerden, this volume, Chapter 6: Setting

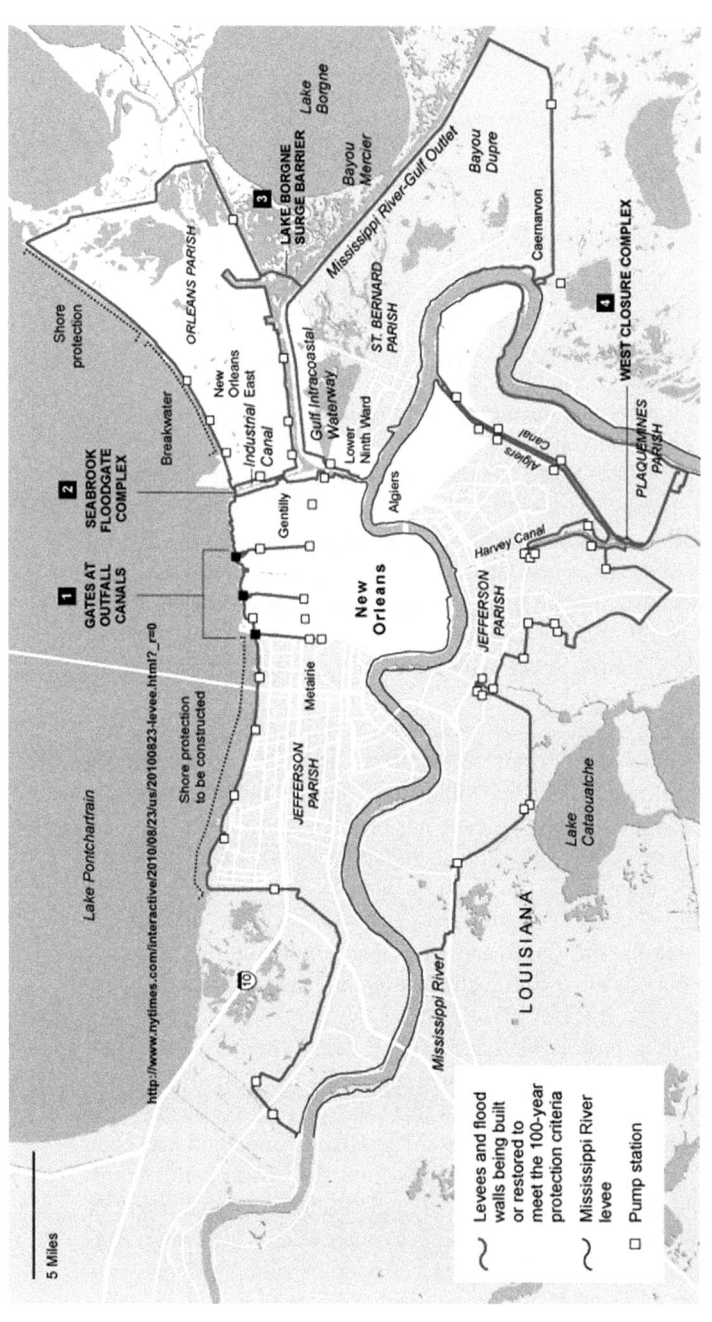

FIGURE 7.1

New Orleans Canals, Levees, and Floodwalls

Source: From http://www.nola.com/hurricane/index.ssf/2013/08/upgrated_metro_new_orleans_lev.html#incart_special-report.

the stage for the catastrophe of Katrina: Environmental degradation, engineering miscalculation, ignoring science, and human mismanagement). A large new area protected by levees opened for housing and neighborhood development. As the Black population of New Orleans grew, they relocated closer to Lake Pontchartrain. Developers drained former swamps to make way for new housing. Draining swampy land resulted in the sinking of this land. With the draining of watery soil, the soil-level sank and the exposed organic matter oxidized and deteriorated. Landthat had sunk several meters below sea level was difficult to protect with levees, storm walls, and pumping stations (Freudenburg et al., 2009).

OIL, CANALS, AND ENVIRONMENTAL DEGRADATION

The migration and displacement of the Black population to eastern New Orleans, and later the migration of low-income immigrants to St. Bernard Parish, were the structural constraints and risk drivers that exposed these populations to a higher risk of tropical storms and hurricanes. Eastern New Orleans and St. Bernard Parish are both closer to the Gulf of Mexico and tidal lakes like Lake Pontchartrain. Except for elevated land near the Mississippi River, much of the land is low-lying, especially after swamps were drained for residential development. Industrial development also took place in these eastern parts of the metro area, and environmental degradation from industrial development is widespread in this area. Once protected by millions of acres of wetlands, this area of predominantly Black and working-class neighborhoods has seen rapid loss of its protective wetlands (van Heerden, this volume, Chapter 6: Setting the stage for the catastrophe of Katrina: Environmental degradation, engineering miscalculation, ignoring science, andhuman mismanagement) .

The discovery of oil in the early 20th century in Louisiana set the stage for economic growth, and massive environmental degradation. In 1901, oil was discovered in Louisiana. Standard Oil (Exxon) built a large refinery in Baton Rouge in 1909, and former Governor Huey Long attracted Texaco to the massive oil deposits in South Louisiana in the 1930s (Marks, 2010b). Extraction of oil accelerated in the 1940s to support the war effort. The oil industry provided a strong economic base for New Orleans and a large number of jobs. However, laying a network of pipes to transport oil hastened erosion of the wetlands. The oil industry cut canals through the wetlands to create an oil extraction infrastructure using small boats. The extraction infrastructure also helped meet the growing demand for a network of canals to transport oil vessels.

By 1928 the Industrial Canal was completed in largely uninhabited land, connecting Lake Pontchartrain to the Mississippi River. The only settlements at that time were Black and Italian immigrant neighborhoods close to the Mississippi River. Italian and other White immigrants would later relocate downriver to St. Bernard Parish. The US federal government financed canal projects benefiting business interests in New Orleans. Several additional canals were built connecting the Industrial Canal to the Gulf, and this resulted in very rapid erosion along the canal banks. Because canals were dredged through marshy soil, erosion widened the canals and consumed thousands of acres of wetlands in eastern New Orleans and St. Bernard Parish (Campanella, 2006).

In the 1930s and throughout the years of the Great Depression, the US federal government greatly expanded available funding for water-related construction projects. During the Great Depression, incipient neoliberal policies aimed to promote economic growth. The projects were believed to contribute to national and local prosperity. Most projects were located in poorer and more rural areas such as eastern New Orleans and St. Bernard Parish, where they were expected to generate little opposition (Freudenburg et al., 2009).

After both the Lower Ninth Ward and St. Bernard Parish were settled, canal building brought additional environmental degradation to these areas. Environmental degradation was facilitated by environmental ideologies that viewed nature as a resource to exploit for economic gain. By 1968, 40 years after the completion of the Industrial Canal, the Gulf Intracoastal Waterway (GIWW) and the Mississippi River Gulf Outlet (MRGO) connected the Industrial Canal to the Gulf of Mexico. The Industrial Canal ran north to south, and separated eastern New Orleans from the rest of the City. The GIWW ran west to east, and the MRGO connected to the GIWW near the Industrial Canal. The MRGO ran toward the southwest, emptying into the Gulf. Building the Industrial Canal/GIWW/MRGO complex occurred after the Lower Ninth Ward and St. Bernard Parish were already settled, representing an environmental injustice to established communities Campanella, 2006; Freudenburg et al., 2009).

Similar to the environmental injustice created by canal development, the building of Interstate-10 through New Orleans negatively affected established low-income neighborhoods. Interstate-10 is on an east-west axis through the Treme', an historically Black neighborhood, and damaged this community by dividing it spatially with a major highway overpass. The highway destroyed an oak-lined avenue that once served as an urban meeting space (Campanella, 2006).

The Lower Ninth Ward is another historically Black neighborhood. The canal complex isolated the Ninth Ward in New Orleans East from the rest of the city, and further split this ward into two segments. Below the Lower Ninth Ward, in St. Bernard Parish, the MRGO created erosion along its banks, literally splitting St. Bernard Parish in two and creating large tracts of open water.

The construction of the MRGO is responsible for the loss of about 30% of the wetlands on the southeast border of Orleans Parish and in adjoining St. Bernard Parish. The canal has no water flow to keep out saltwater, so once the canal reached the Gulf, the MRGO began to provide regular deliveries of saltwater with each high tide. Saltwater intrusion from MRGO killed cypress trees and many kinds of plants that could only survive in less saline conditions (van Heerden, this volume, Chapter 6: Setting the stage for the catastrophe of Katrina:Environmental degradation, engineering miscalculation, ignoring science, and human mismanagement). The soil held in place by these trees and other plants washed away, and wetlands became open water (Freudenburg et al., 2009).

The oil and gas industry further damaged the environment by building wells in the wetlands (Marks, 2010a, 2010b). Between 1937 and 1977, approximately 63 hundred exploratory wells and over 21 thousand development wells were drilled in Louisiana's eight coastal parishes. The energy industry carved a network of canals and pipeline corridors to move drilling barges into the marshes, and to transport oil and gas. The new canals and corridors provided an opening for saltwater to reach and kill freshwater plants. Dredging also created spoil piles along canal banks, preventing normal drainage after heavy rainfalls and drowning plants (Freudenburg et al., 2009).

The broad band of wetlands that historically protected New Orleans against coastal storms began to deteriorate from canal and industrial development. The GIWW and the MRGO widened as land was eroded from the wetlands, and by 2005 much of the wetlands in St. Bernard Parish were degraded or had become open water. By the 1950s the state was losing land at a rate of 10–15 square miles per year. By the late 1960s land was lost at a rate of three to four times faster, with an estimated 45 square miles of coastal wetlands vanishing each year. In the last half of the 12th century, over 1 million of the original 5 million acres of wetland was lost. The 1.9 million square miles lost in the 20th century represents an area nearly as large as Delaware. Since 2000, about 30 square miles are lost each year, which is equivalent to an annual loss of an area the size of Manhattan Island (Marks, 2010a, 2010b; Freudenburg et al., 2009).

When two additional levees were built to protect the Lower Ninth Ward, St. Bernard Parish, and New Orleans East, a dangerous funnel effect was created. The additional levees created a funneling effect for Gulf Water to move into the MRGO and then into the Industrial Canal near the Lower Ninth Ward (van Heerden, this volume, Chapter 6: Setting the stage for the catastrophe of Katrina: Environmental degradation, engineering miscalculation, ignoring science, andhuman mismanagement). The first levee was to protect New Orleans East, which was largely uninhabited at the time. This levee was built to open up New Orleans East to residential development. The second levee encircled the Lower Ninth Ward and part of St. Bernard Parish, protecting already developed areas.

Ironically, because of growing ship dimensions the MRGO was nearly obsolete upon completion in 1968. By 2004 the Mississippi River was able to carry 250 times more traffic than the MRGO. Filling in the MRGO was considered in recent years, but it was estimated that this would take from 15 to 44 years to complete. Even if the MRGO was filled-in completely, the environmental damage in the wetlands was already great enough that a serious risk remains for New Orleans and St. Bernard Parish (Freudenburg et al., 2009).

MIGRATION AND POPULATION DISPLACEMENT SINCE 1960

Like urban areas in rest of the United States, New Orleans has suffered long-term neglect through the abandonment of urban areas by the public, and federal and state government. After the World War II many White urban residents began to leave New Orleans for suburban areas to the west, especially Jefferson Parish. When the federal government began to build interstate highways like Interstate-10 in New Orleans, suburban residents could more easily drive downtown to work. Eventually jobs moved close to workers in the suburbs. The State of Louisiana passed legislation to limit income taxes in New Orleans, forcing it to rely on regressive sales taxes. When the federal government began to distribute tax dollars to the states as block grants, much of this money remained in Baton Rouge, the state capitol, or went to suburban areas of metropolitan New Orleans (Passavant, 2011).

As part of the neoliberal abandonment of cities, many White middle-class families moved west to suburbs adjacent to Orleans Parish. The structural constraints and risk drivers of rapid population growth and displacement began with "White flight." Social service and other disaster-relevant organizations moved to Jefferson Parish and other areas to follow the more affluent White migration (Zakour & Harrell, 2003). Working-class Whites moved downriver to St. Bernard Parish. These areas were more desirable because of low crime and excellent schools. In the 1960s to 1970s, Interstate-10 promoted this expansion. New Orleans East (east of the Industrial Canal) developed in the 1970s because of this interstate highway. The housing development was predominately White in the early years, but became predominately Black when land speculation in the area failed (Freudenburg et al., 2009).

Between 1960 and 1990, most of the White population of New Orleans fled either to the Lakefront, or to Jefferson and St. Bernard Parishes (counties). White populations fled declining public schools, increasing crime rates, and urban infrastructural decay. Whites were attracted to good school districts, safety, suburban lifestyles, reduced congestion of traffic, and a lower cost of living. Whites also resisted school integration, particularly in the Lower Ninth Ward, and they left for St. Bernard Parish (Campanella, 2006). In the eastern section of New Orleans, Creoles and other Black households largely replaced the White population.

When an oil bust occurred in the 1980s, real estate values and development faltered, especially in New Orleans East. This economic crisis hastened the neoliberal segmentation of New Orleans into tourism and consumption zones on the one hand, and zones to contain poor populations on the other (Passavant, 2011). Some of the oldest neighborhoods, located on the Levee of the Mississippi, were gentrified and renovated as tourism grew in these locations. These neighborhoods include the lower Garden District, Faubourg Marigny, Faubourg Treme', Bywater, and the Central Business District. As rents and property taxes rose, poor households and Black households were forced to resettle in areas outside the gentrified neighborhoods and closer to Lake Pontchartrain (Campanella, 2006).

RESULTS OF THE NEGLECT OF NEW ORLEANS

Throughout much of the history of New Orleans the economy of the metro area has experienced frequent crises, and little economic development (Passavant, 2011). This lack of development has led to fragile livelihoods and residence in unsafe locations. In the 1980s, world oil prices crashed, leaving New Orleans in crisis. Poverty rates in the metro area in 2000 were twice the national average, and one-quarter of adults lacked a high school education. In 2000 only 46% of residents owned their homes. Poverty for children was very high before Katrina, and returned to high levels a decade after Katrina. Public transportation has been inadequate, and has been reduced by half since 2005. Another serious social and economic problem is that before Katrina, 120 thousand of the city's residents did not own cars, making evacuation of this population especially difficult.

The poor health status of New Orleans' residents is an additional serious social problem (New Orleans Health Department, 2013). The rate of infant mortality is very high, especially among Blacks, and health indicators are consistently very poor. Overall health indicators are among the lowest in the nation—Louisiana typically ranks 49 or 50 among the states. As of 2000, between 10% and 50% of the population suffers from a disability. Ten percent of the population has special needs, including being elderly (as of 2000). In New Orleans the rate of infant mortality has been consistently high, especially among Blacks.

Disaster mitigation has not been a priority for New Orleans, and in 2000 many levees had critical maintenance problems. In 2000 the complex of ports from New Orleans to Baton Rouge was the center of commerce for the US heartland, making it the busiest port by volume in the United States. About 20% of the nation's energy supply, mainly from offshore oil and gas platforms, passed through this port. Federal funds for civil works projects have mostly benefited navigation interests, and not levee construction and maintenance (Johnson, 2011). Between 2000 and 2005 Louisiana received far more money for Corps of Engineer civil works projects than any other state, or about $1.9 billion. California, with a population seven times larger, received $1.4 billion during the same period. However, most of this money was spent not on levee protection for residents, but on navigation projects that regional politicians saw as offering more immediate payoffs to the economy of Southeast Louisiana (Freudenburg et al., 2009).

VULNERABILITY AND RESILIENCE DURING KATRINA
KATRINA'S NATURAL, TECHNOLOGICAL, AND ORGANIZATIONAL FAILURE HAZARDS

Katrina was triggered and made more severe by a succession of different types of hazards. Initially the Katrina disaster was triggered by the natural hazard of the hurricane. The second stage of disaster was the result of the human-caused hazard of levee and floodwall failures. The final stage of Katrina was the failure of emergency management agencies and government at all levels to effectively respond and protect people. Residents who remained after Hurricane Katrina were only slowly evacuated and rescued from the flooding. Government and some other disaster-relevant organizations engaged in "elite panic." Guns were used to turn New Orleans into a prison and prevent people from evacuating, or even kill people, to protect property. This third type of hazard was fed by racism and the enormity of the storm (Solnit, 2009). Katrina had worsened from a disaster to a humanitarian catastrophe (Fig. 7.2).

FIGURE 7.2

Hurricane Katrina Flooding and Floodwater Depths

Source: From http://www.nytimes.com/interactive/2010/08/23/us/20100823-levee.html?_r = 0.

Because the New Orleans metropolitan area was on the west side (left quadrant) of Katrina, Orleans Parish was spared the strongest winds from the Hurricane. The front, east side (right quadrant), of a hurricane moving north has the highest wind speeds, given the counter-clockwise rotation of hurricanes in the Western Hemisphere. Damage from the winds of Katrina was only moderate. This included broken windows, allowing rain to penetrate the interior of buildings. Some roofs of smaller or nonreinforced buildings were blown off, badly damaging these structures.

Flooding from Katrina's storm surge proved extreme. The hurricane system was very large, and filled the entire Gulf of Mexico. As Hurricane Katrina approached the Central Gulf Coast, its wind speed increased to over 170 mph, a Category 5 hurricane. Because of the extreme wind speed and size of Katrina the hurricane pushed a large volume of water in front of it. The mass of water included water from tidal surges, and a dome of water raised by Katrina's low pressure and very large waves. When Katrina made landfall on the Mississippi Gulf Coast, the surge was over 30 feet, the highest surge ever recorded in a hurricane. As the eye of Katrina approached landfall the hurricane's counter-clockwise winds pushed a large volume of water into Lake Pontchartrain through several of the passes into Lake Pontchartrain. Water surged into the MRGO funnel and then into the Industrial Canal, where the water level rose to over 20 feet (Freudenburg et al., 2009).

When the surge of water from the MRGO and Industrial Canal reached the Lower Ninth Ward, the levee was overtopped. The substandard levees of the Lower Ninth Ward were undermined and collapsed as water from the Industrial Canal and the MRGO flooded this area. A wall of water nearly 20-feet high crashed into the neighborhood. Because of the speed of the floodwaters the death rate was five to seven times greater in the Lower Ninth Ward than the rest of New Orleans (Freudenburg et al., 2009). Levees in New Orleans East, a neighborhood lining the shore of Lake Pontchartrain, were also inferior in construction, and they broke in numerous places, flooding this predominately Black neighborhood.

In the remainder of New Orleans the seawalls in the north of the city, built to channel water as it was pumped out and into the lake, collapsed or were undermined. The seawalls are weaker than levees, which are very wide mounds of soil reinforced with concrete. The seawalls were essentially metal structures a few feet thick that often lack an inverted T shape, surrounded by a thin layer of soil. Numerous levee and seawall failures and leaks poured in massive amounts of water from Lake Pontchartrain, the Industrial Canal, and the Gulf. Eighty percent of the land in New Orleans was flooded, along with much of St. Bernard Parish (van Heerden, this volume, Chapter 6: Setting the stage for the catastrophe of Katrina: Environmental degradation, engineering miscalculation, ignoring science, and human mismanagement).

THE NATURE OF THE HURRICANE KATRINA DISASTER

Many aspects of Katrina's damage involved a high level of surprise. Meteorologists originally forecast Hurricane Katrina to make a westerly turn and miss New Orleans, but around 48 hours before it made landfall, it failed to turn and headed directly toward the city. Partly because of the failure of communications equipment after Katrina made landfall, governmental agencies did not know for several days that the levees along the Industrial Canal and Lake Pontchartrain had failed, and that New Orleans was filling up with lake water. Flooding of the Lower Ninth Ward was especially sudden and without warning.

The flood after Katrina was a long-lasting, and therefore more destructive, event (Campanella, 2006). Because most of New Orleans is at or below sea level, and is surrounded by levees, it was 6 weeks before all the floodwaters could be drained from the city. The draining could not begin until the breaches in the levees were closed and the water level in Lake Pontchartrain returned to mean sea level. During most of this time, city services, utilities, retail establishments, and other services were not available.

THE PATTERN OF DISASTER DAMAGE AND LOSS

Because of the historical pattern of residence of Black people and other minority populations in New Orleans, the lack of mitigation projects in these neighborhoods, and nearby major canals, Black neighborhoods experienced the worst flooding in Katrina. Nearly all predominately Black neighborhoods were flooded in the days after Hurricane Katrina struck. The hurricane and flooding, and the related Murphy Oil Spill, badly damaged St. Bernard Parish, with its many lower income households (Button, 2010).

Disproportionate loss in the Black community was further in evidence when persistent flooding is considered. Sixty percent of residences inhabited by Black people in New Orleans were persistently flooded, while 24% of residences of White people were persistently flooded after Katrina (Campanella, 2006). Persistent flooding is defined as floodwaters remaining until at least September 8, 10 days after the storm. A number of factors helped determine persistent flooding in New Orleans after Katrina. Among these were

- failure of substandard or inadequate levees and floodwalls,
- elevation of land and distance from Lake Pontchartrain, and
- proximity to canals and floodwalls, which channeled the storm surge into the city.

Orleans Parish suffered most from these factors, because of its substandard levees, floodwalls, and canals (van Heerden, this volume, Chapter 6: Setting the stage for the catastrophe of Katrina: Environmental degradation, engineering miscalculation, ignoring science, and human mismanagement). All of New Orleans East lies very near Lake Pontchartrain, and this neighborhood was closest to Hurricane Katrina. The Industrial Canal also separates New Orleans East from the rest of New Orleans, and the Ninth Ward is divided in two by the Intracoastal Waterway or the MRGO (Campanella, 2006).

In the New Orleans metropolitan area, almost 1600 people, most of them older, died in the aftermath of Katrina and the floods that followed (van Heerden, this volume, Chapter 6: Setting the stage for the catastrophe of Katrina: Environmental degradation, engineering miscalculation, ignoring science, and human mismanagement). Because standing floodwater in New Orleans remained for up to 6 weeks, about 80% of all properties in the city were lost due to the flooding. Many built structures collapsed either because of water rushing in from levee or floodwall breaches, or because water created enormous pressure on buildings over many weeks. Because water pressure to extinguish fires was initially very low, fire damaged or destroyed some property and structures.

New Orleans residents relied largely on the tourism and service industries for their livelihoods, and after Katrina jobs for a majority of New Orleans residents no longer existed. Commercial services could not reopen quickly because many workers in this industry did not return from their evacuation. This presented a cycle of job loss in the months after Katrina for New Orleans, particularly for unskilled workers who wished to return to their homes (Kim & Oh, 2014).

An important but lesser known impact of Katrina was the uncovering of contaminated soil on which Black neighborhoods were built in Orleans Parish. Some of the contamination was from lead-based paint, common in older US cities. In other situations, former dumps had been converted into land for residential development. After Katrina, high levels of heavy metals, formaldehyde, and other toxins were found throughout Orleans Parish, particularly in areas that were predominately Black. The EPA and other governmental agencies denied that this was the case, and even when they tested the soil and ground, they failed to release their findings to the public (Freudenburg et al., 2009). Hiding information about chemical contamination was an effort to protect the federal agencies that might be the target of class-action lawsuits (Bullard & Wright, 2009).

KATRINA AS CATASTROPHE

As discussed earlier, the third stage of Katrina involved an inability to evacuate stranded people, and the harsh and largely failed response of government agencies at all levels. This last stage represents the transformation of a disaster into a humanitarian catastrophe. "The most serious looting of all was done not by poor Blacks, but by rich Whites—both in terms of contributing to the risks, in advance of Katrina, and in terms of taking advantage of the recovery afterward" (Freudenburg et al., 2009, p. 136). During the evacuation the practice of subcontracting bus companies to dispatch buses to the disaster scene created unbearable delay and often trauma for survivors, and huge profits for the contractor (Solnit, 2009).

Katrina was the worst case of elite panic in the US history. Elite panic is the fear of social disorder, the poor, minorities, and immigrants. It is the obsession with property crime, and willingness to resort to deadly force. Actions taken during elite panic are based in rumor. Rumors arose of gang shootings involving police or rescue helicopters, but were later retracted. Two and a half days after the storm, on August 31, Mayor Nagin and Governor Kathleen Blanco ordered emergency responders, including police and National Guard, to stop search and rescue to focus on looting. They had chosen to protect property over saving lives (Solnit, 2009).

The 100,000 people who failed to evacuate before Hurricane Katrina suffered disaster-related health problems. Most people that remained were stranded in their homes or in refuges of last resort, such as the Superdome, for days or weeks (Passavant, 2011). With high temperatures often over 100°F and severe damage to utilities such as water, electricity, and natural gas, people experienced new health problems. Persons with a disability, older persons, and young children stranded in the city after the flooding could not access healthcare, and many became more sick before they were rescued. Injuries during the recovery period, when people had returned to the city, were very common. As health problems continued in the months after the flooding, and people were unable to rebuild their homes and return to work or school, rates of anxiety and depression increased (Joseph, Matthews, & Myers, 2013). Many people could not return to their predisaster levels of functioning, and their perceived quality of life continued to deteriorate.

RESILIENCE RESOURCES

ENVIRONMENTAL JUSTICE: RECOVERY BY RACE, INCOME, GENDER, AND AGE

When Hurricane Katrina struck New Orleans in August 2005, some of the most vulnerable populations were also those with little access to resources for response and recovery (Bullard & Wright, 2009). These populations included the poor, Blacks, older people, and women. The expansion of Katrina into a catastrophe limited access to resources needed for resilience (Solnit, 2009). The lack of access to resilience resources was the result of some of the same structural inequalities that had generated high levels of disaster vulnerability. In this section, we discuss four components of resilience: (1) economic, (2) information and communication, (3) social capital, and (4) collective action (Norris et al., 2008). Examination of these resilience components reveals that many resources needed for disaster recovery were lacking or of low quality.

ECONOMIC RESOURCES

Political—economic relationships in socio-ecological systems are an important link connecting vulnerability and resilience. Political and economic relationships provide the continuity between the progression of vulnerability, and networked adaptive capacities after disaster (Wisner et al., 2012). Economic and political inequities have acted to structurally ingrain disaster vulnerability in New Orleans and Southeast Louisiana (see Taylor, 2015). Community disaster resilience is made possible by equity in the distribution of economic resources, and fair distribution of disaster risk and vulnerability (Norris et al., 2008).

Larger structural trends in the political economy of Southeast Louisiana and the nation have created a pervasive vulnerability and fragility among people. For years and decades before Katrina, economic and political inequities favoring community elites made New Orleans more vulnerable to disasters. After Katrina, hierarchies and institutions in New Orleans were often inadequate in disaster response and recovery. Katrina was threatening to elites because power could have devolved to the people, and elites responded violently to threats to the status quo (Solnit, 2009). Those who were poor to begin with recovered from Katrina slowly or not at all. Affluent survivors went into debt and greater vulnerability because of the failure of recovery assistance. The disaster relief system forced survivors to bear the burdens of loss while taking on debt. This was to ensure that the relief industries mobilized to help could profit from the disaster (Adams, 2013).

During the Katrina recovery period, New Orleans experienced the outcomes of many years of neoliberal policies that favored and created instruments of market-driven recovery (Johnson, 2011). Market-oriented strategies of the private for-profit sector were turned into opportunities for making money on the part of companies ostensibly being asked to help people. Banks that had provided mortgages and then processed Small Business Administration (SBA) loans required that residents use their SBA funds to pay off the remainder of their mortgages first. Only then would they release the remaining funds for rebuilding. The working middle-class was impoverished through these debt mechanisms, and they harmed middle-class homeowners and renters (Adams, 2013).

Similar to the case of those who received SBA funding, people receiving Road Home funding sometimes had to use large portions of it to pay the back taxes on properties they could no longer live in. Some recipients found they had to use all of their Road Home funds to pay off mortgages on homes in order to qualify for new loans that would then give them enough to rebuild. This doubled their debt, and many returning residents had to abandon their homes altogether (Adams, 2013).

The company in charge of helping homeowners recover was ICF International. Over the 3 years of ICF International's contract to deliver the Road Home Program, much of what ICF reported as distributed included operational cost within the organization itself. ICF used $6.4 billion to distribute only $1.5 billion in actual relief aid to recipients (Freudenburg et al., 2009). For most homeowners, especially the elderly, the total dollar amount people received was far below what it cost them to rebuild. Many families found the best option was to move away and use what little Road Home money they had received to buy a home in another city (Adams, 2013). The nearly $5 billion that ICF International used for operational costs was far larger than payments to the intended recipients, and ICF's actions were aftermath looting.

In addition to the Road Home Program (Adams, 2013), another case of aftermath looting of taxpayers was the practice of governmental agencies subcontracting out reconstruction jobs that were then subcontracted out to secondary construction companies, who in turn subcontracted out to a third level of companies. Each level of subcontracting kept about half of the money it received from companies or government agencies. The final level of subcontracts paid companies a very small amount of money per repair job. This practice occurred with the Federal Emergency Management Agency (FEMA)s Operation Blue Roof, with debris removal, and with body recovery. In the case of debris removal, of the $500 million allocated by the federal government for this task, only about $65 million was used for debris removal (Freudenburg et al., 2009).

INFORMATION AND COMMUNICATION

Resilient communities have developed enough economically to have an adequate information infrastructure, trusted sources of information, and a responsible media. During disasters, accurate information helps survivors make decisions about their recovery. After a disaster, these capabilities help people to generate narratives of success in disaster recovery. These capabilities are dependent, in part, on a fair distribution of economic resources and disaster risks. Information and communication capabilities are in turn related to a number of social capital variables (Norris et al., 2008; Zakour & Swager, this volume, Chapter 3: Vulnerability-plus theory: The integration of community disaster vulnerability and resiliency theories).

Particularly after Hurricane Katrina and the flooding isolated New Orleans, people who did not evacuate were left with few, if any, sources of information about conditions in the region (Honoré, 2009). Some reporters and members of the media rose to the challenges of Katrina, reporting conditions accurately long before governmental response was fully operational. They accurately portrayed the absence of search and rescue, as well as response.

Some media outlets did not provide a responsible and accurate picture of the first days and weeks after the storm (Honoré, 2009). They provided sensationalist stories that relied on old racist stereotypes of the victims as savage, violent, and even murderous. They reported on rumors as if they were fact, and multiple reports based on the same rumors were seen by the media as corroborated. The theme for many in the media was the breakdown of civilization, and altruism among victims at the Convention Center and throughout New Orleans was ignored. Sensationalist media stories and reports were eventually retracted, but they were retracted quietly a month or more after the storm and initial flooding (Solnit, 2009).

Numerous media sources acted irresponsibly by reporting only stories favorable to elite interests. The strong reaction to Katrina by political and economic elites after Katrina was aimed at preserving a status quo with all of its structural inequalities, and another goal was to protect property at the expense of people. The structural violence of this reaction pressured the media to provide a picture of the Katrina disaster favorable to elites. Information and knowledge were contested in this manner, making decisions by the disaster victims and public in New Orleans more difficult because of the uncertainty that was generated (Solnit, 2009).

SOCIAL CAPITAL

Social support, a type of social capital, had sustained low-income people in New Orleans for generations. Many survivors, especially Black women, lost their social support and infrastructure due to death or relocation of network members (Jones-Deweever, 2011). People who once had a strong social support network and infrastructure, regardless of whether they had preexisting mental-health problems, became extremely distressed. Many stable individuals who had built-in social infrastructures before Katrina felt like isolated persons. Their houses, family social infrastructure, and entire world had disappeared (Adams, 2013).

The altruistic community, made up of disaster survivors who help others recover, emerged to help make up the loss of personal social support networks. Unfortunately the altruistic community is less supportive of poor disaster survivors than high-status survivors. After disasters, older persons are also less likely to receive material aid from the altruistic community unless injuries or other health problems have occurred (Kaniasty & Norris, 2009). Poverty prevented many households from accessing sufficient recovery aid after Katrina, and even sustained help from the altruistic community was lacking.

There was far more mutual aid in the Convention Center and Superdome than was reported by the media or imagined by authorities. Some groups of men were armed, but used their weapons to protect babies, children, and the aged who were stranded in New Orleans. Armed groups prevented assaults on women and girls. They helped to fan older people so that they could survive the extreme heat. When looting occurred in the stores along St. Charles Avenue by "gangs," it was for survival purposes. Organized groups took juice to give to babies and children. They took water, beer, and food for survivors (Solnit, 2009).

Social capital from formal organizations was relatively sparse in rural areas after Katrina. People living in the rural areas of the New Orleans metropolitan area, such as in parts of Plaquemines and St. Bernard Parishes, suffered disproportionately in Katrina. For rural areas, centralized disaster warning, evacuation, and aid systems worked less effectively due to greater distances and transportation problems. These problems arose partly because rural areas have more dispersed populations and lower population densities. In addition, many rural areas in the region have dirt roads and few have superhighways, hindering evacuation (Bassett, 2009).

A number of smaller organizations in New Orleans acted to increase trust, an important component of social capital (Adams, 2013). Based in a radical past (e.g., the Black Panthers), these organizations were able to work quickly and flexibly to reduce conflict during the recovery. Conflict reduction was an important means of building social cohesion. One of these groups was Common Ground, which initially sent medics on bicycles to the West Bank. Common Ground soon set up a clinic in the West Bank that provided medical care. By checking on residents there, and asking if medical or other help was needed, members of Common Ground diffused the explosive atmosphere that had given rise to White vigilantes shooting and killing mostly young Black men (Solnit, 2009).

COLLECTIVE ACTION

The efficacy of collective action by communities and populations in disaster is dependent on the adequacy of the economic, communication/information, and social capital components of resilience. Collective action is similar to the phenomenon of collective agency, or the ability to make and enact decisions to insure a resilient recovery (Norris et al., 2008). After Katrina, collective action was dampened by the severity of the disaster, the structural violence of elites and some relief organizations, and the effects of the forced diaspora of many survivors (Solnit, 2009).

Collective action and collective agency were diminished in Katrina. Without agency, free interaction and social adjustment could not produce a positive disaster community. The scale of the damage from Katrina and the floods that followed was overwhelming to survivors left behind in New Orleans. Mutual aid was not enough to cope with losses, especially since family, neighbors, and friends usually suffered from similar losses. The elite panic of authorities was widespread and disempowering. Victims stuck in public shelters of last resort (i.e., the Superdome) were helpless. Many people evacuated for an extended amount of time. Those who were evacuated from the Superdome and other sites were put on buses and planes, with no knowledge or choice of their destination (Solnit, 2009).

Most returning residents who received support from the Road Home Program noted that the program as a whole probably caused more problems than it solved. This was particularly true in terms of delays in rebuilding. The Road Home Program delayed decisions on whether or not to rebuild (Adams, 2013). A common complaint about the program is that applicants were repeatedly asked to submit the same paperwork. These delays were a kind of structural violence that greatly diminished individual and collective agency.

Consistent with the presence of community social aid clubs, after Katrina, New Orleans witnessed an unprecedented rise in locally organized social movements. Social aid and pleasure clubs have historically been active in the Black community in New Orleans. Over time they had become comprehensive social welfare organizations. They became social clubs that sponsored various celebrations, community activities, insurance funds, labor organizing, and even small businesses (Adams, 2013).

The groups that emerged after Katrina constituted new social movements, and spawned the birth of numerous nonprofits and nongovernmental organizations (NGOs). They became important conduits for volunteer labor, legal support, and financial aid. Some of the NGOs emerged from the ground up. Other organizations emerged as community groups with the explicit goal and purpose of rebuilding neighborhoods with volunteer support and donations from churches and individuals (Solnit, 2009).

Many organizations not only helped people recover, but also revisited the past context of structural inequality and social injustice in New Orleans. They focused not just on rebuilding, but on a socially "just" rebuilding. Not-for-profit organizations undertook political action to hold those responsible for unsafe conditions accountable. They promoted the development of a genuine participatory and open-ended approach to recovery (Adams, 2013).

The growth of nonprofit grass-roots organizations in post-Katrina New Orleans offers a good example of an important shift in American neoliberal policies. These community organizations have fueled the growth of political movements committed to local sociopolitical action aimed at rectifying inequalities and injustices. Grass-roots organizations have created a new playing field in the world of disaster response and preparedness. The community organizations have, in many cases, given voice to local political concerns and effective social activism (Adams, 2013).

SUMMARY AND CONCLUSIONS
THE POLITICAL ECOLOGY OF KATRINA AND SOUTHEAST LOUISIANA

The level of disaster vulnerability increased for decades and generations before Katrina occurred, and vulnerability disproportionately affected the poor, people of color, and women. "The death and destruction of ... Hurricane Katrina can be viewed as due in part to economically and socially inscribed practices associated with the capital and commodity flows that created and sustained them" (Oliver-Smith, 2009, p. 24). Katrina reveals the structural inequalities of race, class, age, and gender. The unequal distribution of vulnerability is visible throughout the ecology of Southeast Louisiana. Inequalities are inscribed in coastal zones, wetlands, and seascapes of Southeast Louisiana. These aspects of the natural world are associated with the technologies of levees, seawalls, canals, spillways, and pumping systems governing the flow of water in the region (Robbins, 2012).

The issue of environmental justice helps explain why a much higher proportion of land with Black residences flooded in the metropolitan area, compared to any other racial and ethnic group (Bullard & Wright, 2009). Over the last three centuries, Black people were moved or migrated to swampy areas at the "back-of-town" that were nuisance flood-prone areas near canals. St. Bernard Parish, where working-class Whites had migrated, has long been outside of the main system of levees in Orleans and Jefferson Parishes. Jefferson Parish has a larger tax base than either Orleans or St. Bernard Parishes, and enjoys a newer, more uniform, and more effective levee system. Both race and class have distributed low-status people to low-lying land with inadequate and aging levee protection, next to major canals and waterways (Freudenburg, This Volume, 2009).

Environmental injustices can likewise occur when social "bads" are distributed to established minority neighborhoods or communities. By 1794 the Carondelet Canal cut through the Tremé neighborhood, occupied largely by Creoles and other people of color. The canals built in the eastern part of New Orleans, and the small channels constructed for the thousands of oil wells, caused massive erosion of the wetlands protecting Southeast Louisiana (van Heerden, this volume, Chapter 6: Setting the stage for the catastrophe of Katrina: Environmental degradation, engineering miscalculation, ignoring science, and human mismanagement). Building the Industrial Canal/GIWW/MRGO complex took place after the Lower Ninth Ward and St. Bernard Parish was already settled, greatly increasing the vulnerability of these two areas.

Katrina is partly about the ways unequal power has shaped our world, and our beliefs about the world (Solnit, 2009). Unequal power has helped determine which set of knowledge is accepted as truth, and which competing knowledge has been rejected as false. The stories of powerful elites, who claim their projects have inconsequential effects on the natural world, have prevailed. The warnings and stories of danger and environmental destruction have been rejected or ignored (Freudenburg et al., 2009; Robbins, 2012).

Although we are better able to predict the environmentally damaging effects of economic development projects, this ability has lagged behind the rate of advance of new and damaging technologies. Technological advances and their use in environmental projects have created wealth for the few, and greater disaster vulnerability for the many and their communities. Human technology has advanced to the point that environmental harm can continue to grow over years and generations after projects are initiated and completed (Freudenburg et al., 2009).

IMPLICATIONS FOR VULNERABILITY-PLUS THEORY

In this section, we examine the implications for Vulnerability-plus (V+) theory, as described by Zakour and Swager (this volume, Chapter 3: Vulnerability-plus theory: The integration of community disaster vulnerability and resiliency theories). First, using an historical perspective, we discuss the causal chain consisting of root causes, structural constraints, and unsafe conditions. We describe how the natural, technological, and organizational failure hazards interact with unsafe conditions to produce the Katrina disaster. Then we look at resilience resources as intervening variables between the disaster and the triggering hazards. The nature and quality of these networked resources represent the level of disaster resilience after Katrina. Recovery after Katrina appears to be uneven, and results both in survivor wellness, and in continued community dysfunction.

Root causes

The neoliberal policies of the last 40 years have been a root cause for the decline of cities like New Orleans. White middle-class families have chosen to move to suburban areas for better schools and lower crime. The state government of Louisiana has prohibited the income tax in New Orleans, and has supported suburban areas through the distribution of block grants. As part of a massive highway-building project the federal government built Interstate-10 through New Orleans. The new interstate highway allowed suburban workers to commute to their jobs in downtown New Orleans. Eventually, jobs moved to the suburbs, along with service organizations and amenities. Supported by neoliberal policies, this trend has resulted in an increasingly meager tax base for cities like New Orleans, as well as substantial social problems including both economic and health crises (New Orleans Health Department, 2013; Passavant, 2011).

The neoliberal policies of the federal government facilitated the construction of major canals in New Orleans to promote economic growth (Johnson, 2011). The growth machine in New Orleans was able to obtain US federal tax dollars for the construction of the Industrial Canal, the GIWW, and the MRGO. Canal building benefitted only a few well-connected project sponsors. The construction projects redistributed damage to the natural environment of the Lower Ninth Ward and St. Bernard Parish. They made eastern New Orleans and St. Bernard Parish much more vulnerable to storm surges by funneling water to these two areas (Freudenburg et al., 2009).

Racist ideologies existed from the founding of New Orleans, and included the belief that Blacks were inferior to Whites and were subhuman. Blacks were owned as slaves for 140 years in New Orleans, and were further oppressed by Jim Crow Laws (Campanella, 2006). In the 20th century racist ideologies continued to justify the economic and political marginalization of Black people in Louisiana.

Structural constraints, dynamic pressures, and risk drivers

The root cause of racist ideology justified and promoted the structural constraint, or risk driver, of rapid population change and displacement of Black people. Beginning before the Civil War, Whites relegated slaves and people of color to shantytowns in the swampy areas outside of the French Quarter. After the Civil War, cheap rents and low-cost land attracted former slaves, including migrants from outside New Orleans, to these swampy areas. Intensified segregation forced Black populations off the Natural Levee of the Mississippi, and onto marshy land nearer Lake Pontchartrain. Because of racist covenants, elevated land near Lake Pontchartrain was off-limits to Black people. Throughout the 21st century, Blacks continued moving onto low-lying land away from the river and Lakefront, and to eastern areas of New Orleans (Campanella, 2006).

After the World War II, because of diminishing investments by government and a shrinking tax base, New Orleans developed health crises, high rates of poverty, and other social problems (Passavant, 2011). Millions of dollars allocated to the metropolitan area for levee building and maintenance were spent on navigation projects for the Mississippi River (Freudenburg et al., 2009). These structural constraints, combined with rapid population growth and displacement, interacted with root causes to produce unsafe locations, especially for Black and poor residents of New Orleans.

Unsafe locations

The series of canals, and damage to wetlands surrounding eastern New Orleans and St. Bernard Parish, left these areas and their populations highly vulnerable to surge waters from Gulf hurricanes. Hurricane Katrina channeled water into the Industrial Canal through the funnel consisting of the MRGO and levees meant to protect St. Bernard Parish and the Lower Ninth Ward (van Heerden, this volume, Chapter 6: Setting the stage for the catastrophe of Katrina: Environmental degradation, engineering miscalculation, ignoring science, and human mismanagement). The water level reached 20 feet in the Industrial Canal, and easily overtopped the substandard 13-foot high floodwall in the Lower Ninth Ward. The water from this funnel combined with lake water in the Industrial Canal, and overtopped levees in New Orleans East.

Characteristics of Katrina

Partly because of the characteristics of Hurricane Katrina, the storm created severe damage and losses in New Orleans. Hurricane Katrina was massive, its flooding was sudden, and the disaster that followed was long-lasting. Katrina was especially large, filling up most of the Gulf of Mexico. This allowed the hurricane to push a massive wall of water into Southeast Louisiana and Mississippi (Freudenburg, 2009).

The levee failures that accompanied Katrina's surge were surprising because they were not reported until several days after Katrina made landfall, and were initially unknown to the residents who remained. Flooding in the Lower Ninth Ward was particularly surprising and traumatic because of the sudden collapse of the levee next to the Industrial Canal. The levee system had numerous weak links and gaps, and failed to keep out Katrina's surge waters (van Heerden, this volume, Chapter 6: Setting the stage for the catastrophe of Katrina: Environmental degradation, engineering miscalculation, ignoring science, and human mismanagement). The levees helped hold in floodwater, and the pumps in New Orleans took up to six weeks to empty all of this water into the lake.

The pattern of damage and loss from Katrina revealed that some populations in New Orleans were highly vulnerable. The loss of life, and property damage, disproportionately affected older people, Blacks, and low-income people. Of the 1600 people who died in Katrina in Southeast Louisiana, most were over the age of 65. In Orleans, Jefferson, and St. Bernard Parishes (counties), about 60% of residences inhabited by African-Americans, and 24% of homes inhabited by White residents, remained flooded 10 days after Katrina made landfall (Campanella, 2006). Much of St. Bernard Parish, which is predominately White and working class, either flooded persistently or became uninhabitable because of the Murphy Oil spill triggered by Hurricane Katrina (Button, 2010).

Economic resources

The neoliberal policies and governance that had increased the level of disaster vulnerability for low-income, Black, and older people harmed their chances for a resilient recovery. A market- and insurance-driven approach to disaster recovery disadvantaged low-income people, rural households, and older people (Adams, 2013). These populations were less likely to have any homeowner or renter insurance, and were less likely to have adequate insurance or important options such as flood or hurricane coverage. Older homeowners were often on fixed incomes. Many low-income households lived in older, nonengineered, housing in low-lying sections of the city. These individuals were mostly renters, and they lost all of their possessions in Hurricane Katrina. The Road Home Program helped only homeowners (not renters) and gave uninsured homeowners only 70% of what insured homeowners received from their insurance carriers (Bassett, 2009).

Economic inequality and poverty, supported by structures of domination and neoliberal policies, left many people both vulnerable to disaster, and without resources for recovery. Relief organizations were able to profit from governmental funds allocated to help survivors. The Road Home Program, the SBA, and other relief agencies were delayed in helping survivors, and profited from governmental funds rather than distributing all of their funds to survivors (Adams, 2013).

Information and communication resources

Information and communication resources were of mixed quality after Katrina (Honoré, 2009). The media in New Orleans and throughout the United States portrayed the disaster and response after Katrina with uneven accuracy. Just as some in the media blindly supported the claims of the Growth Machine in the years leading up to Katrina, some media members were inaccurate and sensationalistic in their reporting during the disaster. Some in the media adopted the racist ideologies that were root causes of disaster vulnerability, and they exaggerated stories of violence and crime committed by Blacks. Some of the media supported elite panic efforts to protect property rather than lives (Solnit, 2009).

Social capital resources

Katrina's flooding, and the evacuation of stranded survivors, damaged social capital and social support networks. The heroism of the altruistic community in Katrina was substantial, but under-reported by the media. The natural, technological, and other hazards of Katrina devastated strong support networks by harming and dispersing network members. Some nonprofit organizations such as Common Ground worked to strengthen social cohesion and reduce conflict during the immediate aftermath of Katrina. Many social movements arose from existing institutions like social aid and pleasure clubs, reconstituting social support and providing recovery aid to survivors (Adams, 2013; Solnit, 2009).

Collective action resources

The nature of the Katrina disaster weakened collective action. The scale of the disaster, elite panic, the massive involuntary diaspora after Katrina, and delays in relief (e.g., the Road Home Program) dampened collective action. Conversely, social movements and the responses of community organizations strengthened collective action. During the recovery period, locally organized social movements began an unprecedented rise to strengthen collective agency. Community-based organizations helped people recover through a focus on socially just rebuilding (Adams, 2013).

SUBSTANTIVE IMPLICATIONS OF KATRINA

Understanding the cultural and social uniqueness of New Orleans is important for a successful and resilient recovery. Any recovery plan for New Orleans will fail if it does not take the socio-cultural and historical context into account (Morris & Kadetz, this volume, Chapter 10:Culture and resilience: How music has fostered resilience in post-Katrina New Orleans). An understanding of context, especially inside knowledge of the community, is critical for a resilient recovery. At the same time, the way in which local elites have increased the community's disaster vulnerability is not unique to New Orleans. The story of its progression to vulnerability is applicable to many communities across the United States and globally (Freudenburg et al., 2009). Ecological change that unduly harms some while benefiting others raises questions about alternative ways of doing things (Adams, 2013; Robbins, 2012). Ecological degradation challenges the hidden costs of environmental change from levee construction, oil extraction, and canal building projects (van Heerden, this volume, Chapter 6: Setting the stage for the catastrophe of Katrina: Environmental degradation, engineering miscalculation, ignoring science, and human mismanagement).

Changing root societal causes can be a powerful strategy for disaster risk reduction, but it is very difficult (Zakour & Gillespie, 2013). Addressing root causes, including the exploitation of natural resources such as oil, and the projects to increase efficiency of shipping, will cause reductions in those sectors of the regional economy. This will reverberate in the employment sector, eliminating the capital and jobs needed for establishing an economic base. These effects on the employment sector will only increase poverty and decrease the socioeconomic status of the population. If environmentally sustainable urban planning had been implemented to reduce the footprint of New Orleans, the human rights of several hundreds of thousands of displaced people to return to their community would have been severely limited (Campanella, this volume, Chapter 2: Settlement shifts in the wake of catastrophe, Oliver-Smith, 2009).

The experiences of New Orleans show that canal and levee development projects tend to be obsolete soon after completion, but their damage to socioecological systems is long-lasting. Although their economic benefits were quite modest, these projects have created a great deal of ongoing environmental damage. The time it took for major canal and lock projects to be conceived, politically supported, funded, and built was many decades. Over these decades, ship size would increase in terms of depth and width, and many ships became too large to use the new canals and locks. The Industrial Canal barely escaped obsolescence when it was completed in 1928. The MRGO was obsolete upon delivery in 1968, although it continues to damage the local environment (Freudenburg et al., 2009).

POLICY IMPLICATIONS

Scientists and others in the public need to challenge mistaken conclusions, especially the assumption that sensible policy choices are not politically feasible. For every dollar spent on disaster mitigation, society saves an average of $4. This means that investments in environmental protection and building practices that adapt to our surroundings pay major dividends. Wiser approaches to risk management make rational sense regardless of political party or perspective (Freudenburg et al., 2009).

A first more general suggestion is part of a heuristic approach to disaster resilience and future vulnerability:

> *Emergency managers, those involved in disaster risk reduction, and community members should continually learn from reconstruction after major disasters such as Katrina.*

The process of resilience and recovery needs to include a component of learning. Through learning, recovery can continually improve, and lessons learned from other disasters can contribute to this improvement. According to Oliver-Smith (2009) "Perhaps our most important task is to discover and implement those aspects of reconstruction that feasibly, within the limit of action permitted by existing political—economic structures, can reduce both environmental degradation and vulnerability to hazards" (p. 24).

A second important policy suggestion regarding environmental justice is:

> *Make income equality a priority, among populations in urban, suburban, and rural areas, and among states and regions.*

The federal and state governments in the United States must be directly responsible for guaranteeing the resources for resilient disaster recovery. Environmental justice principles affirm that the importance of addressing structural inequalities of race, poverty, and gender (Marks, 2010a). Root causes are powerful leverage points, but difficult to change (Oliver-Smith et al., 2016; Zakour & Gillespie, 2013). Despite this difficulty an important step that can immediately be taken is to transfer a larger share of tax funds from the Outer Continental Shelf oil extraction from federal to Louisiana or coastal parish coffers. This step would not damage the oil industry or jobs from this industry.

A third suggestion calls for changes in the FEMA requirements for floodplain development:

FEMA should modify its requirements so that levees protect land from 500-year floods, and FEMA should stop allowing new floodplain development in areas not protected by levees.

FEMA guidelines allow nearly unlimited development in floodplains, as long as levees initially protect land from 100-year floods (Campanella, this volume, Chapter 2: Settlement shifts in the wake of catastrophe). Once FEMA has declared an area that has 100-year protection, it does not matter if phenomena such as subsidence, or deterioration of levees from lack of maintenance, reduces protection below the 100-year level (Freudenburg et al., 2009). Paradoxically, levee failures are responsible for roughly one-third of all flood disasters in the United States. Levees cause land to settle, sink, and subside, increasing flood levels when they actually fail. The false confidence that people gain from levee construction can lead to building on floodplains and increased vulnerability to flooding (Freudenburg et al., 2009).

A fourth policy suggestion is that

Energy corporations, and consumers of cheap oil and gas, need to pay for restoring the wetlands.

The energy industry and energy consumers have failed to pay for the destruction of the Louisiana's protective wetlands or for the series of disasters in the first decade of the 21st century. The interdependent but highly unequal relationship between Louisiana workers and energy companies has never been addressed, even though it has shifted risk and environmental damage onto the victims. Energy companies, largely headquartered outside of the state, have accumulated capital (Marks, 2010a) and should return some of this money to Louisiana for wetlands restoration.

Closely related to the previous suggestion is that

Community development and organization approaches need to be used to encourage a participatory approach to disaster prevention and reconstruction.

Oil companies have never been held accountable for the loss of wetlands in coastal Louisiana (Marks, 2010a). Profits from oil and gas production, and massive state subsidies for the energy industry, have allowed the industry to extract huge surpluses from the people of Louisiana. The oil industry and consumers of cheap gasoline have outsourced environmental damage and other costs to Louisiana. "In prevention and in reconstruction people can address those conditions which bring about disaster by undertaking political action to … hold those responsible for vulnerable conditions accountable.… The challenge thus becomes the development of policy that supports a genuine participatory and open-ended approach" (Oliver-Smith, 2009, p. 25).

REFERENCES

Adams, V. (2013). *Markets of sorrow, labors of faith. New Orleans in the wake of Katrina.* Durham, NC: Duke University Press.

Bassett, D. L. (2009). The overlooked significance of place in law and policy: Lessons from Hurricane Katrina. In R. D. Bullard, & B. Wright (Eds.), *Race, place, and environmental justice after Hurricane Katrina: Struggles to reclaim, rebuild, and revitalize New Orleans and the Gulf Coast* (pp. 49–62). Boulder, CO: Westview Press.

Bullard, R. D., & Wright, B. (Eds.), (2009). *Race, place, and environmental justice after Hurricane Katrina: Struggles to reclaim, rebuild, and revitalize New Orleans and the Gulf Coast.* Boulder, CO: Westview Press.

Button, G. (2010). *Disaster culture: Knowledge and uncertainty in the wake of human and environmental catastrophe.* Walnut Creek, CA: Left Coast Press.

Campanella, R. (2006). *Geographies of New Orleans. Urban fabrics before the storm.* Lafayette, LA: Center for Louisiana Studies, University of Louisiana at Lafayette.

Cutter, S. L. (1995). Race, class and environmental justice. *Progress in Human Geography, 19*(1), 107–118.

Freudenburg, W. R., Gramling, R. B., Laska, S. B., & Erikson, K. T. (2009). *Catastrophe in the making: The Engineering of Katrina and the disasters of tomorrow.* Washington, DC: Island Press.

Heynen, N., Kaika, M., & Swyngedouw, E. (2006). Urban political ecology: Politicizing the production of urban natures. In N. Heynen, M. Kaika, & E. Swyngedouw (Eds.), *In the nature of cities: Urban political ecology and the politics of urban metabolism* (pp. 1–19). New York: Routledge Press.

Honoré, R. L. (2009). *Survival: How a culture of preparedness can save you and your family from disasters.* New York: Atria Books.

Johnson, C. (2011). Introduction: The neoliberal deluge. In C. Johnson (Ed.), *The neoliberal deluge: Hurricane Katrina, late capitalism, and the remaking of New Orleans* (pp. xvii–l). Minneapolis, MN: University of Minnesota Press.

Jones-Deweever, A. (2011). The forgotten ones: Black women in the wake of Katrina. In C. Johnson (Ed.), *The neoliberal deluge: Hurricane Katrina, late capitalism, and the remaking of New Orleans* (pp. 300–326). Minneapolis, MN: University of Minnesota Press.

Joseph, N. T., Matthews, K. A., & Myers, H. F. (2014). Conceptualizing health consequences of Hurricane Katrina from the perspective of socioeconomic status decline. *Health Psychology, 33*(2), 139–146.

Kaniasty, K., & Norris, F. H. (2009). Distinctions that matter: Received social support, perceived social support, and social embeddedness after disasters. In Y. Neria, S. Galea, & F. H. Norris (Eds.), *Mental health and disasters* (pp. 175–200). Cambridge: Cambridge University Press.

Kim, J., & Oh, S. S. (2014). The virtuous circle in disaster recovery: Who returns and stays in town after disaster evacuation? *Journal of Risk Research, 17*(5), 665–682.

Marks, B. (2010a). Louisianans, oil & petro-addiction. *Against the Current, 147,* 4–6.

Marks, B. (2010b). The Gulf disaster: No end in sight. *The Indypendent, 153,* 6–7, June 23.

Norris, F. H., Stevens, S. P., Pfefferbaum, B., Wyche, K. F., & Pfefferbaum, R. L. (2008). Community resilience as a metaphor, theory, set of capacities, and strategy for disaster readiness. *American Journal of Community Psychology, 41,* 127–150.

New Orleans Health Department (2013). *Child and family health in New Orleans: A life course perspective of child and family health at a neighborhood level.* Retrieved from: https://www.nola.gov/getattachment/Health/Data-and-Publications/Child-and-Family-Health-in-New-Orleans-December-2013.pdf/

Oliver-Smith, A. (2009). Anthropology and the political economy of disasters. In E. C. Jones, & A. D. Murphy (Eds.), *The political economy of hazards and disasters* (pp. 11–28). Lanham, MD: Altimira Press.

Oliver-Smith, A., Alcántara-Ayala, I., Burton, I., & Lavell, A. M. (2016). *Forensic investigations of disasters (FORIN): A conceptual framework and guide to research (IRDR FORIN Publication No. 2).* Beijing: Integrated Research on Disaster Risk.

Passavant, P. A. (2011). Mega-events, the Superdome, and the return of the repressed in New Orleans. In C. Johnson (Ed.), *The neoliberal deluge: Hurricane Katrina, late capitalism, and the remaking of New Orleans* (pp. 87–129). Minneapolis, MN: University of Minnesota Press.

Robbins, P. (2012). *Political ecology. A critical introduction* (2nd ed.). West Sussex, UK: John Wiley and Sons.

Solnit, R. (2009). *A paradise built in hell. The extraordinary communities that arise in disaster.* New York: Penguin Books.

Taylor, M. (2015). The political ecology of climate change adaptation: Livelihoods, agrarian change and the conflicts of development *(Routledge explorations in development studies).* New York: Routledge/Earthscan.

Wisner, B., Blaikie, P., Cannon, T., & Davis, I. (2004). *At risk: Natural hazards, people's vulnerability and disasters* (2nd ed.). New York: Routledge.

Wisner, B., Gaillard, J. C., & Kelman, I. (2012). Framing disaster: Theories and stories seeking to understand hazards, vulnerability and risk. In B. Wisner, J. C. Gaillard, & I. Kelman (Eds.), *The Routledge handbook of hazards and disaster risk reduction.* London: Routledge. (Chap. 3, pp. 18-34).

Zakour, M. J., & Gillespie, D. F. (2013). *Community disaster vulnerability: Theory, research, and practice.* New York: Springer.

Zakour, M. J., & Harrell, E. B. (2003). Access to disaster services: Social work interventions for vulnerable populations. *Journal of Social Service Research, 30*(2), 27–54.

CHAPTER

The resilience in the shadows of catastrophe: Addressing the existence and implications of vulnerability in New Orleans and Southeastern Louisiana

Regardt J. Ferreira[1] and Charles R. Figley[1,2]

[1]*Tulane University, New Orleans, LA, United States* [2]*University of the Free State, Bloemfontein, South Africa*

CHAPTER OUTLINE

Introduction 193
    Vulnerability Defined 194
    Resilience Defined 196
The New Orleans and Southeastern Louisiana "*Catastrophe*" 198
The New Orleans Vulnerability and Resilience Paradigm 200
Causal Process: *The New Orleans and Southeastern Louisiana "Catastrophe"* 201
    Root Causes 203
    Dynamic Pressures 204
    Unsafe Conditions 205
Predictors of Social Vulnerability in Louisiana: A Multilevel Analysis 207
Addressing Vulnerability in New Orleans and Southeastern Louisiana 208
Conclusion 209
References 210
Further Readings 213

INTRODUCTION

This chapter aims to examine the progression of vulnerability in Southeastern Louisiana that led to Hurricane Katrina being one of the deadliest and costliest disasters in the history of the United States. By addressing the root causes of social vulnerability, the impact of future disasters on the psychosocial well-being of individuals, households, and communities can be mitigated and can become more resilient in the wake of potential hazards (e.g., floods, hurricanes, etc.), that could result in large-scale disasters.

The study field of disaster resilience has become prevalent in what has been characterized as a "collective surge in science, policy, and practice" in the disaster field, along with other concepts of risk and vulnerability (Tierney, 2015). Vulnerability and resilience are closely related, but have key distinctions. The following section provides definitions of vulnerability and resilience, followed by a discussion on the vulnerability and resilience paradigm in New Orleans and Southeastern Louisiana.

VULNERABILITY DEFINED

Extreme natural events are not regarded as disasters until a vulnerable group of people is exposed to such an event. The concept of vulnerability was coined in the early 1980s as a way to reduce losses from disasters. A renewed emphasis on vulnerability emerged in the early 2000s. The most recent work on vulnerability is disconnected from a Marxist perspective, but still focuses on changes within a system, making it consistent with social work values and practices (Gillespie & Danso, 2010).

Particular social groups are more prone to damage, loss, and suffering in the context of differing hazards. Wisner, Blaikie, Cannon, and Davis (2004) provide key variables explaining the variations of impact. One key variable is class, which includes different levels of wealth within a community. Class levels can include occupations of the target system, caste, ethnicity, gender, disability, health status, age, and immigration.

Vulnerability is the reflection of the current state of the individual and his collective social, physical, economic, and environmental conditions. These conditions are shaped recurrently by behavioral, attitudinal, cultural, socioeconomic, and political influences on individuals, families, communities, and countries. The causes of vulnerability are not only natural conditions, but also processes relating to social, political, and economic environments. This process is edified with the Pressure and Release (PAR) model. Gillespie (2010) identifies causes of vulnerability including

- a lack of access to information, knowledge, and technology;
- weak or nonexistent political representation or power;
- limited social capital;
- building age and quality;
- frail and physically limited individuals; and
- type, quality, and age of infrastructure and lifelines.

Vulnerabilities are structural and situational in nature (Gillespie, 2010). This is attributed to the way lives of different groups of people are structured and shaped by structural patterns based on politics, economics, environmental management practices, race and class relations, the gender-based division of labor, and other factors. Social status and situational or context-specific living conditions that vary over time might also shape vulnerability. The main concept grounding vulnerability is social and especially distributive justice. Within this conceptualization, the market value of individuals and populations is inversely related to their level of disaster vulnerability (Zakour, 2010).

Where hazards appear to be directly related to loss of life and damage to property, the social, economic, and political origins of the disaster remain as possible root causes. In the simplest form, vulnerability is created by these social, economic, and political processes that determine how hazards affect people in different ways and varying degrees of intensity (Wisner et al., 2004).

A subtype of vulnerability is social vulnerability. According to Cutter (2006), "social vulnerability is partially a product of social inequalities—those social factors and forces that create the susceptibility of various groups to harm, and in turn affect their ability to respond, and bounce back (resilience) after the disaster." Social vulnerability is most often described using the individual characteristics of people, including age, race, health, income, type of dwelling unit, and employment. Social vulnerability is the product of social inequalities that influence or shape susceptibility of various groups to harm, and that also shapes their ability to respond to adversity (Cutter, Borhuff, & Shirley, 2003).

Social vulnerability in many instances is ignored, and only becomes visible to the broader society in times of adversity. According to Wisner (2001), vulnerable groups include

- residents of group-living facilities,
- the frail elderly,
- persons with a physical or mental disability,
- renters,
- poor households,
- female-headed households,
- ethnic minorities,
- recent residents,
- large households, and
- neighborhoods with large concentrations of children, youth, the homeless, and tourists.

Publicly available census data on some of the socially vulnerable groups identified by Wisner (2001) can be used as a barometer for determining social vulnerability in a community.

According to Cutter et al. (2003) there is a general consensus within the social science community about some of the major factors that have an influence on social vulnerability. These include:

- lack of access to resources, which includes information, knowledge, and technology;
- limited access to political power and representation;

- social capital, including social networks and connections;
- beliefs and customs;
- building stock and age;
- frail and physically limited individuals; and
- type and density of infrastructure and lifelines.

Cutter et al. (2003) indicates that social vulnerability variables found within disaster management literature are age, gender, race, and socioeconomic status. Other characteristics identify special needs populations or those that lack the normal social safety nets necessary in disaster recovery, such as the physically or mentally challenged, immigrants, the homeless, transients, and seasonal tourists. The quality of human settlements (housing type and construction, infrastructure, lifelines) and the built environment are also important in understanding social vulnerability, especially as these characteristics influence the potential of economic losses, injuries, and fatalities from hazards (Cutter et al., 2003).

For operational purposes, it is more useful to work with broad definitions and commonly understood characteristics. Cutter et al. (2003) agree with this approach, since there are numerous meanings of vulnerability and resilience in the hazards literature. An operational definition of vulnerability within the context of this chapter can be described as individuals, households, and communities that are more susceptible to harm or strife when faced with an adverse situation (e.g., Hurricane Katrina), who would be impacted to such an extent that recovery from the adverse situation would take a significant amount of external resources.

RESILIENCE DEFINED

Miller et al. (2010) traced the origin and research traditions that embrace either resilience and vulnerability and found considerable overlap in their respective origins in ecological and social theory. They suggest that this largely explains the continuing differences in approach to social—ecological dimensions of change. We return to this issue toward the end of this chapter.

Resilience is complex and multifaceted. To identify key factors in disaster resilience on an individual (micro) and community (macro) level, it is important to understand what factors contribute to the vulnerability and resilience of the individual and the community (Goodman et al, 1998; Norris, Stevens, Pfefferbaum, Wyche, & Pfefferbaum, 2008). A generalist resilience definition within the context of individuals and human systems is "Resilience is the ability of individuals or human systems to absorb stressors and return to their original state when that stressor is lifted without creating permanent damage or harm" (Hobfoll, Stevens, & Zalta, 2015).

The resilience definition of Hobfoll et al. (2015) indicates that resilience can be approached from both an individual and community level and is dependent on more than one particular factor. Similar to Hobfoll et al. (2015) is Goldstein and Brooks' (2006) definition of individual resilience describing it as the "ability to meet life's challenges and pressures with confidence and perseverance" (Brooks & Goldstein, 2002). Resilient individuals demonstrate strength in four key psychological constructs: (1) social competence, (2) problem solving, (3) autonomy, and (4) a sense of purpose. Aspects of social competence include social support resources, communication and responsiveness, empathy, and forgiveness. Problem-solving skills may be evidenced by an individual's ability to use emotion-focused or problem-focused coping styles (as opposed to avoidant coping).

According to Ungar (2008), resilience is defined both as the *opportunity* and *the capacity* of individuals to navigate their way through life and adapt to the various psychological, social, cultural, and physical demands and resources available. Communities can make this more or less difficult. Social factors posessing the power to increase individual resilience include the immediate family environment, caring relationships outside the family, and opportunities to participate and contribute within social networks. Community organizations have the potential to confer resilience to the individual. With regard to the power of protective systems in the development of resilient youth, Bernard (2004) identifies the importance of caring, respectful, reciprocal social relationships within school and community programs. An examination of factors that contribute to individual resilience should include measures of program quality as well as program availability in a community.

Community resilience is a common feature of complex systems that include cities, communities, and ecosystems. These systems continually evolve through cycles of growth accumulation, crisis, and renewal. In order to describe resilience on the macro level, an holistic approach is needed (Norris, et al., 2008). In establishing the various factors relating to community disaster resilience, it is vital to first gain an understanding of resilience from a community context. The concepts of resilience and vulnerability are opposite sides of the same coin, but both are relative terms. Some scholars argue that the opposite of vulnerability is resilience (Adger, Hughes, Folke, Carpenter, & Rockström, 2005).

Resilience on a macro level is not latent and can be increased or decreased. Vulnerability should be taken into account as a determining factor that can result in an increase or decrease of resilience on the macro level. From a practical standpoint, community resilience can be increased when a community implements risk reduction measures preparing for the possibility of disaster impact, has the response capacity to minimize the impact of a disaster, and has disaster recovery measures in place (Ronan & Johnston, 2005). For a community to increase its level of resilience toward disaster events, it is imperative that socioeconomic resources are developed, community competence is increased, and distribution of information and social capital is fostered (Norris et al., 2008). By strengthening economic resource levels and equities, diverse communities will increase their economic resilience. Ensuring that there is a form of collective action and decision-making skills can increase community competence. Collective efficacy in the community empowers communities with decision-making capabilities.

Information distribution in the community is essential, and in order for this to be effective, infrastructure is needed. Information should come from trusted sources that are associated with positive narratives. Lastly, social capital can increase social support and social participation, and allow for bonds, roots, and commitments to be formed in the community. An operational definition of resilience within the context of this chapter can be described as the capacity to withstand an adverse event with as little possible disruption of daily activities.

THE NEW ORLEANS AND SOUTHEASTERN LOUISIANA "*CATASTROPHE*"

New Orleans is situated in Louisiana, a state in the Southern United States with a rich cultural history. From 1686 to 1790, France and Spain colonized and governed the lower Mississippi River Valley. The region is named after Louis XIV, who was King of France in 1682. In the early 1800s, the US government purchased land to the west of the Mississippi River. Historically, it is regarded as the biggest land acquisition in American history. The acquisition of land granted the US government the ability to expand westward for settlement and trade, and secured borders against any possible threat. The Louisiana Purchase expanded trade, and allowed for a trade route to be established along the Mississippi River (Baker, 2011; Fortier & McLoughlin, 1913; Kelman, 2006).

Louisiana has a rich, colorful, historical background, recognized for its rich multicultural and multilingual heritage. Strong influences from Spanish, French, Native American, and African cultures have resulted in a fusion of historical and cultural backgrounds (Bates & Swan, 2007). There is a rich mixture of people residing in Louisiana (Table 8.1). Current-day residents of Louisiana include the original American Indian inhabitants, as well as descendants of German, Spanish, French, English, Irish, Italians, Acadians, Africans, and West Indians (Louisiana, 2012; U.S. Census Bureau, 2015).

The New Orleans Metro area pre-Katrina had a fairly socially vulnerable demographic composition (Table 8.2). One-fifth of the city's population were

Table 8.1 Population and Racial Composition of Louisiana Jurisdictions

	Louisiana	N.O. Before 2005	N.O. After 2005
Total	4,670,724	484,674	384,320
White (%)	63.4	26.6	31.2
Black (%)	32.5	66.7	58.8
Asian (%)	1.8	2.7	3.0
Hispanic or Latino (%)	4.8	0	5.5
Native Am./Eskimo (%)	0.7	0.2	1.5

Table 8.2 Demographic Variables

	N.O. Before 2005 (%)	N.O. After 2005 (%)
Children below 5 years	20.0	18.0
Adults age 65 and older	20.0	18.0
Elderly living alone	34.2	32.0
Owned housing units	46.0	46.0
Earned less than $25 K per year	46.4	37.1

children below 5 years and older adults aged 65 years and older, 34.2% of the elderly lived alone, less than half of the city owned their housing units, and 46.4% of individuals earned less than $25,000 per year (Plyer, Shrinath, & Mack, 2015). New Orleans Metro (Orleans Parish) post-Katrina in 2015 had a total population of 384,320, and the demographics are somewhat different (Table 8.2). Comparatively between White and Black there is a complete switch between the two racial groups compared with New Orleans and Louisiana (Plyer et al., 2015; US Census Bureau, 2015).

New Orleans is no stranger to disasters. It is common knowledge that the Mississippi River is the largest river in the United States (Kammerer, 1990). The river system drains an estimated 41% of the continental United States, with a watershed area of around 1,245,000 square miles, resulting in the Mississippi being the third largest watershed of any river in the world (Independent Levee Investigation Team, 2006). The Mississippi River has changed the landscape and psyche of New Orleans and Louisiana residents with some communities experiencing annual losses due to flooding from the river (Kelman, 2006). The Mississippi River has resulted in a series of levees being erected around the city and the region, serving as a protection barrier from possible flooding.

Some of the earliest reported disasters in Louisiana can be traced back as far as 1718. Over the following 300 years the Mississippi River has caused several flooding events (Independent Levee Investigation Team, 2006). Hurricanes strike the coastline of Southeastern Louisiana with a mean frequency of two hurricanes every 3 years. From 1759 to 2000 more than 172 hurricanes have caused devastation and loss of life (Bates & Swan, 2007; Federal Emergency Management Agency, 2015; Independent Levee Investigation Team, 2006). The most notable of these are shown in Table 8.3.

A number of disasters have had a lasting impact on the human psyche of New Orleans and Southeastern Louisiana. The horrible yellow fever epidemic of 1853 took the lives of nearly ten thousand residents in New Orleans. Historians report that the disaster could have been averted, if it were not for the class and racial segregation in the city. The city was crippled and brought to its knees with the poor being left behind and the wealthiest citizens fleeing the city from the epidemic (Kelman, 2006; Lafayette Cemetery Research Project New Orleans, 2012). Louisiana has gained notoriety for train wrecks, structural fires, structural

Table 8.3 The Most Notable Hurricanes That Have Affected Louisiana

Year	Hurricane Name
1812	Great Louisiana Hurricane
1831	Great Barbados Hurricane
1893	Record Hurricane
1915	Grand Isle Hurricane
1965	Hurricane Betsy
1969	Hurricane Camille
1998	Hurricane Georges
2004	Hurricane Ivan
2005	Hurricane Katrina
2005	Hurricane Rita
2008	Hurricane Ike

disasters, river and maritime accidents, and industrial chemical spills. The Mississippi River has claimed a number of steamboats and passengers, with the river being notorious for accidents over the past two centuries (Beitler, 2007; Kelman, 2006; The Times-Picayune, 2010).

THE NEW ORLEANS VULNERABILITY AND RESILIENCE PARADIGM

The city of New Orleans provides a great example of a system that poses both vulnerability and resilience traits. From a pre-Katrina standpoint, many will regard the city as very vulnerable, whereas post-Katrina the use of the concept resilience has been synonymous for some. Resilience provides a capacity for adaptation that helps the impacted community mitigate and adapt to changed circumstances after a disturbance, instead of just returning to the exact same state as before the disturbance (Laska, 2012). Through maladaptation strategies a community can destabilize resiliency, creating increased vulnerability. A prime example of a maladaptation strategy is when the government enacted policies to promote economic resilience in the region. This effort conflicted with ecological resilience when developers damaged local ecosystems, as with the building of the Industrial Canal. This, in turn, led to Hurricane Katrina's destruction of wetlands around New Orleans, leading to decreased resilience and increased vulnerability (Freudenburg, Gramling, Laska, & Erikson, 2009).

Resilience for a majority of those who remained behind when Katrina struck was significantly lowered because they had no means of transportation (Eisenman, Cordasco, Asch, Golden, & Glik, 2007; Litman, 2006). Policymakers, including the Mayor of New Orleans, should have been aware of this vulnerability for those living in economically blighted areas of the city. Those vulnerable communities had a large percentage who suffered from mental disorders and suffered from medication withdrawal because of elevated health and mental health difficulties (Rhodes et al., 2010),

Post-Katrina, Orleans Parish is still regarded as being more socially vulnerable compared with the surrounding parishes of St. Charles, Jefferson, Plaquemines, and St. Bernard. Disasters tend to precipitate predisaster situations. In reviewing social vulnerability data compared between pre- and post-Katrina, there are clear differences to be observed between the parishes in Southeastern Louisiana. By using income as an indicator of social vulnerability, it is clear that there are differences to be observed. In regards to household income for parishes (counties), St. Tammany households had an annual income of $63,210, Jefferson households with $46,961, compared to Orleans that had an annual household income of $35,504 (The Data Center, 2016).

According to Plyer et al. (2015), data indicate that Orleans Parish had a very poor population, with slow job growth and extremely high levels of poverty. Post-Katrina the disparities in Orleans Parish, compared to the surrounding parishes and nationally, have increased significantly, with employment and income disparities between minorities and nonminorities increasing. For example, the household income for White households in Orleans Parish is comparable to national averages for White households. The median income for Black households in Orleans Parish is 20% lower compared to the national average for Black household income.

According to Plyer et al. (2015), data indicate that the economy of Orleans Parish has rebounded for some, and in some instances it is better than pre-Katrina levels. It should be noted however that there are several social and environmental trends that may be stressed and test resilience when the area is faced by the next big disaster.

CAUSAL PROCESS: *THE NEW ORLEANS AND SOUTHEASTERN LOUISIANA "CATASTROPHE"*

Hurricane Katrina was one of the worst natural disasters in the history of the United States. The effects of the disaster lead to a number of structural failures that caused further damage to the City of New Orleans (Brunsma, Overfelt, & Picou, 2007; Independent Levee Investigation Team, 2006). Levee failures in New Orleans caused extensive flooding that inundated the city for weeks. Secondary effects of Hurricane Katrina resulted in a natural-technological or "natech" disaster. Dangerous hazmat releases resulted from toxic oil and chemical spills. The hazmat releases are referred to as "toxic gumbo." "Toxic gumbo" is bacteria infested hazardous floodwaters that engulfed the region. Louisiana experienced 10 major oil spills with an estimated 134 minor oil spills as a result of

Hurricane Katrina. According to estimates, eight to nine million gallons of oil were spilled as a result of Katrina being not only a natural disaster but also a technological disaster. As a natural-technological disaster, Katrina caused the third worst oil spill in the history of the United States (Brunsma et al., 2007). Empirical evidence suggests that a number of factors have contributed to the high number of disasters in Southeastern Louisiana.

A causal approach is needed to identify factors that have caused and are currently causing social vulnerabilities within New Orleans and Southeastern Louisiana. Application of the PAR model (Wisner et al., 2004) provides an opportunity to delineate the processes causing Hurricane Katrina to be such a large-scale and unmanageable disaster. Vulnerability-generating processes is a key element within the PAR model. Vulnerability in New Orleans and the surrounding communities in Southeastern Louisiana is dissected to gain a better understanding of what caused individuals and communities in the region to experience repetitive losses over time.

The PAR model is employed here to establish the causal chain associated with vulnerability-generating processes that resulted in Hurricane Katrina. The PAR model provides a foundation for conceptualization and measurement of vulnerability. A multilevel modeling approach to social vulnerability measurement proposed by Ferreira (2013) supplements the measurement of social vulnerability in the Southeastern Louisiana Region.

The premise of the PAR model is that the source of disasters falls more in the social realm than in the purely natural realm. A comparative perspective offers the possibility to understand an event via a scale of causation (Anderskov, 2004; Wisner et al., 2004). The PAR model is grounded in the idea that a disaster can be described as the crossing of two forces (Anderskov, 2004). On one side is a process generating vulnerability and on the other side is exposure to hazards. A visual metaphor for this process is suggested by a "nutcracker", for the pressure placed on people intensifies from either side over time—both from exposure to vulnerability and from the impact of the hazard. In the case of induced pressures on either side of the PAR model, the risk of a possible disaster increases.

The PAR model is divided into the three interrelated and causal phases of disasters, known as the progression of vulnerability. These three phases/links are (Wisner et al., 2004) (1) root causes, (2) dynamic pressures, and (3) unsafe conditions.

The hazard side of the PAR model consists of various hazard types. A complete outline of the factors addressed in the model is to be found in Fig. 8.1 (Wisner et al., 2004, p. 50).

Accordingly, the progression of vulnerability, with Hurricane Katrina *as a case study,* will be examined by means of applying the PAR model to Orleans Parish (New Orleans). There are three main elements that are addressed in this discussion; root causes, dynamic pressures, and unsafe conditions.

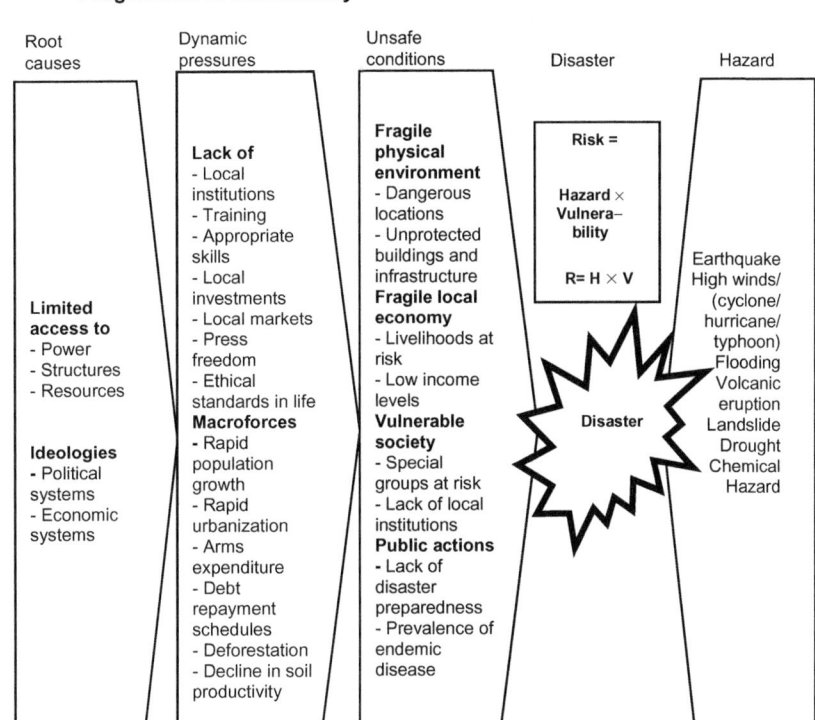

FIGURE 8.1

PAR Model: The Progression of Vulnerability (Wisner et al., 2004).

ROOT CAUSES

Root causes refer to a set of widespread and general, interrelated processes within a society. Root causes are regarded as distal in the context of one or more factors, arising from a distant center of economic or political power (Wisner et al., 2004). Examples of a spatial distal element of root causes playing a role in a communities' functioning include distant economic or political power structures that contribute to an increase in vulnerability. Another distal factor is temporal distance, that relates to events in history and refers to a decline in welfare that can be exacerbated by history. The final distal factor linking root causes is the "distance" between cultural assumptions, ideologies, beliefs, and social relations in the everyday existence of individuals (Wisner et al., 2004). The root causes of vulnerability in the communityinclude economic, demographic, and political processes.

These factors also affect the allocation and distribution of resources to different groups of people.

Factors relating to root causes resulting in high levels of vulnerability are complex. With regard to Orleans Parish and Southeastern Louisiana, the root causes can be attributed to colonization, slavery, racial segregation, and marginalization of certain minorities (Bates & Swan, 2007). Louisiana has had a fair share of political ideologies and an economic system affecting its residents negatively. Marginalization of certain racial groups started as early as 1751, when there was constant strife reported between the colonists and the natives (Fortier & McLoughlin, 1913). The Natchez and the Chickasaws were often in conflict with the colonists.

One of the first recorded disasters in New Orleans where the needs of locals were not met was with the yellow fever epidemic of 1853. Response from authorities was lackluster and caused the deaths of nearly ten thousand residents. Residents were left to fend for themselves during the epidemic. The poor and vulnerable were left behind, while the rich left New Orleans to escape the effects of the epidemic (Kelman, 2006; Lafayette Cemetery Research Project New Orleans, 2012).

Almost 150 years later, history was repeated by Hurricane Katrina. Portions of the community were able to leave the region in time, while others had no means of evacuation. New Orleans, at the time, had 127,000 residents without vehicles (Van Heerden & Bryan, 2007). Lack of governance and political will from authorities seem to be among the many causes of such high levels of vulnerability in the state of Louisiana (Van Heerden & Bryan, 2007). Issues of authority and management are extremely complex during times of disaster (Dynes & Rodriguez, 2007). This is evident from the yellow fever epidemic and Hurricane Katrina.

The high levels of vulnerability are also attributed to the marginalization of minorities in Louisiana (Bates & Swan, 2007). Root causes have led to an increase in social vulnerability and are partially due to the current political and economic system that has benefited a small group of individuals and marginalized several others in New Orleans and Southeastern Louisiana.

DYNAMIC PRESSURES

The second link in the PAR model is dynamic pressures. This link refers to the processes that are "translating" the effects of root causes, both in a temporal and spatial form, into unsafe conditions. Temporal and spatial features are more contemporary or immediate than conjunctional manifestations of general underlying political, social, and economic patterns (Wisner et al., 2004). Dynamic pressures within the context of Southeastern Louisiana channel the root causes into particular forms of unsafe conditions (Wisner et al., 2004).

Dynamic pressures relating to an increased sense of social vulnerability in Louisiana are a result of the lack of appropriate skills and the lack of local investments in marginalized areas of Orleans Parish. In many instances, there has been a lack of appropriate skills, as well as distrust in the community when it comes to evacuating for disasters in Louisiana (Elder, et al., 2007; Van Heerden & Bryan,

2007). These issues were evidenced with Hurricanes Ivan, Katrina, and Ike. Some citizens have not been able to conceptualize the extent of damage an event such as a hurricane might do to the environment. Apart from the lack of appropriate evacuation skills among citizens, there is a serious lack of disaster preparedness among local officials and residents (Van Heerden & Bryan, 2007). Another problem associated with the lack of appropriate skills in the region has been the focus on terrorism prevention instead of disaster preparedness by state and federal agencies in light of the events of 9/11 (Tierney & Bevc, 2007).

The lack of investment in local infrastructure in Louisiana is highlighted by the failure of the Levee system during Hurricane Katrina (Kelman, 2006). The 2012 Report Card for Louisiana's Infrastructure supports this (Movassaghi, 2012). Louisiana's entire current infrastructure received the highest grade of a B − for dams. Levees and bridges received grades of C − and D +, respectively. The grade points can be described as representing marginally crumbling infrastructure (Movassaghi, 2012). The crumbling infrastructure, especially the levees, was exposed during Hurricane Katrina, and ultimately can be regarded as one of the main reasons the city flooded.

A macroforce within the dynamic pressure link increasing vulnerability in the region is the constant threat to the wetlands in the region (Burby, 2006; Gordon, Buchanan, Singerman, Madrid, & Busch, 2011). Wetlands are an essential part of the ecosystem, serving not only as a buffer from storm surges, but also as a purification system for water in Louisiana (for a full discussion see Chapter 6: Setting the stage for the catastrophe of Katrina: Environmental degradation, engineering miscalculation, ignoring science, and human mismanagement (this volume)). The biodiversity of the wetlands in Southeastern Louisiana is under constant threat from residential development, chemical spills, and the mining of cypress trees (Gordon et al., 2011; Van Heerden & Bryan, 2007).

UNSAFE CONDITIONS

The third link of the PAR model is unsafe conditions. The vulnerability of a population being expressed in time and space in conjunction with hazards translates into unsafe conditions. Living within hazardous locations, the inability to afford safe housing and shelter, lack of government protection, and having to engage in dangerous practices to sustain livelihoods are results from root causes and dynamic pressures (Wisner et al., 2004). These factors are all present within New Orleans and Southeastern Louisiana. Unsafe conditions are dependent upon the initial level of well-being of the people; and the interaction of the level of well-being between regions, microregions, households, and individuals varies. When referring to unsafe conditions, no single element, especially technical and apolitical determinants of people's vulnerability, should be regarded separately from the entire range of factors and processes that tend to create a vulnerable state (Wisner et al., 2004).

New Orleans has a history of being faced with adversity. The region has been placed at constant risk with a fragile local economy and physical environment. The fragile local economy had been struggling before the impact of Hurricane

Katrina, coupled with an ailing US economy. The aforementioned unsafe conditions placed lower income level residents in the region at a higher level of social vulnerability, thereby creating a more vulnerable society.

With the intersection of the progression of vulnerability and the high prevalence of hazards, New Orleans and Southeastern Louisiana have the misfortune of being one of the worst disaster-affected areas in the United States. The main causes of the high prevalence of disasters in New Orleans are due to areas being inhabited that are not suitable for humans, dependence on assistance, marginalization of certain groups, and a lack of political will among politicians and administrators to bring about change. The combination of these factors on the one hand, and the high prevalence of hazards in Southeastern Louisiana on the other, has created one of the most vulnerable regions in the United States.

The progression of vulnerability of New Orleans and Southeastern Louisiana is depicted in Fig. 8.2.

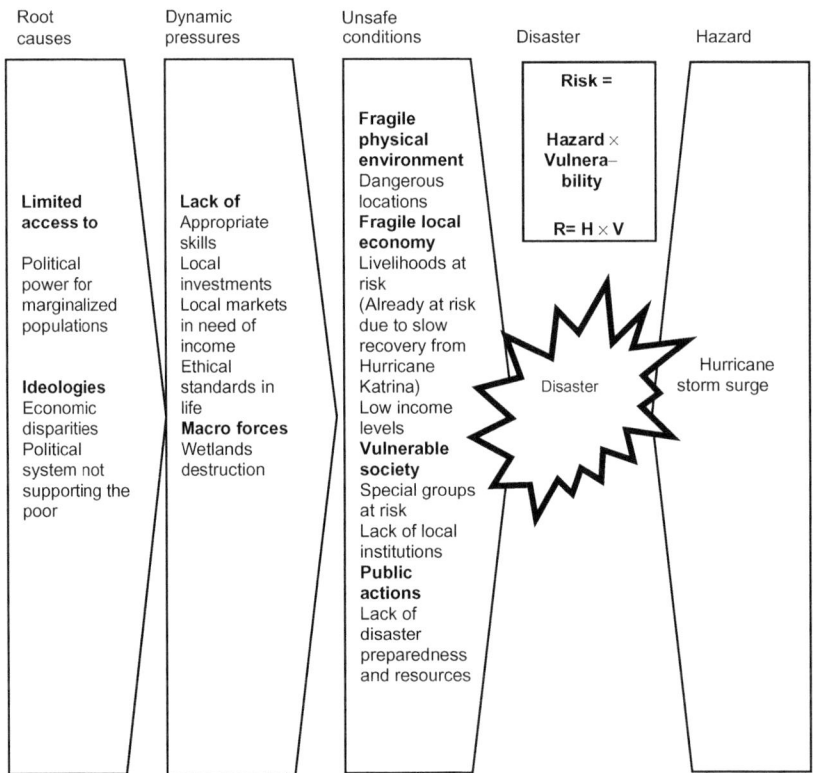

FIGURE 8.2

PAR Model for New Orleans and Southeastern Louisiana.

The PAR model for New Orleans and Southeastern Louisiana presented in Fig. 8.2 highlights several vulnerabilities pre-Hurricane Katrina. The progression of vulnerability associated with the PAR model for New Orleans and Southeastern Louisiana clearly illustrates that the root causes resulted in the impact of Hurricane Katrina being as devastating as it was to the region. The socioeconomic disparities in the region linked with several dynamic pressures (e.g., corruption, lack of political will and appropriate skills) and the fact that the region is unsafe (e.g., physical environment, lack of disaster preparedness) created one of the largest and, in many regards, most unmanageable of disasters.

PREDICTORS OF SOCIAL VULNERABILITY IN LOUISIANA: A MULTILEVEL ANALYSIS

A study was undertaken by Ferreira (2013), to develop an understanding of the predictors of social vulnerability in individuals nested within Louisiana communities. The Behavioral Risk Factor Surveillance System and 14 other community-level data sources were used. The model investigated the influence of parish disaster history, operational resilience, and socioeconomic resilience on individual social vulnerability. Individual social vulnerability was categorized according to *demographics* (eg., belonging to a minority groups, older than 65 years, female, presence of children in household, high-school diploma), *livelihood* (eg., household income less than $50,000), *social support* (eg., living alone), *societal protection* (eg., health care access), and *well-being* (eg., self-rated health and mental health status).

The research design of the study was a multilevel repeated cross-sectional design with a three-level nested structure. Using a representative sample of 34,685 individuals from 2004 to 2010, nested in 56 Louisiana parishes, the trend study allowed for an understanding of the factors that predict individual social vulnerability in Louisiana.

Overall, the results indicate that there were differences between parishes and their levels of individual social vulnerability; individual social vulnerability decreased from 2004 to 2010 and several statistically significant predictors of social vulnerability were identified. Statistically significant community-level predictors of individual social vulnerability were lack of educational attainment ($p < 0.001$), communities with less access to a household phone ($p < 0.001$), community poverty ($p < 0.05$), and community unemployment ($p < 0.05$). Statistically significant two-way interactions included number of disasters ($p < 0.05$) and total population per square mile ($p < 0.05$), and number of disasters and number of physicians per 100,000 population ($p < 0.05$). The predictors of social vulnerability from this study correspond with social vulnerability indicators as identified by Wisner (2001) and Cutter et al. (2003).

ADDRESSING VULNERABILITY IN NEW ORLEANS AND SOUTHEASTERN LOUISIANA

The impact of future disasters on individuals, families, and communities in New Orleans and Southeastern Louisiana is inevitable. The vulnerability process associated with Hurricane Katrina, both predisaster and postdisaster, provides opportunities to learn, improve, and lessen the impact of future disasters. The impact can be lessened through fostering the progression to safety and resilience for individuals, families, and communities by addressing the root causes that create social vulnerability.

This section provides important initiatives for community practitioners and policymakers to work proactively in strengthening individuals, families, and communities pre- and postdisaster. Guided by the Progression of Safety model (the reverse of the PAR model) (Wisner et al., 2004) and past empirical findings (Ferreira, 2013), recommendations are aimed at decreasing vulnerability in Orleans Parish and the surrounding parishes.

The processes of reducing the level of disaster vulnerability should be guided by addressing root causes and reducing pressures in order to achieve safe conditions. The main factors that contribute to vulnerability—namely social, economic, and political mechanisms (dynamic pressures) that translate root causes into unsafe conditions—can sometimes be blocked, changed, or even reversed. This section aims to highlight possible recommendations, that can ensure a safer and risk-free environment.

Advocating for the most socially vulnerable (e.g., the elderly and marginalized minorities) and providing opportunities to reduce the pressures associated with vulnerability can result in safer conditions. Individuals in poverty tend to be more vulnerable, in part due to unemployment and a possible lack of education. It is accepted that poverty alleviation will not happen immediately, but better educational opportunities can lead to increased employment opportunities. And with better employment, poverty can be reduced. It is therefore important that communities work toward creating sustainable employment opportunities for individuals. A starting point for creating access to education would be to create work-study programs in which individuals could receive a subsidized education stipend, associated with possible job placement. This will ensure that individuals also receive practical experience, and will be ready for the job market upon completion of their studies. Access to basic health services is imperative. Community decision-makers should ensure that all individuals have access to basic health services. Health services need to be made available to all in need, irrespective of background. Healthy citizens increase community well-being and lessen the financial burden on other resources within the community.

Creating holistic support systems for individuals that are living alone is important. Support systems (e.g., community groups or neighbors) are essential during times of disaster. Community practitioners can provide the guidance and tools for individuals to connect with faith-based, social, and community groups. Receiving emotional and physical support during a time of disaster is important for all, but especially for individuals who are social vulnerable. Older adults tend to be more isolated during disasters. By implementing outreach programs, the needs of isolated groups, such as the older population, are attended to postdisaster. In practice, implications on a macrolevel are just as important as on a microlevel. Macropractice implications are intended to increase community competence, communication, social capital, and economic growth. If practitioners build on these suggestions, vulnerability can be decreased in communities. With a decrease in vulnerability, communities will be able to reorganize and recover from the impact of disasters in a shorter time.

It is not always feasible to improve education from the grass roots level in a community. Instead, communities should make it more rewarding for highly educated individuals to move to, and live in a particular community with less educated individuals. Providing housing incentives and tax relief are possible solutions for attracting such individuals. By "recruiting" highly educated residents in areas with lower levels of education, pockets of low-income areas can be prevented. The concept of promoting mixed-income areas should be fostered at a macro level. Ensuring that communities focus on diversified employment opportunities is also very important. When disaster strikes it tends to have a cyclical effect on employment. Communities that rely on one source of employment are more susceptible to the impact of a disaster. This vulnerability can have a ripple effect on the community.

Promoting intergenerational living within communities will decrease social vulnerability on a micro- and macro-level. This would be especially beneficial in areas where there are limited resources available. Intergenerational living provides support networks for people who live alone, improving access to resources for individuals and families.

Access to communication resources is essential during times of disaster. Having access to communication resources improves disaster awareness and preparedness. Tax incentives and breaks should be provided to communications companies, to ensure that everybody has access to either a landline or to mobile communication.

CONCLUSION

This chapter aimed to examine the progression of vulnerability in southeastern Louisiana that led to the psychosocial consequences of Hurricane Katrina, including some of the recovery post-Katrina. As illustrated in this chapter, addressing the root causes of vulnerability and, in particular, social vulnerability, provides a roadmap for both minimizing the psychosocial costs by directing postdisaster assistance quickly to the most socially vulnerable.

The PAR model allows for the assessment of the causal processes that resulted from Hurricane Katrina. With the significant increase in disasters worldwide, there has been a greater sense of awareness among the public and private sectors to be better prepared for disasters.

In quantifying disaster-related themes with a vulnerability prediction model, we are able to understand who the most vulnerable are, how to better prepare communities for disasters, and what makes communities disaster-resilient. The PAR model linked with Ferreira's (2013) findings of social vulnerability predictors in Louisiana, help with the conceptualization of vulnerability generating processes in any given disaster context. The model also provides the opportunity to identify the socially vulnerable populations and sectors that are susceptible to disasters in any given region.

REFERENCES

Adger, W. N., Hughes, T. P., Folke, C., Carpenter, S. R., & Rockström. (2005). Social-ecological resilience to coastal disasters. *Science, 309,* 1036–1039.

Anderskov, C. (2004). Anthropology and disaster—An analysis of current trends within anthropological disaster research, and an attempt to construct an approach that facilitates theory building and applied practices—analyzed with vantage point in a case-study from the flood-prone Mudardara District in Mozambique. Retrieved from <http://www.anthrobase.com/Txt/A/Anderskov_C_03.htm#21_Research%20Framework> Accessed May 15, 2008.

Baker, L.F. (2011). Louisiana purchase. In *The encyclopedia of Arkansas and culture, The Central Arkansas Library System*. Retrieved from <http://www.encyclopediaofarkansas.net/encyclopedia/entry-detail.aspx?entryID = 2383>.

Bates, K. A., & Swan, R. S. (2007). *Through the eye of Katrina: Social justice in the United States*. Durham, NC: Carolina Academic Press.

Beitler, S. (Producer). (2007). *The ninth Mississippi River disaster in three weeks*. Available from http://www3.gendisasters.com/louisiana/52/hermitage%2C-la-steamer-paris-c.-brown-wreck%2C-jan-1889?page = 0%2C0.

Bernard, M. (2004). *Resiliency: What we have learned*. San Francisco, CA: WestEd.

Brooks, R., & Goldstein, S. (2002). *Nurturing resilience in our children: Answers to the most common parenting questions*. New York, NY: McGraw-Hill.

Brunsma, D. L., Overfelt, D., & Picou, J. S. (Eds.), (2007). *The sociology of Katrina: Perspectives on a modern catastrophe*. Lanham, MD: Rowman & Littlefield Publishers.

Burby, R. J. (2006). Hurricane Katrina and the paradoxes of government disaster policy: Bringing about wise governmental decisions for hazardous areas. *The Annals of the American Academy of Political and Social Science, 604*(1), 171–191.

Cutter, S. (2006). *The geography of social vulnerability: Race, class and catastrophe*. Retrieved from <http://www.understandingkatrina.ssrc.org/cutter/> Accessed April 13, 2008.

Cutter, S., Borhuff, B., & Shirley, W. (2003). Social vulnerability to hazards. *Social Science Quarterly, 2*(84), 242–243.

Dynes, R. R., & Rodriguez, H. (2007). Finding and framing Katrina: The social construction of disaster. In D. L. Brunsma, D. Overfelt, & J. S. Picou (Eds.), *The sociology of Katrina: Perspectives on a modern catastrophe*. New York: Rowman & Littlefield Publishers, Inc.

Eisenman, D. P., Cordasco, K. M., Asch, S., Golden, J. F., & Glik, D. (2007). Disaster planning and risk communication with vulnerable communities: Lessons from Hurricane Katrina. *American Journal of Public Health, 97*, 109–115.

Elder, K., Xirasagar, S., Miller, N., Bowen, S. A., Glover, S., & Piper, C. (2007). African Americans' decisions not to evacuate New Orleans before Hurricane Katrina: A qualitative study. *American Journal of Public Health, 97*, 124–129.

Federal Emergency Management Agency. (2015). *Disaster declarations by year*. Retrieved from <https://www.fema.gov/disasters/grid/year> Accessed August 17, 2015.

Ferreira, R.J. (2013). *Predictors of social vulnerability: A multilevel analysis*. Retrieved from <http://ir.library.louisville.edu/cgi/viewcontent.cgi?article = 1434&context = etd> Accessed August 15, 2015.

Fortier, A., & McLoughlin, J. J. (1913). *Louisiana*. Catholic Encyclopedia. New York, NY: Robert Appleton Company.

Freudenburg, W. R., Gramling, R., Laska, S., & Erikson, K. T. (2009). Disproportionality and disaster: Hurricane Katrina and the Mississippi River-Gulf Outlet. *Social Science Quarterly, 90*(3), 497–515. <http://onlinelibrary.wiley.com/doi/10.1111/j.1540-6237.2009.00628.x/abstract> Accessed August 25, 2015.

Gillespie, D. F. (2010). Vulnerability: The central concept of disaster curriculum. In D. F. Gillespie, & K. Danso (Eds.), *Disaster concepts and issues* (pp. 3–14). Alexandria, VA: CSWE Press.

Gillespie, D. F., & Danso, K. (2010). *Disaster concepts and issues: A guide for social work education and practice*. Alexandria, VA: Council on Social Work Education Press.

Goldstein, S., & Brooks, R. B. (2006). *Handbook of resilience in childhood*. New York: Springer Science and Business Media, Inc.

Goodman, R., Speers, M., McLeroy, K., Fawcett, S., Kegler, M., Parker, E., et al. (1998). Identifying and defining the dimensions of community capacity to provide a basis for measurement. *Health Education & Behavior, 25*, 258–278.

Gordon, K., Buchanan, J., Singerman, P., Madrid, J., & Busch, S. (2011). *Beyond recovery: Moving the Gulf Coast toward a sustainable future* (pp. 1–74). Washington, DC, Center for American Progress & Oxfam America. <http://www.oxfamamerica.org/explore/researchpublications/beyond-recovery-moving-the-gulf-coast-toward-a-sustainable-future/>.

Hobfoll, S. E., Stevens, N. R., & Zalta, A. K. (2015). Expanding the science of resilience: Conserving resources in the aid of adaptation. *Psychological Inquiry, 26*(2), 174–180. Available from http://dx.doi.org/10.1080/1047840X.2015.1002377.

Independent Levee Investigation Team. (2006). *Investigation of the performance of the New Orleans flood protection systems in Hurricane Katrina on August 29, 2005*.

Kammerer, J.C. (1990). *Water Fact Sheet: U.S. Geological Survey, Department of the Interior—Largest Rivers in the United States* (pp. 1–2). U.S. Geological Survey, Department of the Interior, Reston, VA.

Kelman, A. (2006). *A river and its city: The nature of landscape in New Orleans*. Berkley, CA: University of California Press.

Lafayette Cemetery Research Project New Orleans (Producer). (2012). *The yellow fever epidemic in New Orleans-1853*. Available from http://www.lafayettecemetery.org/yellowfever1853_page1/the-yellow-fever-epidemic-in-new-orleans---1853-page-2.

Laska, S. (2012). Dimensions of resiliency: Essential resiliency, exceptional recovery and scale. *International Journal of Critical Infrastructures*, *8*(1), 47−62. <http://www.inderscienceonline.com/doi/abs/10.1504/IJCIS.2012.046552> Accessed September 24, 2015.

Litman, T. (2006). Lessons from Katrina and Rita: What major disasters can teach transportation planners. *Journal of Transportation Engineering*, *132*(1), 11−18.

Louisiana. (2012). *About Louisiana*. Baton Rouge. Retrieved from http://louisiana.gov/Explore/About_Louisiana/.

Miller, F. H., Osbahr, E., Boyd, F., Thomalla, S., Bharwani, G., Ziervogel, B., et al. (2010). Resilience and vulnerability: Complementary or conflicting concepts? *Ecology and Society*, *15*(3), 11. Retrieved from http://www.ecologyandsociety.org/vol15/iss3/art11/.

Movassaghi, K. (2012). *Report card for Louisiana Infrastructure, 2012* (pp. 1−58). Reston, VA. American Society of Civil Engineers.

Norris, F. H., Stevens, S. P., Pfefferbaum, B., Wyche, K. F., & Pfefferbaum, R. L. (2008). Community resilience as a metaphor, theory, set of capacities, and strategy for disaster readiness. *American Journal of Community Psychology*, *41*, 127−150.

Plyer, A., Shrinath, N., & Mack, V. (2015). *The New Orleans index at ten; measuring Greater New Orleans' progress towards prosperity*. Retrieved from <https://s3.amazonaws.com/gnocdc/reports/TheDataCenter_TheNewOrleansIndexatTen.pdf> Accessed January 15, 2016.

Rhodes, J., Chan, C., Paxson, C., Rouse, C. E., Waters, M., & Fussell, E. (2010). The impact of Hurricane Katrina on the mental and physical health of low-income parents in New Orleans. *American Journal of Orthopsychiatry.*, *80*, 237−247.

Ronan, K. R., & Johnston, D. M. (2005). *Promoting community resilience in disasters: The role for schools, youth and families*. New York, NY: Springer.

The Data Center. (2016). *Who lives in New Orleans and metro parishes now?* Retrieved from <http://www.datacenterresearch.org/data-resources/who-lives-in-new-orleans-now/> Accessed August 7, 2016.

The Times-Picayune (Producer). (2010). *Chemical spill in Norco closes schools, forces residents out*. Retrieved from <http://www.nola.com/politics/index.ssf/2010/04/st-charles.html> Accessed March 23, 2011.

Tierney, K. (2015). *Resilience and the neoliberal project discourses, critiques, practices—and Katrina*. American Behavioral Scientist 0002764215591187. Retrieved from <http://abs.sagepub.com/content/early/2015/06/17/0002764215591187> Accessed November 24, 2015.

Tierney, K., & Bevc, C. (2007). Disaster as war: Militarism and the social construction of disaster in New Orleans. In D. L. Brunsma, D. Overfelt, & J. S. Picou (Eds.), *The sociology of Katrina: Perspectives on a modern catastrophe* (pp. 33−50). New York: Rowman & Littlefield Publishers, Inc.

Ungar, M. (2008). Resilience across cultures. *British Journal of Social Work*, *38*(2), 218−235. Retrieved from http://dx.doi.org/10.1093/bjsw/bcl343.

U.S. Census Bureau. (2015). *State and county quickfacts*. Retrieved from <http://quickfacts.census.gov> Accessed October 30, 2015.

Van Heerden, I., & Bryan, M. (2007). *The storm: What went wrong and why during Hurricane Katrina—The inside story from one Louisiana scientist*. New York: Penguin Group (USA) Inc.

Wisner, B. (2001). *'Vulnerability' in disaster theory and practice: From soup to taxonomy, then to analysis and finally tool*. Paper presented at the International Work-Conference Disaster Studies of Wageningen University and Research Centre in June 2001. Wageningen: Wageningen University.

Wisner, B., Blaikie, P., Cannon, T., & Davis, I. (2004). *At risk, natural hazards, people's vulnerability and disasters* (2nd ed.). New York: Routledge. Available from http://www.preventionweb.net/files/670_72351.pdf.

Zakour, M. J. (2010). Vulnerability and risk assessment: Building community resilience. In D. F. Gillespie, & K. Danso (Eds.), *Disaster concepts and issues* (pp. 15–33). Alexandria, VA: CSWE Press.

FURTHER READINGS

Blaikie, P., Cannon, T., Davis, I., & Wisner, B. (1994). *At risk: Natural hazards. People's vulnerability, and disasters*. London: Routledge.

Picou, J. S., Marshall, B. K., & Gill, D. A. (2004). Disaster, litigation, and the corrosive community. *Social Forces, 82*, 1493–1522.

Problematizing vulnerability: Unpacking gender, intersectionality, and the normative disaster paradigm

Paul Kadetz[1] and Nancy B. Mock[2]

[1]*Drew University, Madison, NJ, United States* [2]*Tulane University, New Orleans, LA, United States*

CHAPTER OUTLINE

Introduction	215
Problematizing the Essentializing of Female Vulnerability in Disaster Research	218
Fitting One Size of Vulnerability to All	218
Vulnerabilities in the Context of New Orleans	220
Framework for Understanding Urban Vulnerability in New Orleans	220
The Intersectionality of Gendered Vulnerability in New Orleans	222
Financial Vulnerability	223
Housing Vulnerability	223
Health Care, Education, and Transportation	224
Political Economy and Neoliberal Vulnerabilities	225
Other Intersectionalities of Gendered Vulnerability in New Orleans	226
Conclusion	228
References	229
Further Reading	230

INTRODUCTION

Census data profile New Orleans as one of the nation's most impoverished cities in one of the nation's most impoverished states. "Nearly twice as many people in pre-Katrina New Orleans were below poverty level than nationally (24.5% vs. 13.3%). [And] in New Orleans, as in most of the South, race and gender often are interrelated with poverty" (NCCROW, 2008, p.12). Even before Hurricane Katrina, multidimensional and multigenerational poverty in New Orleans exhibited a substantial-gendered component. The inner city poverty trap is characterized by single parent households headed by women of color. Indeed, this cycle of intergenerational poverty, perpetuated in the inner city, was well-understood as a major source of ongoing social vulnerability in New Orleans. The levels of poverty in post-Katrina New Orleans have remained high and above the national average, revealing a recovery paradigm that has reconstructed poverty traps, as opposed to launching vulnerable populations on resilience trajectories. But, poverty alone cannot explain vulnerability. For, "differential vulnerability grows out of historically situated inequalities that limit access of some to secure housing, adequate incomes, food supplies and legal rights" (Bolin, Jackson, & Crist, 1998, p. 37). In crises, social exclusions thwart access to resources and options out of vulnerability. Thus, sociocultural systems that privilege certain groups over others ultimately produce the disaster vulnerability of these excluded groups. We believe that the story of Hurricane Katrina exposes the complex interactions between gender, race, class, age, and vulnerability.

Although the complexity of the contextual factors relating gender to vulnerability in disaster research is identified in research concerning low-income countries, this paradigm has been sorely lacking in disaster research in the United States. The UN Human Rights Committee—a body of independent experts entrusted with the task of monitoring the implementation of the International Covenant on Civil and Political Rights—when examining the report submitted to it by the United States, expressed its concerns "about information that the poor, and in particular African-Americans, were disadvantaged by the rescue and evacuation plans implemented when Hurricane Katrina hit the United States, and continue to be disadvantaged under the reconstruction plans" (UN, 2006, p. 63). The committee recommended the United States "review its practices and policies to ensure the full implementation of its obligation to protect life and of the prohibition of discrimination, whether direct or indirect, as well as of the United Nations Guiding Principles on Internal Displacement, in matters related to disaster prevention and preparedness, emergency assistance and relief measures. In the aftermath of Hurricane Katrina, the State party should increase its efforts to ensure that the rights of the poor, and in particular African-Americans, are fully taken into consideration in the reconstruction plans with regard to access to housing, education and healthcare" (UN, 2006, pp. 63–64). What can explain this egregious equity gap in US disaster policies and interventions?

Bolin et al. (1998) argue that the normative "applied and managerialist" paradigm employed in the United States ignores the role of social inequality and, thereby, affords no room for an accurate assessment of the complexity of vulnerability, particularly with regards to gender. The disaster management discourse in the United States has been overly rationalistic and technocratic; embracing a narrative of a mythical "level playing field," rather than targeting important social differences and developing equitable social infrastructure. US disaster research tends to portray disasters as predictable and orderly and thereby prevention and recovery are depicted in the same formulaic and reductive manner. Chapter 4, A systems approach tovulnerability and resiliencein post-Katrina New Orleans (this volume) concerning complexity illustrates how complex social events, such as disasters and their aftermath, can never be accurately portrayed in any manner other than via a complex systems approach. Furthermore, the US approach to disasters favors technological solutions, while ignoring the importance of the given social world and the "effects of historically embedded human practices that are implicated in creating the 'natural disaster' in a complex array of social activities situated in time and place" (Bolin et al., 1998, p. 27). And yet, in the context of the United States, "Women are far more likely than men to be household heads who rely on their own low incomes or modest government assistance, so the national *shortage of affordable housing* can become a personal crisis" (Enarson, 2010, p. 135).

Victims of disasters are commonly portrayed as a hegemonic group in which predisaster social vulnerability is irrelevant. "The functionalist and positivist theoretical orthodoxy that has prevailed in U.S. disaster research is, in large part, driven and supported by the public policy-emergency management orientation of federal funding agencies. The concern with making disaster research an objective and quantitative 'science' has resulted in a neglect of the experiences of those whose lives and livelihoods are caught up in the disaster, and this neglect is most pronounced in the case of women" (Bolin et al., 1998, pp. 28–29). Much is lost in this approach. The risk and vulnerabilities of disasters are "distributed in ways that reflect the social divisions that already exist in society" (NCCROW, 2008, p. 11). In other words, disasters reflect socially constructed vulnerabilities.

Addressing the complexity of gendered vulnerability is imperative in disaster research, both in terms of prevention and intervention, as well as for building pathways to resilience in recovery. An analysis of gendered vulnerability in Hurricane Katrina and the absence of a strategy to build resilience to reduce gender inequality during disaster recovery provides an illustrative case example in this chapter. But, we can problematize this ordering of vulnerability within discrete social categories.

Many of the challenges in vulnerability and disaster literature can be broken down to the appropriateness of the level of analysis and the disaggregation of dependence on multiple and universalized social determinants of vulnerability. Vulnerability cannot be fully understood from highly aggregated assessments in which, for example, gender and linkages to other intersectionalities of social inequality are ignored. Nor, can gender be employed as a monolithic category. The concept of essentialism posits that every entity is deterministically imbued with an essence that is particular to that thing or being and necessary for its function and identity. Thereby, essentialism obliterates the impact of the social and cultural over what might be considered the irrefutably biological; or the sole impact of nature to the detriment of nurture. Essentialism in disaster research and intervention "that simply compare disaster responses between men and women routines [sic] avoid any indepth analysis of the social-structural inequalities (economic, political, legal occupational, familial, ideological, cultural) underlying and variously producing observed gender differences. As a consequence, important questions about the social dynamics of gendered experiences in disasters are left unmasked" (Bolin et al., 1998, p. 29). Typically there is "a narrow range of women's characteristics deemed relevant for social analysis [...] women tend to be treated as universal constructs undifferentiated by class, ethnicity, culture, age, race and national origin. Social heterogeneity and inequality are frequently obscured or oversimplified" (Bolin et al., 1998, p. 34). Hence, essentialized dichotomous categories such as male/female assume, for example, that all females should be considered to have the same vulnerability to disasters regardless, of specific contextual cofactors including: income, education, race/ethnicity, age, disability, sexual orientation, and/or if the woman is a single parent. Highly aggregated data from essentialized social categories do not provide the necessary information needed for targeting disaster vulnerability and providing interventions for those most in need in any given population.

Furthermore, vulnerability is ultimately determined by the given social context. Both gender and race are social constructions that are frequently misrepresented as biologically predetermined. As social constructions, gender and race are completely dependent on context. Gender and race then may more accurately be understood as reflections of a given society's values, beliefs, justifications, and behaviors that create and perpetuate hierarchy, distinction, exclusion, and ultimately inequality. In other words, gender and race are the reified boundaries of Othering [i.e.,"The perception or representation of a person or group of people as fundamentally alien from another, frequently more powerful, group" (OED, 2017).], which serve to protect the dominant social elite and their status from all Others.

This chapter examines the intersections of gender, race, and class in the creation of postdisaster vulnerability in New Orleans. While pre-Katrina-gendered vulnerability was stark, the lack of focus on the reduction of disparities—a great opportunity for recovery efforts—provides important lessons for other contexts. In this chapter, we integrate the literature and secondary data analysis, most notably the Katrina Index and the American Community Survey, to examine the nature of urban inner city poverty and the lack of deliberate rights-based recovery efforts aimed to reduce it.

PROBLEMATIZING THE ESSENTIALIZING OF FEMALE VULNERABILITY IN DISASTER RESEARCH
FITTING ONE SIZE OF VULNERABILITY TO ALL

Are all women to be considered more vulnerable than men in disaster contexts? Can disasters not endanger the well-being of boys and men? In this chapter the term *gender* points to "the range of 'socially constructed' roles, behaviours, attributes, aptitudes and relative power associated with being female or male in a given society at a particular point in time" (Esplen, 2009, p. 2). Hence, gendered vulnerability is not exclusive to females. Depending on the social context, certain groups of men may be considered more vulnerable than certain groups of women. For example, Enarson (2010, p. 129) offers: "women have specific needs in late pregnancy, their domestic work is generally discounted, and they are at increased risk of abuse post-disaster; [however] men are at risk in male dominated relief occupations, may be feel it 'unmanly' to ask for counseling, and suffer the effects of unmanageable stress or substance abuse." Thus, it is not accurate to assume that all women are, by default, more vulnerable than men in disasters. Vulnerability can be better understood as an outcome of multiple layers of alterity (or othering), inequality, and social exclusion. Researchers have rarely examined "women of different cultures, classes and ethnicities and the heterogeneity of their disaster experiences" (Bolin et al., 1998, p. 32).

Essentializing female vulnerability is a form of objectification in which women are perceived as one massive Othered group, rather than as individual subjects. Men and women really belong to multiple groups from which they form identities. Baumann notes: "Attributions of culture and community can clearly not be reduced to one factor alone" (1996, p. 5). She offers an illustrative case example of the multidimensionality of identity in a predominantly South Asian London neighborhood: "The vast majority of all adult Southallians saw themselves as members of several communities, each with its own culture. The same person could speak and act as a member of the Muslim community in one context, in another take sides against the Muslims as a member of the Pakistani community, and in a third count himself part of the Punjabi community that excluded other Muslims, but included Hindus, Sikhs, and even Christians" (Baumann, 1996). Hence, a static representation of the Self or Other cannot account for the changes one must adopt in order to survive in any social environment. Nor can a rigid approach account for the fact that people adopt several group identities between which they freely move. Identity changes as needs and contexts change. This fluidity of movement between multiple social identities can be understood as a mechanism of survival in a social landscape that requires fluid navigation. Vigh (2008) describes this as a capacity to "navigate" the myriad needs, necessities, and obstacles of social worlds. Identity is a plastic and dynamic (political−economic) strategy (Cohen, 2004) "formed by internal organization and stimulated by external pressures to defend" sociocultural, as well as "economic and political interests" (Banks, 1996, p. 35). Thereby, it would be erroneous to try to analyze or represent this fluidity within the rigid boundaries of a closed, linear, simple systems approach (see Chapter 4: A systems approach tovulnerability and resiliencein post-Katrina New Orleans (this volume) for a full discussion of the need for complex systems thinking for disaster research and interventions).

Yet the complexity of subjectivities is often disregarded in disaster research with a shorthand of gender stereotyping in which women are reduced to nurturers and being female equals vulnerability. Butler concisely illustrates the issue of representing the complexity of the subject for which "elaborate predicates of color, sexuality, ethnicity, class, and able-bodiedness invariably close with an embarrassed 'etc.' at the end of the list [...] these positions strive to encompass a situated subject but invariably fail to be complete" (1990, p. 143). Women (and men) differ by "age, ability, social class, ethnicity, race, education, occupation, marital status, sexual orientation and identity. Neglect of the specific needs of those whose lives do not fit the heteronormative ideal, single mothers as well as lesbian, bisexual, and transgender women, results in unique hardships for these groups and points to the need to disaggregate the at-risk population to include, not only 'women', but the diversity of women" (NCCROW, 2008, p. 6).

VULNERABILITIES IN THE CONTEXT OF NEW ORLEANS

The intersectionality of vulnerabilities in New Orleans is specific to the context of New Orleans. In the specific context of pre- and post-Katrina New Orleans, women, as a group, *do* appear to exhibit greater vulnerabilities than men in several sectors. When Hurricane Katrina struck the Gulf Coast "one in four women residing in the City of New Orleans lived below the poverty line; more than half (56%) of families with children were headed by women and two-fifths of these lived in poverty; over a third (35%) of African American women in Louisiana were officially poor, the worst record in the region and nation; and over half (61%) of the poor people over [age] 65 in the City of New Orleans were women" (Enarson, 2010, p. 125).

New Orleans makes an interesting case study in part, because nearly half of New Orleans households (41%) were headed by women pre-Katrina compared to the national average of 18.8%. Post-Katrina this number dropped by 10% (NCCROW, 2008). The burden of maintaining some semblance of "normal" day-to-day family life fell largely on the shoulders of women. Women also have emerged as civic leaders to organize and lead collective actions for local, state, and national level renewal and reform in response to the structural crisis and political vacuum left in the wake of the storm. Thus, it is not surprising that a post-Katrina study of 1043 adults found women to be "2.7 times more likely than men to have Post-traumatic Stress Disorder and 1.3 to 2 times more likely than men to have an anxiety or mood disorder other than PTSD" (NCCROW, 2008, p. 9).

"Access to information, resources, legal rights, and entitlements" can determine who will be able to avoid or cope with a disaster more successfully (Bolin et al., 1998, p. 37). In general, New Orleans women were found to typically have less access to resources, such as transportation, immediate cash or savings, secure housing and employment, and less control over decision-making and economic resources.

FRAMEWORK FOR UNDERSTANDING URBAN VULNERABILITY IN NEW ORLEANS

Fig. 9.1 depicts a framework for understanding the link between urban infrastructure, disaster policies, and vulnerability to shocks. This framework, built from the sustainable livelihoods framework of the Department for International

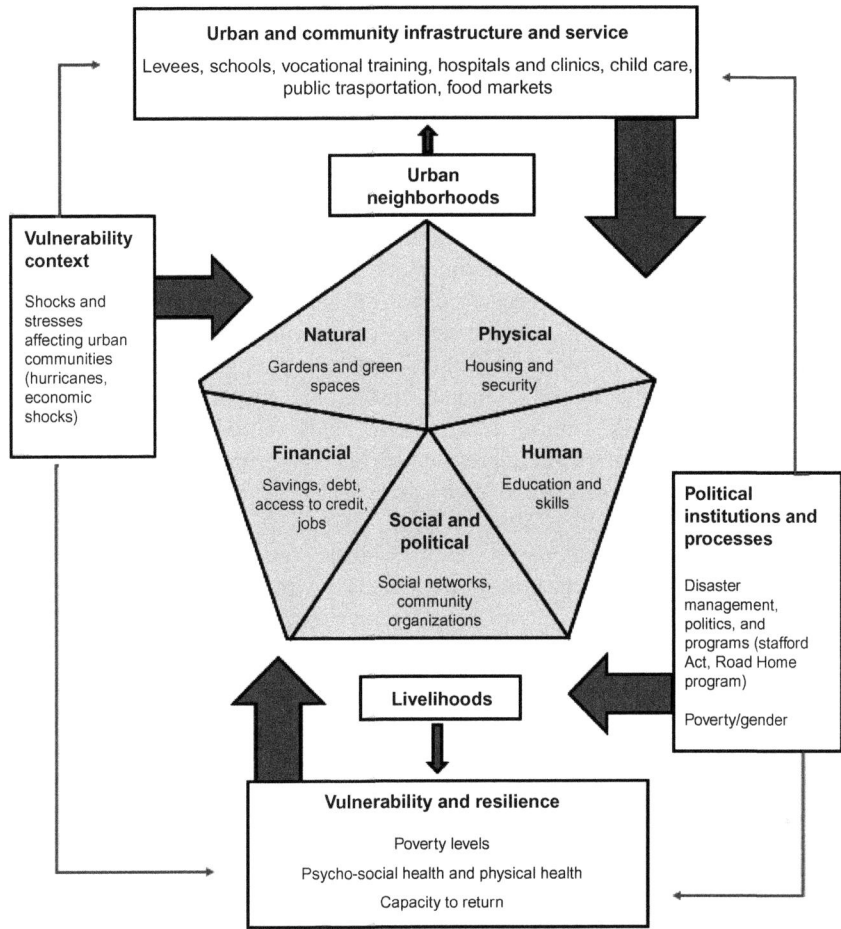

FIGURE 9.1

Sustainable Livelihoods Framework for Understanding Urban Vulnerability in New Orleans

Source: Adapted from DFID's Sustainable Livelihood Guidance Sheets; Rakodi, C., & Lloyd-Jones, T. Livelihoods, a people-centered approach: Earthscan, p. 9.

Development of the UK (DfID), established the centrality of livelihood, and the ability to provide the inputs necessary for households to survive and thrive. We have adapted this model for an urban context. The adapted model recognizes the centrality of infrastructure and services for the creation of resilience pathways in urban settings. It establishes the critical linkages between services available in neighborhoods, livelihood, and well-being outcomes. Urban environments present unique challenges to resilience-building efforts. While natural, physical, social, human, and financial assets are key vulnerability and resilience determinants everywhere and at every level of scale for social ecological systems, a key element for understanding vulnerability and resilience in urban settings is the degree to which urban systems can provide the necessary sources of production of services: schools, hospitals, public transportation, and childcare; all of which can ensure livelihood success of its residents.

Fig. 9.1 provides a framework for understanding how the Katrina recovery process did not sufficiently incorporate the intersectionality of gender, class, and race into a poverty eradication strategy. Intersectionality promotes an understanding that human beings are shaped by the interaction of different social locations (e.g., "race"/ethnicity, indigeneity, gender, class, sexuality, geography, age, ability, migration status, religion). These interactions occur within a context of connected systems and structures of power (e.g., laws, policies, state governments and other political and economic unions, religious institutions, media). Thus, individuals have multi-dimensional identities that intersect with multi-dimensional forms of othering and power relations in aggregated ways. "Power operates by disciplining people in ways that put people's lives on paths that make some options seem viable and others out of reach" (Collins & Bilge, 2016, p. 9). Hence, identity categories both adopted by and placed on individuals "gain meaning from power relations of racism, sexism, heterosexism, and class exploitation" (Collins & Bilge, 2016, p. 7). Through such processes, interdependent forms of privilege and oppression shaped by colonialism, imperialism, racism, homophobia, ableism, and patriarchy are created. Thus, intersectionality provides a three-dimensional understanding of othering and can be conceptualized as the intersection of multiple -isms and -phobias and other forms of social exclusion, usually of a dominant group in the dominant society; although this can change in different social groups and subgroups. As a lens for disaster interventions, intersectionality problematizes simple, essentialized groupings of race, class, or gender, in relation to their need/assets and vulnerability/resilience, for more accurate complex understandings.

THE INTERSECTIONALITY OF GENDERED VULNERABILITY IN NEW ORLEANS

In comparing pre-Katrina (2000) and post-Katrina (2008) populations there is a 41% reduction (from 45,183 to 26,819) in the number of single mothers in New Orleans. A 57% reduction from 23,131 to 9883 in the number of single mothers who are living in poverty. And a 56% reduction from 33,675 to 15,118 in the number of single African-American mothers. The reduction of poor African-American women from the population could be reflected in the 37% reduction in black female poverty in New Orleans from 2000 to 2008. Within this same time frame there was a 0.001% and 9% reduction in the number of poor white and Hispanic women, respectively. It is unclear if this reduction is an outcome of people deciding to leave New Orleans or being internally displaced and unable to return due to compromised social welfare across sectors post-Katrina.

FIGURE 9.2

Race, Gender, and Wages from 2005 to 2007

Source: *From American Community Survey 2005, 2006, 2007, United States Census Bureau.*

FINANCIAL VULNERABILITY

The heterogeneous population of the City of New Orleans provides a pertinent example of the need for a more complex analytical framework, such as, for example, multilevel analysis, in order to more accurately understand vulnerability in complex emergency and humanitarian research/interventions. At the time of Hurricane Katrina, 25.9% of women in New Orleans were living below the poverty line, compared to 20% of men (Institute for Women's Policy Research (IWPR), 2010). According to our research, prior to Hurricane Katrina, 65% of all single female households with children under 18 years of age were classified as poor, comprising more than 80% of public housing tenants. And female householders represent 20% or more of many low-income neighborhood families. If we assess the aggregate average annual income for males and females in New Orleans, we find that the average income for males is consistently higher than the average for females from 2005 to 2008. In general, males earned 18% more than females in 2005; 38% more in 2006; 29% more in 2007; and 27% more in 2008 (Fig. 9.2).

HOUSING VULNERABILITY

As discussed throughout this book, vulnerability is significantly affected by access to the resources of social welfare and social networks. Although social welfare is even more essential postdisaster, need does not always guarantee access. In the case of post-Katrina New Orleans, institutional support for the vulnerable was severely altered across several sectors, particularly housing. The lack of access to "thousands of public housing units helps explain why 83% of single mothers were unable to return to their communities two years after the storms"

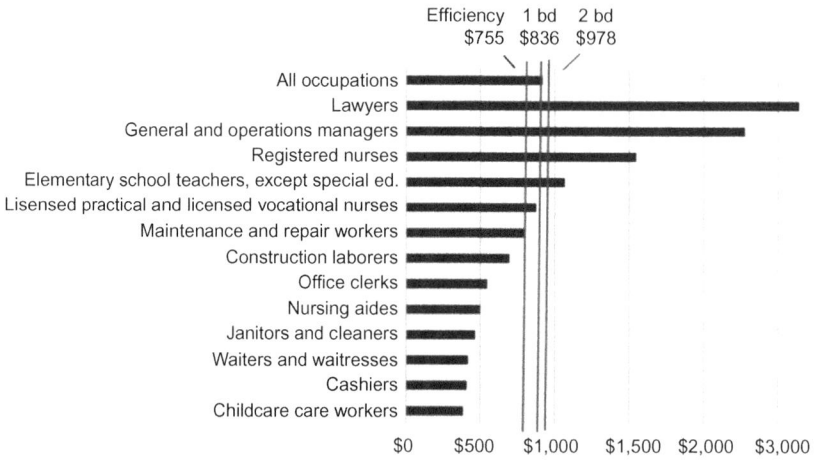

FIGURE 9.3

Affordable Monthly Rent by Occupation, With 2007 Fair Market Rents

(Enarson, 2010, p. 135). As mentioned, more than 80% of public housing in New Orleans was comprised of single female parent households. Yet, of the 4812 inhabitable units in the four major public housing complexes pre-Katrina, only 1410 remained postdisaster (or a 71% decrease in subsidized housing). However, the majority of these units are specifically reserved for mixed-income families or the elderly; clearly not targeting poor single female parent households (IWPR, 2010). In tandem with reductions in subsidized housing, rents increased by an average of 39% from 2005 to 2008, with the greatest increase occurring post-Katrina from 2005 to 2006 (Fig. 9.3). Women are disproportionately in need of low-income housing, but a majority of displaced single female parent families could not find affordable housing in post-Katrina New Orleans.

Five years after Hurricane Katrina the percentage of African-American females in New Orleans decreased 24.2%, compared with before Katrina (IWPR, 2010). And yet, many vulnerable women remain in post-Katrina New Orleans. Of the total households that are headed by women in New Orleans: 88% of households receive certificates and tenant-based vouchers, 77% are in public housing, and 85% are in HUD programs. The lack of affordable housing, and the obstacles to renting, buying, and renovating a home post-Katrina, have contributed to local women's vulnerability to housing discrimination, forced evictions, and displacement, and also to the disruption of social support networks (Willinger, 2008, p. 8).

HEALTH CARE, EDUCATION, AND TRANSPORTATION

However, for the vulnerable who were able to find affordable housing in post-Katrina New Orleans, their vulnerability has been exacerbated via lack of access

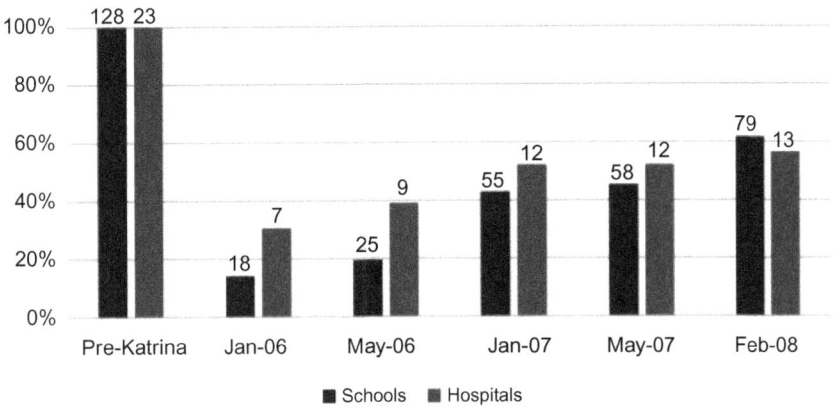

FIGURE 9.4

Functioning Schools and Hospitals During Early Recovery

Source: From Katrina Index.

to several other sectors. For example, Charity Hospital, which was the primary and often only source of health care for the poor of New Orleans, never reopened post-Katrina. Furthermore, many of the public elementary schools were closed and replaced by private Charter schools post-Katrina. Overall, there were 128 schools and 23 hospitals pre-Katrina, and 3 years after Katrina only 79 (39% reduction) schools and 13 hospitals (44% reduction) were functioning (Fig. 9.4).

Similarly, public transportation is imperative for the poor. Buses are the primary means of public transportation in New Orleans. Pre-Katrina there were a reported 368 operational buses in New Orleans, but 3 years later, there were a total of only 76 buses operating; representing an 80% reduction in accessible transportation (Fig. 9.5).

Childcare would be particularly important for single mothers. Operational childcare facilities in New Orleans were reduced from 275 pre-Katrina to 117 post-Katrina, for a 58% reduction in available childcare (Fig. 9.5). According to the Greater New Orleans Community Data Center, as of June 2008 only just 117 childcare facilities were operating post-Katrina.

POLITICAL ECONOMY AND NEOLIBERAL VULNERABILITIES

Klein (2007) argues that elites take advantage of the aftershock of disasters to roll out marked neoliberal policies and practices, that result in the rapid privatization of former public goods and the dismantling of the welfare state. This "disaster capitalism" benefits from the misery of others. This trend can clearly be identified across housing, education, transportation, and education sectors, and in the vulnerable who were excluded from these sectors and thereby, rendered far more vulnerable upon their return or simply kept out of the new New Orleans. Women are

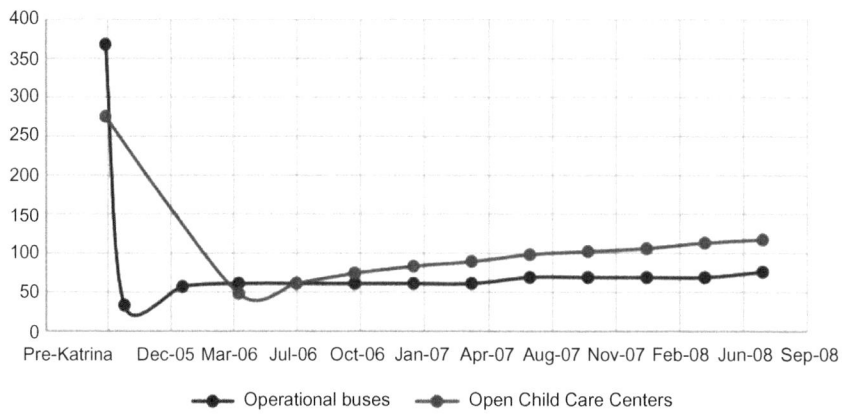

FIGURE 9.5

Operational Buses and Child Care Centers in New Orleans

Source: *From New Orleans Regional Transit Authority, Agenda for Children, and Louisiana Department of Social Services Bureau of Licensing.*

often particularly marginalized in these contexts. However, even in the absence of a disaster, neoliberalism fosters individual and community vulnerabilities, particularly in terms of dismantling social cohesion, which can markedly thwart any recovery effort. Coburn (2000) identifies how neoliberalism and neoliberal development will create inequity among individuals and foster competition that hinders the development of social cohesion. The literature on disasters in low-income settings identifies how neoliberalism can compromise local social structures "including gender inequality and family relations, generating differential vulnerability to disaster in the process" (Bolin et al., 1998, p. 36).

OTHER INTERSECTIONALITIES OF GENDERED VULNERABILITY IN NEW ORLEANS

Poverty exacerbates vulnerability in disasters "because it limits resources that in turn allow for more opportunities for escape or survival when escape is impossible" (IWPR, 2010, p. 1). The relationship between poverty and vulnerability is even more pronounced in postdisaster recovery and reconstruction. When vulnerability is studied as a function of gender *and* race/ethnicity, marked variation is identified especially in terms of annual income. Within-gender group variation demonstrates on average a higher gap in average annual income between white and African-American females than between white and African-American males, both before and after Hurricane Katrina. For example, over the 4-year period from 2005 to 2008, white females earned on average 39%, 50%, 39%, and 51% more than African-American females, respectively (Fig. 9.2). The widest gap in incomes in

New Orleans is between white males and African-American females during this period with a difference of 43%, 67%, 58%, and 51%, respectively. And when comparing white males and African-American males, the gap is on average 28%, 46%, 42%, and 36%, which was nearly twice as high (post-Katrina) as that between white males and white females. Hence, both within- and between-group variation demonstrates that females cannot be assessed statistically as an essentialized group, especially in terms of their marked income differences. The fact that these differences were markedly exacerbated post-Katrina may also be an indication of the increased vulnerability for women of color.

However, income variation alone does not provide adequate information in order to distinguish gendered vulnerability. We have demonstrated that post-Katrina recovery efforts did not sufficiently address goods and services needed to lift the most vulnerable into resilience trajectories. These services are major determinants of livelihoods in urban settings. In the United States, the urban inner city represents a particularly vulnerable component. In general, pre-Katrina, structural inequality made it virtually impossible for many to return or survive in post-Katrina New Orleans. This is evidenced across all of the sectors analyzed.

Numerous other criteria need to be examined to assess the intersectionalities of gendered vulnerability. For example, women generally comprise a larger proportion of the elderly population in New Orleans. "African American women over-65, although still larger in number than White women over 65, decreased as a percent of the over 65 female population" (NCCROW, 2008, p. 7). Although the reported proportion of female and male mortality post-Katrina was nearly identical (49.4 and 50.6, respectively), 60% of reported deaths post-Katrina were over the age of 65 (IWPR 2010). Pregnancy is also an important consideration in assessing gendered vulnerability, particularly due to limitations in mobility. Furthermore, gender-based violence represents a category that is extremely problematic in disasters. Loss of domicile and protective mechanisms can exacerbate postdisaster gender-based violence. However, gender-based violence is often ignored in disaster research. Post-Katrina gender-based violence recorded in Mississippi increased from 4.6/100,000/day pre-Katrina to 16.3/100,000/day (or a 72% increase) exactly 1 year later.

Spatial targeting of vulnerability is imperative for effective preparedness, prevention, and recovery. However, data collection in terms of normative geographic or administrative boundaries, such as census tracks, can prove problematic. Although census tracts are often the most readily available sources of demographic data, in actuality, these depict an average of factors contributing to vulnerability for a given (often arbitrary) geographic designation. Hence, for example, more vulnerable subgroups within the designated geographic group may be rendered invisible by this methodology (Fig. 9.1). To redress the inherent problems with arbitrary geographic and administrative spatial targeting, we recommend using the local designations of a given community. New Orleans had one of the highest nativity rates in the United States in which generations often lived within a given community. Hence, community self-definition in this context could be valuable not only for defining the community, but also for involving the community organically in recovery processes, as well. Hence, local knowledge of communities and employing mechanisms for engaging communities effectively would be beneficial in redressing vulnerability.

CONCLUSION

Gender inequality and other forms of exclusion have remained largely unexplored in US disaster research. "The concern with policy and technocratic control has contributed to a relative neglect of the complexities of social inequalities except in the limited context of how they might create problems in disaster management" (Bolin et al., 1998, p. 29). In this chapter, we have argued that the essentialized depiction of vulnerability and "disaster victims," while ignoring the predisaster social factors that engender and perpetuate inequality and exclusion, serves to exacerbate the problem of vulnerability, rather than properly address it.

Clearly, in order to address gendered and intersectional vulnerability in disasters, all of the complex factors contributing to vulnerability and inequality would need to be assessed. But this would require analysis in disaster preparedness, recovery, and reconstruction that is far more specific than current aggregated groupings of sex, gender, and race/ethnicity. The complexity of gendered vulnerability through more complete data collection would provide a means to disaggregate gender via a series of criteria that includes income, age, ethnicity, marriage status, sexual orientation, as well as other specific criteria, such as education, religion, and HIV-status. However, ideally these groups really need to be even further disaggregated to even more specific levels of analysis and then analyzed in combinations to understand how collectively they can impact one another.

"A well-grounded understanding of the structures of vulnerability is the first step toward meaningful mitigation. Knowing to what extent women, and other historically marginalized groups, have particular vulnerabilities to various kinds of natural and technological hazards facilitates the development of appropriate assistance programs, ones that don't exacerbate social inequalities" (Bolin et al., 1998, p. 43). In order to more accurately identify vulnerability for both pre- and postdisaster intervention, it is imperative to carefully assess the appropriate level of analysis. For example, even a household level of analysis may be too aggregated a level to accurately depict factors contributing to gendered vulnerability. Examples of the need for sublevel analyses are demonstrated in food security studies. For example, despite more than 60 years of nutritional interventions for the worst intractable chronic malnutrition in the Western hemisphere, in indigenous Guatemala, the success of any nutritional program will remain negligible as long as the household level of analysis is ignored; which would reveal a common issue with males receiving the largest portions of meals, while female children and mothers receive the smallest portions (Kadetz, 2009).

It is imperative that disaster research and interventions address "the complex and heterogeneous nature of classes, cultures, ethnicities and genders [and thereby avoid] "sweeping generalizations that ignore diversity" (Bolin et al., 1998, p. 42). Assessments that include social inequality in the production of vulnerability must be prioritized. Disaggregated assessments, such as risk mapping, will help determine the geography of inequality and vulnerability and how best to target interventions. Building equity in recovery interventions can also be fostered through vulnerability programming and education, particularly in regards to gender and the building of capacity among vulnerable groups, especially through including women in emergency management. A response and recovery emphasis on strengthening local capacity of the most vulnerable in any group is essential. Assuring that gender is addressed in all disaster-related policies will help to more successfully address the issues of gender and disaster vulnerability. Increasing the social capital and social cohesion of women in the period before the event is essential to strengthening the resilience of women and other vulnerable segments of society. Thus, in conclusion, the issues of inequity in disaster research and interventions may well result from a paradigm that needs to better differentiate between and within genders and their myriad intersectionalities.

REFERENCES

Banks, M. (1996). *Ethnicity: Anthropological constructions*. London: Routledge.
Baumann, G. (1996). *Contesting culture: Discourses of identity in multi-ethnic London*. Cambridge: Cambridge University Press.
Bolin, R., Jackson, M., & Crist, A. (1998). Gender inequality, vulnerability and disasters: Issues in theory and research. In E. Enarson, & B. H. Morrow (Eds.), *The gendered terrain of disaster—through women's eyes* (pp. 27–44). Westport, CT: Praeger.
Butler, J. (1990). *Gender trouble: Feminism and the subversion of identity*. New York: Routledge.
Coburn, D. (2000). Income inequality, social cohesion and the health status of populations: The role of neo-liberalism. *Social Science & Medicine*, *51*(1), 135–146.
Cohen, A. (2004). *Urban ethnicity*. London: Routledge.
Collins, P. H., & Bilge, S. (2016). *Intersectionality*. Cambridge: Polity Press.
Enarson, E. (2010). Gender. In B. D. Phillips, D. S. K. Thomas, A. Fothergill, & L. Blinn-Pike (Eds.), *Social vulnerability to disasters* (pp. 123–154). Boca Raton, FL: CRC Press.
Esplen, E. (2009). *BRIDGE gender and care overview report*. Brighton: BRIDGE/IDS.
Institute for Women's Policy Research (IWPR). (2010). *Women, disasters and Hurricane Katrina*. Fact Sheet IWPR#D492.
Kadetz, P. (2009). *Government, NGO, and community factors affecting malnutrition in twelve indigenous communities of Lake Atitlan, Guatemala: An assessment for sustainable solutions*. Technical Report for The Royal Geographical Society.
Klein, N. (2007). *The shock doctrine: The rise of disaster capitalism*. New York: Henry Holt.

NCCROW. (2008). *Katrina and the women of New Orleans.* Tulane University. Newcomb College Center for Research on Women (NCCROW).

Oxford English Dictionary Online (OED). September 2017. Oxford University Press. <http://www.oed.com/viewdictionaryentry/Entry/5788>. Accessed 06.06.17.

UN. (2006). *Report of the Human Rights Committee* (Vol. 1). New York: United Nations Press, A/61/40.

Vigh, H. (2008). Crisis and chronicity: Anthropological perspectives on continuous conflict and decline. *Ethnos, 73*, 1.

Willinger, B. (Ed.), (2008). *Katrina and the women of New Orleans.* New Orleans, LA: Newcomb College Center for Research on Women, Tulane University.

FURTHER READING

Gramsci, A. (1994). In F. Rosengarten (Ed.), *R. Rosenthal (Trans.).* Letters from prison. New York: Columbia University Press.

Scott, J. C. (1985). *Weapons of the Weak: Everyday Forms of Peasant Resistance.* New Haven: Yale University Press.

Disaster Resilience

PART III

CHAPTER 10

Culture and resilience: How music has fostered resilience in post-Katrina New Orleans

James R.G. Morris[1] and Paul Kadetz[2]
[1]Stephen F. Austin State University, Nacogdoches, TX, United States
[2]Drew University, Madison, NJ, United States

CHAPTER OUTLINE

Introduction	233
Background: Supporting Resilience From the Outside	234
Understanding Resilience and Designing the Research	236
The Context for This Research	236
Questions to be Answered	238
Methodology	239
Study Design	239
Sample	239
Data Collection	240
Data Analysis	240
Ethics	241
Locating Resilience in Musical Performance	241
Risk Factors, Stress, and Mental Health	241
Protective Factors and Assets	247
Social Support	247
Connections to Community and Mentoring	248
The Impact of Music Performance on Performers and Audience Members	249
Hobfoll's Conservation of Resources Theory	251
Conclusion	251
References	253

INTRODUCTION

Resilience has often been depicted as a positive social quality that is either already embedded in a community or is somehow quietly disseminated from outside the community through interventions that incorporate community participation. The idea that resilience is dynamic and can be engendered from within the community is underexamined. That resilience can be generated, fostered, and perpetuated through engagement with cultural outputs is considered even less frequently. This chapter illustrates the central importance of a community's generation of cultural outputs, specifically that of music, for individual and community resilience in the aftermath of a disaster.

BACKGROUND: SUPPORTING RESILIENCE FROM THE OUTSIDE

In the normative disaster recovery discourse, resilience has often been portrayed as bestowed upon a community from the outside. Although numerous NGOs sprang up post-Katrina, often filling the void left by inaccessible federal programs, few were specific to the needs of musicians. The New Orleans Musicians Clinic (NOMC), established in 1997, is a nonprofit health care clinic for musicians, Mardi Gras Indians, Social Aide and Pleasure Club members, and other allied members of the music, arts, and cultural community of New Orleans. The combination of need and opportunity that existed in New Orleans allowed the NOMC to position itself to the funding public as an established and stable nonprofit that was in existence prior to Hurricane Katrina. The proclaimed goal of the NOMC was to address the important and unmet needs of the vulnerable bearers of local traditions.

Sweet Home New Orleans (SHNO) was a small start-up nonprofit that was conceived, created, and implemented post-Katrina. It was created to fill the void at the NOMC for displaced members of the cultural community who were actively trying to return to New Orleans. SHNO acted as a clearinghouse for available assistance for members of the cultural community who were actively engaged in the traditions of New Orleans before Katrina. SHNO would not have come to fruition were it not for a confluence of human and environmental factors that existed post-Katrina. SHNO was created when twelve disparate, and often competing, agencies across the United States came together with the intent of directing funds that were coordinated through a case management process that addressed current and future unmet needs, resources, and sustainability. Its goal was to return these tradition bearers back into their neighborhoods after hurricane Katrina so they could rebuild their homes and lives, and continue the intergenerational transmission of New Orleans culture. It was believed that future generations who perpetuate and embody the traditions of New Orleans would be lost post-Katrina, if permanently displaced to Atlanta, Houston, and other points across the country. The future of these traditions was considered to be at stake, due to the unlikelihood of recreating these unique cultural conditions anywhere other than in the communities in which they originated. Even in the transient 21st century, place is central to culture and identity.

Prior to the development of SHNO, members of the cultural community who were seeking assistance had to contact an agency, make an appointment with a grant manager, explain their situation, get approved, and help coordinate interventions. However, the coordination of paperwork when requesting assistance from several organizations post-Katrina was difficult at a time when most residents were displaced and living with family or friends out of town. Survivors did not have access to vital paperwork, and were often functioning under a burden of posttraumatic stress. SHNO staff reported this situation as an opportunity to apply their knowledge, values, and skills toward an engine of change.

With ample input from agency partners and community members, SHNO was crafted to fit the needs of both displaced clients and their agency partners. SHNO liaised with this broad base for support, as witnessed in the church-based work crews of college students on their spring break who would come with hammers and crowbars to "gut houses". It was also manifested in the referrals that SHNO made to Neighborhood Housing Services of New Orleans, and the partnership established with the local Habitat for Humanity, for those seeking home ownership. Thus, completing a SHNO intake was the equivalent of simultaneously seeking assistance from over a dozen agencies.

SHNO facilitated the return of a significant proportion of the cultural community of New Orleans. The process of relocating clients back into New Orleans was complex and costly. Many musicians were unable to successfully return on their own. Clients saw disruptions in their ability to earn money and find meaning in their lives while displaced. However, clients who came through the SHNO process were believed to be more able to empower themselves in their return to New Orleans. SHNO's success was measured by the number of people who could return to New Orleans to promote and engage in the traditions of the city.

However, the economic reality of the "new New Orleans" was harsher toward many in the arts and culture community. The recovery process for the city was slow and uneven. Those neighborhoods on higher ground tended to have wealthier residents who were better insured. Those neighborhoods on lower ground, often where cultural traditions were borne and bred, tended to have lower incomes, with homes of older construction that were often insufficiently insured or uninsured. When the storm came through and flooded the city, many musicians were among the hardest hit. This reality was coupled with an influx of out-of-town contractors who prioritized bidding and winning jobs over the completion of work. Those areas with homes inhabited by occupants who were still employed and funded by insurance company pay-outs, were more likely to be rebuilt, and to be rebuilt first.

The financial crisis of 2008 hampered the ability of funding bodies to make grants available, because the growth of their endowments was impacted. By 2008, some of the luster and appeal New Orleans enjoyed as a cause in the immediate aftermath of the Katrina disaster had faded, and public sentiment and the dollars associated with humanitarian aid had shifted to other desperate situations across the globe. These changes undermined the long-term stability of SHNO and in August 2013, SHNO closed its doors. The founders of the organization had never intended to create a permanent organization, especially if this resulted in resources being diverted from those in need. SHNO had fulfilled its mission in assisting the cultural community of New Orleans to help itself to return and recover after Hurricane Katrina. In this way, SHNO helped foster the resilience of these communities.

UNDERSTANDING RESILIENCE AND DESIGNING THE RESEARCH

This research seeks to understand the factors that enabled some musicians to return to New Orleans and create new forms of art, reflective of the human experience. Among numerous factors affecting resilience, community resilience can be understood in terms of resource accessibility. According to Hobfoll's Conservation of Resources (COR) theory "resource loss [is] disproportionately weighted in comparison to resource gain". COR theory further posits that to prevent resource loss or establish resources, other resources must be invested (Hobfoll & Lilly, 1993, p. 128). Hence, it is important to identify who has what level of access to which resources, and who can avoid which situations in order to best enable future generations to better face future disasters. The stories of musicians provide an understanding of the resiliency they experienced as musicians in New Orleans, as well as displaced New Orleanians.

THE CONTEXT FOR THIS RESEARCH

Disaster mental health literature has addressed the human psychic reactions to disasters, including hurricanes (Kaiser, Sattler, Bellack, & Dersin, 1996), with particular recent attention to Hurricane Katrina (Gabe, Falk, McCarty, & Mason, 2005). This literature identifies the unmet needs of those who survive such a disaster (Zottarelli, 2008) and suggests techniques for a broad range of coping strategies for those affected (Norris, Perilla, Riad, Kaniasty, & Lavizzo, 1999). While much is known about the events surrounding Hurricane Katrina and the subsequent flooding of the city of New Orleans, the impact on musicians in the city is less well known. Disasters severely affect the physical structures of the environment, but they also have lasting effects on

- the rate of repopulation (Freedy, Saladin, Kilpatrick, Resnick, & Saunders, 1994),
- stress (Hobfoll, 1989; Norris et al., 1999),
- mental health (Ursano, Fullerton, & Terhakopian, 2008),
- household finances (Oppenheimer, 2008),
- community (Corley, 2010; Raeburn, 2007), and
- culture (Patterson, Weil, & Patel, 2010; Williams & Spruill, 2007).

With so many adverse outcomes of disasters, one counter-intuitive finding is that increased community resilience to stress is also a product of disaster exposure (Bonanno, Galea, Bucciarelli, & Vlahov, 2006; Jensen, 2005; Linley & Joseph, 2004).

The post-Katrina diaspora of musicians and tradition bearers left an irreplaceable gap in the continuity of the transmission of cultural traditions (Pais & Elliott, 2008). Those who were able to return encountered a much different city than the one they left (Zottarelli, 2008). Many were not able to return to their homes, and those who did were faced with the multiple factors of trying to rebuild their homes while facing a lack of employment (Zottarelli, 2008). Moreover, those returning faced issues ranging from the complete loss of community support structures, the lack of health care infrastructure, and myriad unmet mental health issues stemming from losing their former lives, to living in a disaster area crippled by slow and often disorganized recovery efforts (Freedy et al., 1994). Although confronted with these and other myriad challenges, many have survived. In fact, some have thrived, rebuilding their homes, and piecing together their former lives to something newer and more resilient (Le Menstrel & Henry, 2010). The story of these resilient individuals and groups is examined here in order to better understand community strengths and vulnerabilities that can be leveraged to overcome disasters.

Musicians in New Orleans were among the most disadvantaged, the most exposed to the disaster, and the most vulnerable (Elliott & Pais, 2006). The music composed and played by New Orleans' musicians, especially those who experienced the Katrina disaster, engendered healing (Kish, 2009). Their ability to capture and express the pain and sorrows of disasters facilitated coping mechanisms, social cohesion, and community healing (Jenson, 2005). By engaging an affected community in a social and culturally relevant context of healing, musicians and the music they play enable the construction of new social support networks to be created and "new versions of home and community in which music offers itself as a vital means of disaster recovery" (Kish, 2009, p. 671).

Musicians are such an integral cultural, social, and economic part of New Orleans that they were considered to be a barometer of how the city was recovering from Katrina (Raeburn, 2007). The musicians responded to displacement, exile, and homecoming; the changes and growth of their musical repertoire; and the symbiotic relationship they share with their cultural environment. Through examining their response to the disaster, a better understanding of the recovery of New Orleans, and disaster recovery in general, can be gained (Le Menstrel & Henry, 2010). Katrina exposed the city's disaster vulnerability due to the physical, social, cultural, economic, and structural disparities embedded in the history of New Orleans (Kish, 2009). These vulnerabilities have existed in varying forms since the city was first established and have been integrated into the unique self-identities of New Orleanians on both sides of Canal Street.

The intergenerational transmission of culture is evident in the unique cultural exchanges of "second lines". and brass band funerals. Second lines are comprised of the people who informally join in with the (Main line) musicians in a parade or other procession. According to Regis (2001, p. 755), "The distinctive interaction between the club members, musicians, and second liners produces a dynamic participatory event in which there is no distinction between audience and performer"(Coclanis & Coclanis, 2005). The social and cultural meaning of such events is deeply rooted in the history and neighborhoods of New Orleans. While these events are major economic engines of the tourism industry, they are both symbolic and concrete remnants of cultural expressions rooted in the historical structural violence of slavery, segregation, and racism (Turley, 1995). The intergenerational transfer of cultural traditions is an integral aspect of New Orleans music, and is crucial in maintaining the continuity of New Orleans culture and traditions for future generations (Raeburn, 2007; Rowell, 2007; SHNO, 2010).

Residents of New Orleans experience a deep bond between place and identity. Le Menstrel and Henry (2010) interviewed musicians who had been displaced after Hurricane Katrina and found a deep-seated connection to New Orleans as motivation to return and rebuild. These musicians felt an obligation to return, not only for themselves, but also as a moral obligation to the future of the city and its music. The authors reported that musicians were willing to forgo more stable and lucrative employment, lower crime rates, and better schools, in order to return to create, perform, and teach music. The level of predisaster attachment to New Orleans in a postdisaster environment is a significant factor in the decision-making process of musicians in regard to returning to rebuild their lives and careers, as well as their self-identity as a New Orleans musician.

The contributions of New Orleans to music, art, food, culture, and architecture are rivaled by few cities in the New World (Coclanis & Coclanis, 2005; Sakakeeny, 2006). The cultural contributions of New Orleans are very real, and steeped in traditions of the neighborhoods (Smith, 1994). Those who bear these traditions are a conduit from the past and will serve as a bridge to the future (Jenson, 2005; Le Menstrel & Henry, 2010). Le Menstrel and Henry propose that if the cultural traditions of brass bands, school bands, and progressive new music return to New Orleans following Hurricane Katrina, the city's cultural future can hope to be saved.

QUESTIONS TO BE ANSWERED

This research seeks to identify factors that have enabled New Orleans musicians to cope with the disaster of Hurricane Katrina. Overall, this research seeks to determine how disaster-affected persons migrate from vulnerability to resiliency. The questions to be answered in this study include

- What are the risk factors of being a musician?
- What stress and/or mental health issues are associated with being a musician?
- What resources are critical for musicians to thrive and what factors of resilience were most useful for New Orleans musicians following Hurricane Katrina?

- What are the sources of social support for musicians in New Orleans?
- How were they able to establish stability once displaced from New Orleans?
- What factors enabled them to return to New Orleans and rebuild their lives once back in the city?
- How were musicians able to return to their careers in New Orleans and move on, despite hardships?
- What impacts do performing music have for musicians, and listening to music have on musicians and audiences?
- How do connections to the New Orleans music community affect musicians?
- What role does mentoring play in the transmission of culture for musicians?
- Finally, what needs to be done in the future to better enhance resilience and reduce vulnerability for musicians and other cultural bearers in disasters?

METHODOLOGY
STUDY DESIGN

This study utilizes a qualitative, ethnographic, semistructured interview process with a targeted and snowball sampling of New Orleans musicians recognized by their peers for being particularly resilient. This qualitative study is a recursive Variable-Generating Activity (Figley, Cabrera, & Speciale, 2011) using ethnographic video interviews.

SAMPLE

A peer nomination process was initiated from a database of musicians. Musicians could nominate any number of peers, including themselves. Participants were nominated for their perceived resilience by their peers from a list of musician contacts from the 501(c) 3 nonprofit organization, New Orleans Musicians Foundation (NOMAF). By interviewing these community "role models," sources of stress, strategies for coping, insights into factors impacting resilience, and other insights could be identified for other musicians working in New Orleans and beyond. A recruitment email was submitted to NOMAF's list of 596 musician contacts requesting participation in this project. Ten musician informants were secured for this research. All participants were age 18 or over, and have lived and worked in New Orleans before and after Hurricane Katrina. All agreed to be interviewed and were interviewed between February 1, 2013 and February 28, 2013. Eight of the 10 musicians interviewed evacuated in anticipation of Hurricane Katrina, with one musician staying in New Orleans for the duration of the storm and aftermath, and one musician was on tour in South America. Of the eight who evacuated, five went to Texas, one went to Baton Rouge, Louisiana, one to Tennessee, and the last went to New York City. Eight of the 10 musicians have returned to New Orleans full-time. All informants were living and working in New Orleans as professional musicians prior to Hurricane Katrina.

All musicians sustained some level of damage to their homes through the course of Hurricane Katrina. Two musicians sustained a full loss of home and contents. Three musicians sustained heavy damage that included flood damage necessitating substantive repairs. Three musicians sustained moderate damage that included street flooding necessitating substantial repairs. Three musicians sustained relatively minimal damage to their home with no or little street flooding, necessitating moderate repairs. Common repairs included house gutting, new roofs, interior repair, or replacement of sheet rock due to water damage, and issues associated with damage from trees or windblown debris breaking windows and fences. One informant's house was in the Lower Ninth Ward, in the path of a major breech of the levee of an industrial canal, and was pushed off its foundation. The entire neighborhood has yet to recover and the informant has since settled in an adjacent neighborhood.

DATA COLLECTION

The 10 musicians who garnered the most nominations for perceived resilience were contacted for their participation in in-depth, semistructured interviews concerning informant's lived experiences and perceptions before, during, and after Hurricane Katrina. The Variable-Generating Activity (VGA) video analysis protocol generated appropriate items for the interviews. Videotaping the interviews facilitated the collection of nonverbal, facial, postural, and body language forms of communication. It also allowed for appropriate cultural interpretation of the nonverbal, which would have been lost or misinterpreted if just a transcript alone was used (Ekman, 1993; Krauss, Chen, & Chawla, 1996). As no research tool existed based on the lived experiences of musicians in a postdisaster context, the use of ethnographic interviewing is an appropriate precursor in the development of such a tool, for the lived experiences of the musicians can now be identified for future inquiry (Campbell & Russo, 2001). The interview instrument that emerged was reviewed by informants to improve wording and reduce redundancy.

DATA ANALYSIS

The data collected were analyzed using triangulation techniques, according to the VGA protocol, in order to generate a series of statements (variables) based on the lived experiences of musicians before, during, and after Hurricane Katrina. The VGA process utilizes multiple viewings of video interviews by different researchers and is ideal for studying new sample groups, because it provides contextual information about that population in order to generate hypotheses that can be systematically investigated (Figley et al., 2011).

ETHICS

All informants completed and signed consent forms authorized by the Tulane University Internal Review Board (IRB) for this project. Informants were told their participation in the study was entirely optional, that they can withdraw from the study at any time without consequence, and that there will be no benefit or penalty for their choice to participate or refrain from participating in this study. All musicians consented to donate their videotaped interviews to the Tulane University Traumatology Institute as Oral Histories and for future research.

LOCATING RESILIENCE IN MUSICAL PERFORMANCE

This study explored the resilience of musicians in New Orleans. The primary research question is: "What are the factors that have enabled resilient musicians in New Orleans the ability to cope with the outcomes of Hurricane Katrina?" In order to address this question, 10 secondary research questions were developed that are consistent with the research literature associated with postdisaster mental health and the social psychology of musicians in New Orleans. These questions were consolidated into the following themes of the informants' experiences.

RISK FACTORS, STRESS, AND MENTAL HEALTH

Income instability manifests itself in the lives of the informants through lack of pay for performances, irregular employment, and the effects these have on self-esteem and maintenance of a stable home life. Each manifestation acts in concert as a source of stress. The items relating to income instability as a risk factor are listed in Table 10.1, items 1–10.

Health issues for the informants include lack of health care and health care options, a profession that is not conducive to a healthy lifestyle, and diminished capacity post-Katrina to address health and mental health needs. Resilient musicians reported taking their health seriously, proactively maintaining their health, and seeking access to health care when in need (Table 10.1, items 11–20 and 43–47).

Exposure to drugs and alcohol was described by informants as readily available in their work environment and a particular temptation for musicians, because many musicians are given free drinks while working in clubs or at parties (Table 10.1, items 21–24).

The remaining factors concerning stress and mental health issues (Table 10.1, items 27 and 30) include

- performance anxiety (Table 10.1, items 25 and 26),
- competition among musicians (Table 10.1, item 28),
- desire for appreciation (Table 10.1, item 29), and
- management of personalities in a band (Table 10.1, item 31).

Table 10.1 New Orleans Resilient Musicians Scale (Norms)

Item #	Items
Risk Factors, Stress, and Mental Health	
1	Not getting gigs can affect your self-esteem.
2	Job instability is a source of stress for musicians.
3	Stress over money is common for many musicians.
4	The club owners need to realize that musicians are working a job too and need to receive fair pay for work.
5	Musicians are underpaid.
6	Musicians do not always get proper compensation or recognition for their work.
7	Financial stress is a significant risk factor for musicians.
8	Unstable finances are a risk factor for musicians.
9	The instability of gigs and income is a risk factor for musicians.
10	Being a musician can make it very difficult to have a stable home life.
11	Musicians' physical health can be at risk by not always being able to eat healthy and not getting the proper amount of sleep.
12	Musicians' physical and mental health can be greatly impacted by the everyday stressors.
13	Musicians can be stressed when having to play music under poor conditions in which they would rather not be playing.
14	Musicians are at risk due to the lack of health care.
15	Mental health resources for musicians in New Orleans is limited.
16	Being a musician can have a negative impact on a person's physical health.
17	Musicians may not always pay proper attention to their health.
18	Hurricane Katrina was very stressful for musicians and their families.
19	Benefits and health care are limited resources in the music industry.
20	Low self-esteem can be a real issue for musicians.
21	Musicians are exposed to drugs, alcohol, and smoky environments.
22	Being a musician brings with it a risk of exposure to drugs and alcohol.
23	Drugs and alcohol are a risk factor for musicians.
24	Musicians often work in high-risk environments.
25	Performance anxiety is a stress that all musicians deal with on some level.
26	Performing can generate a lot of fear for musicians.
27	After Katrina, a lot of musicians felt very stressed out because they were missing New Orleans.
28	Competition among musicians is a significant source of stress.
29	The desire for appreciation stresses musicians.
30	Being a musician can cause stress in a person's home life.
31	If a musician is in a band, trying to manage all the personalities can be stressful.

(Continued)

Table 10.1 New Orleans Resilient Musicians Scale (Norms) *Continued*

Item #	Items
Protective Factors and Thriving	
32	One of the benefits of being a musician is getting to do something that you love.
33	Love for their art can help musicians survive hard times.
34	It is important to "feel" the music not just formally learn it.
35	Resilient musicians are very dedicated to their artistry.
36	A strong drive to perform can help a musician to be resilient.
37	Social support is a protective factor for musicians.
38	Being a musician in New Orleans is a special thing.
39	Knowing that they are important to the community can be a protective factor for musicians.
40	It is important to surround yourself with people that are interested in what you are doing.
41	Resilient musicians have a network of supporters that they can rely upon.
42	Faith in God can be very important to helping a musician be resilient.
43	Having some sort of physical exercise is important.
44	Maintaining a good diet is important.
45	Maintaining good health is important for musicians.
46	Access to medical care is important.
47	In order to thrive, musicians must take care to attend to their health.
48	Resilient musicians must have a cool head to stay away from negative influences that they may be exposed to.
49	Resilient musicians maintain a positive outlook.
50	A resilient musician maintains a positive initiative and attitude.
51	Resilience is one day at a time.
52	A resilient musician is one who does not throw in the towel because of adversity.
53	Resiliency means that you can acclimate to the conditions in your environment.
54	To succeed a musician needs to be flexible and do what it takes.
55	To be resilient a musician needs to continue to learn and grow artistically.
56	A musician should be themselves and not try and model themselves after anyone else.
57	Having a steady source of income is important.
58	It is important to get paid appropriately for a gig.
59	It is important for fans to support musicians.
60	It is important to live within your financial means.
61	It is important to plan and prepare.

(Continued)

Table 10.1 New Orleans Resilient Musicians Scale (Norms) *Continued*

Item #	Items
Social Support	
62	Communication between musicians can provide positive social support.
63	It is important that musicians help each other.
64	It is important for musicians to encourage each other.
65	Musicians often get strong social support from other musicians and musicians' organizations.
66	The musical community in New Orleans is very supportive.
67	The social aid and pleasure clubs provide funds for needy people in the community.
68	The SHNO was helpful after Katrina.
69	The NOMC and Tipitina's Co-op are important resources for New Orleans Musicians.
70	There has been more social support for the musicians post-Katrina than pre-Katrina; for example, Sweet Home, the New Orleans Musicians Relief Fund (NOMRF), and the Musicians Clinic.
71	The NOMC has been a beneficial resource for musicians' health care.
72	A stable family is a form of support and very important for musicians to maintain.
73	Strong traditional music families can provide support systems within the music culture and community.
74	Personal relationships (family, friends) are a very important form of social support for musicians.
75	Family support plays an important role in the livelihood of musicians.
76	Musicians receive social support from varied sources including family, church, and community members.
77	Community cohesion as a social support can help musicians through difficult times.
78	Mentoring is one type of social support that could be very beneficial to musicians.
79	Social support for others can be very important in helping a musician to continue to pursue their career.
80	Resources to assist musicians in dealing with some of their risk factors (i.e., drugs, money issues, etc.) can help musicians to be more resilient.
81	Social support can help a musician maintain a positive frame of mind.
82	Social support can be helpful in assisting musicians with financial stresses.
83	Social support organizations (e.g., MusiCares) can play an important role in helping musicians recover after a disaster.
84	Social support from other musicians can have a positive psychological impact.
85	Social support organizations (e.g., the Musicians Clinic) can be very helpful to musicians.
86	The people of New Orleans and the support of other musicians are what makes it special.

(Continued)

Table 10.1 New Orleans Resilient Musicians Scale (Norms) *Continued*

Item #	Items
Psychological Impact of Music	
87	Performing allows musicians to express their thoughts and feelings.
88	By performing music you can channel emotions and feelings to the audience.
89	Attending others musical performances can help a musician to experience the crowd response more intimately.
90	Attending other musicians' performances has the ability to inspire and help one grow.
91	Hearing a live performance stirs emotions.
92	Performing can give musicians a sense of making a contribution to those who hear them.
93	Performing makes musicians happy because they are making the audience happy.
94	As a musician you get to travel and see other parts of the world. You are rewarded by giving of yourself and seeing other people smile.
95	Playing music can energize you.
96	Playing music and doing what you love has the ability to keep you young and sane.
97	A successful performance can bring a musician a great deal of satisfaction and pleasure.
98	Attending other musicians' performances brings a feeling of enjoyment.
99	Attending performances of other talented musicians can be very exciting.
100	Listening to the music of others is a joy because we love the music.
101	After Katrina being able to play music again was one of the only things that felt normal.
102	Post-Katrina, musicians felt they were bringing life back to the city.
103	Performing after Katrina was a healing experience.
Connection to Community and Mentoring	
104	Being part of the New Orleans musical community brings exposure to many talented artists.
105	Being connected to the New Orleans music community can give young children in the community the chance of a different direction that they may not have had otherwise.
106	Mentoring is very important and can have a huge impact on your success as a musician.
107	Having positive mentors is important for the professional and personal development of young musicians.
108	It is important for older musicians to pass along the culture to younger musicians, so that the next generation does not lose the art and culture that is New Orleans.
109	Older musicians in New Orleans are very open to helping younger musicians.
110	The church was important in exposing me to music as a child.

(*Continued*)

Table 10.1 New Orleans Resilient Musicians Scale (Norms) *Continued*

Item #	Items
111	Church, school, and community are all-important to the music culture and can lead children in the path of having a music career.
112	Becoming a musician can be a struggle without an encouraging mentor.
113	Mentors have the ability to change one's life forever.
114	In New Orleans there is a support system among musicians unlike any other place.
115	Connection to the musical community can provide creative and emotional support for a musician.
116	Performing in a supportive environment like New Orleans can help a musician to be resilient.
117	New Orleans has culture that is unlike any other part of the United States.
118	New Orleans represents a very unique and positive environment for musicians.
119	There is more camaraderie between musicians in New Orleans as opposed to other cities.
120	The passing of cultural traditions within families is very important in New Orleans.
121	Family members can be particularly positive mentors.
122	Coming from a music-influenced background/family can provide you with the ability to learn about music and the industry.
123	Musicians in New Orleans enjoy each other. They enjoy playing and working with one another.
124	Musicians who have experienced New Orleans often find that they become very connected to the city.
125	There is a new appreciation for New Orleans music and culture since Katrina.
126	The bond between musicians has gotten stronger since Katrina.
127	Remembering your community and your culture can help you to go on after a tragedy.

This Scale is composed of statements by musicians interviewed during 2013. Each statement is a self-description. The purpose of the Scale is to better understand musicians' resilience. Each statement was answered according to informant's identification with a statement on a Likert scale from 0 ("Not like me") to 4 ("That's totally me").

These findings are consistent with previous research concerning

- risk factors of musicians (Dobson, 2011; Elliott & Pais, 2006; Fredrickson & Rooney, 1988; Roset-Lobet, Rosines-Cubells, & Salo-Orfila, 2000; Winick, 1959),
- stress (Bourque, Siegel, Kano, & Wood, 2006; Norris et al., 1999),
- mental health (Bourque et al., 2006; Jenkins, Laska, & Williamson, 2007; Kaiser et al., 1996; Lee, Shen, & Tran, 2009; Nigg, Barnshaw, & Torres, 2006), and
- access to health resources (Elrod, Hamblen, & Norris, 2006; Freedy et al., 1994; Linley & Joseph, 2004).

The findings illustrate that New Orleans musicians are subject to the occupational stressors of finding, retaining, and being compensated for work without sufficient access to physical and mental health services that could potentially ameliorate the negative impact of the stressors.

PROTECTIVE FACTORS AND ASSETS

Factors supporting musician resiliency were grouped under the theme of "protective factors and assets". These factors are satisfaction and dedication to one's craft, social support, health, and personal characteristics. Satisfaction and dedication to music kept musicians engaged in making and performing music (Table 10.1, items 32−36). Social support was considered to be important by informants and is identified in connections to friends, community, and spirituality (Table 10.1, items 37−42). Personal characteristics identified by informants as being affiliated with resiliency are a positive attitude, avoiding negative influences, perseverance in the face of challenges, and the ability to adapt to situations and grow artistically (Table 10.1, items 48−56). Finances were identified as a protective factor for informants, and include income stability and ability to manage finances in an unstable financial situation (Table 10.1, items 57−61).

These findings are consistent with previous research on creativity and flexibility (Dobson, 2011), social support, optimism and perseverance (Le Menstrel & Henry, 2010), and resilience (Bonanno et al., 2006; Jensen, 2005; Linley & Joseph, 2004). These findings illustrate that New Orleans musicians derive great benefit and resilience from the creative work they do, and are dedicated to their craft despite their profession's risk and stress factors. Social support networks provide emotional, psychological, spiritual, and community support, which all serve to protect musicians on a daily basis. Resilience is fostered by personal characteristics of perseverance, flexibility, and positivity while avoiding negative influences, all which reportedly serve to keep these informants active and able to continue performing, despite previously discussed threats to their self-esteem, ability to find and keep work, and exposure to high-risk situations. In addition, the ability to manage finances seems to protect musicians from irregular income and work patterns, which appear to be major threats to resilience. Resilience may also be reflected (and engendered) in informants' ability to access significant and diverse personal, psychological, health, social support, and financial management resources in order to successfully navigate the potential negative impact of life as a musician.

SOCIAL SUPPORT

This thematic category examines the sources and impact of social support for informants. Friendships and engagement with other musicians and inclusion in the musical community were identified as important sources of social support, particularly for the specific issues associated with being a musician in post-Katrina New Orleans (Table 10.1, items 62−66).

Involvement in organizations that are specific to the needs of the music and cultural community in New Orleans was singled out for their importance to informants for their social support. These include Social Aide and Pleasure Clubs, New Orleans Musicians Clinic (NOMC), Tipitina's Office Co-op, Sweet Home New Orleans (SHNO), The New Orleans Musicians Relief Fund, and Sweet Relief. It was also identified by informants that more social support from these and other organizations has occurred post-Katrina than previously (Table 10.1, items 67−71). Other sources of support identified include family, friends, fans, and one's community neighborhood. Informants reported that social support played a helpful and important role in facilitating their ability to continue their musical career (Table 10.1, items 72−86).

As mentioned, musicians face numerous issues associated with their lifestyle and profession and must be able to access limited resources in order to achieve stability in their home and work lives. Social support from other musicians serves to provide a positive psychological impact for a musician and can support the ability to locate employment. Organizations directed toward musicians (such as NOMC and SHNO) assist with financial support, access to medical care, and with other risk factors that are specific to musicians. These organizations have played an important role in fostering resiliency post-Katrina. Social support from family, friends, and the community facilitates the maintenance of a stable and supportive personal life that, in turn, will assist in the musicians professional lives and the cultural life of the city. These sources of social support and the functions they provide serve to insulate musicians from the exacerbated vulnerabilities many encounter when an unanticipated shock, such as a disaster, occurs. Without these support buffers, musicians would be exposed to the ongoing vulnerabilities associated with being a musician.

These findings are consistent with previous research concerning the sources of social support for post-Katrina musicians (SHNO, 2010) and the impacts of social support (Le Menstrel & Henry, 2010; Rowell, 2007; Voltmer, Schauer, Schroeder, & Spahn, 2008).

CONNECTIONS TO COMMUNITY AND MENTORING

This theme further examines the impact of the New Orleans music community on informants and the role of mentoring in the transmission of culture. Benefits of engagement with the musical community include the enjoyment that comes with playing and working with other local musicians, and an increased connection to, and appreciation of, New Orleans as a musical city (Table 10.1, items 123−127). Young musicians identified the benefits of personal and professional development and the intergenerational transmission of culture in their interactions with older musicians (Table 10.1, items 104–109). Social support is enhanced through the unique systems of support for musicians in New Orleans, which include music community relationships and mentors. The influence of musical families was specifically identified as an important source of social support and listed as being a source of positive mentorship that enables transmission of cultural traditions, as well as the life skills associated with being a professional musician (Table 10.1, items 110–122).

These findings are consistent with previous research concerning

- cultural traditions (Smith, 1994),
- cultural transmission (Bonner, 2010; Jenson, 2005; Turley, 1995),
- the value of mentoring (Raeburn, 2007; Rowell, 2007),
- the use of social support networks for professional musicians (Coulson, 2010),
- and the development of community networks (Le Menstrel & Henry, 2010; SHNO, 2010; Weil, 2010).

These findings illustrate that community involvement is a crucial source of support for musicians professionally employed, as well as for young musicians beginning to learn occupational skills associated with becoming a professional musician. In addition, mentoring and community have meaningful contributions to the understanding and appreciation of the historical and cultural basis of New Orleans music for young musicians, which in turn can impact the likelihood of the continuance of music education and can ground future music production in the context of the evolving sounds of New Orleans.

Through the development of traditional brass bands and second lines as viable economic engines of the city's tourism industry, real and viable employment opportunities exist for young musicians steeped in the cultural context of the genre, in addition to fostering proficiency in the technical skills of music production. The perpetuation of these cultural outputs provides economic opportunities, albeit in the informal economy, in a city with civic issues concerning education, crime, and unemployment for segments of an already at-risk population. Community networks and mentoring significantly impact the future of music in New Orleans, as increased attention to New Orleans music and culture post-Katrina have led to the increased commercialization of the culture in lieu of the authentic manifestations of culture.

THE IMPACT OF MUSIC PERFORMANCE ON PERFORMERS AND AUDIENCE MEMBERS

In this theme the social impact of music for both performers and audience is examined. As a cultural language and a central component of ritual, music can serve to strengthen group identity and group purpose, foster sharing and trust, and ultimately enhance social cohesion, all of which are fundamental for building community resilience. Music can be considered more of an emotional, rather than a cognitive, language. The emotions communicated by musicians facilitate engagement; among themselves, with the audience, and among the audience (Table 10.1, items 87−93).

Although the importance of engaging in and observing ritual for the perpetuation and survival of a group has been well-documented in the anthropological literature, the specific value of contemporary cultural performance for group resilience can benefit from further examination. The conception of performative culture as a conduit for community healing dates to the Greek idea of catharsis (κάθαρσις), or the emotional purging and subsequent purification and renewal of a group, resulting from artistic expression that potently engages the group's emotions. Informants describe their engagement and interaction between audience and performer as energizing and providing a particular source of positive emotions. Informants also identified that their performance of music kept them feeling youthful and mentally stable (Table 10.1, items 94−100). Post-Katrina performances were reported by informants as engendering a feeling that they were "bringing life back to the city." They also reported a sense of community healing through music, possibly through instilling a sense of normalcy (Table 10.1, items 101−103).

These findings are consistent with the literature examining the role of music in community healing (Stige, 2017), for establishing a connection with audience members and developing the musical repertoire (Kish, 2009), and for performance as an opportunity for a musician to engage in their normal behavior (Le Menstrel & Henry, 2010). Informants consistently articulated the importance of the symbiotic nature of engagement between audience and performer. This relationship is described as delicate, although powerful, and manifested itself consistently in post-Katrina performances as a function of expression of emotion concerning the destruction of the city; the struggle to rebuild; and the interruption of seemingly mundane aspects of life in an urban environment after a disaster.

Informants reported that by being New Orleanians who had experienced Hurricane Katrina, their ability to identify with the struggles of their audience, and vice versa was enhanced. Thereby, the capability for their performances to reach their audiences and heal can be compared to the healing capabilities of a shaman that results from the shaman's own suffering and the liminal, near-death, experiences they share with their patients. This symbiotic expression of emotion (or "suffering with") further promoted connections over shared trauma, and thereby, further facilitated the potential for healing. Hence, the voice of the musicians through their music was also the voice of the audience. Musicians describe their experiences on stage as being a very authentic and cathartic experience shared with the audience. The psychological impact of both performing and listening to music for musicians, reported to be positive and emotional, may be understood as an important, culturally relevant means of postdisaster healing.

HOBFOLL'S CONSERVATION OF RESOURCES THEORY

The findings of the study are consistent with Hobfoll and Lilly's (1993) COR theory. This integrated model of stress resulting from trauma examines the resources of

- object,
- personal characteristics,
- conditions,
- energy, and
- social support resources.

These five resources were then compared to the classification of five themes based on informant experiences:

- Risk Factors, Stress, and Mental Health,
- Protective Factors and Assets,
- Social Support,
- Community Connections and Mentoring, and
- Impact of Music on Performers and Audience.

In general, resources were affected negatively by the presence of risk factors and positively by the presence of protective factors.

CONCLUSION

Culture has seldom been portrayed as an integral factor in the recovery and community rebuilding efforts of disasters. The emphasis in the recovery literature has been predominantly on the rebuilding of physical infrastructure with technological solutions, rather than the rebuilding of social infrastructure and the mining of cultural assets, which ultimately is essential for community resilience. This research also corroborates the importance of place for identity. In addition to the centrality of the cultural environment, informant identity and self-esteem were drawn from their physical and social environments. When removed from their familiar and supportive environment, New Orleans musicians were deprived of protective factors that allow them to engage in the historical, traditional, community, personal, social, and professional support networks, all critical to their successful recovery. When compounded with the significant stresses endemic of being a musician in New Orleans, including issues with finances, future employment, stress, and high-risk behaviors, musicians are at further risk for negative consequences to health, employment, and mental health. Resilience is identified in the qualities of perseverance, maintaining a positive outlook in times of struggle, and reliance on the community and collaborations to reduce risk factors.

Although surviving a disaster can increase resilience, the process of recovery puts great strain on personal and psychological resources and resource management. Informants who were able to thrive after Katrina reported being able to do so because of their ability to manage resources. On returning to New Orleans, many musicians were able to access resources from their social support networks, enhancing their social capital. They were also able to access new and existing musician-specific organizations (e.g., NOMC and SHNO). These organizations not only provided access to services for health care, musical instruments, and rebuilding, but also provided a de facto social support network of musician assistance organizations dedicated to providing services to bring back musicians.

Also significant as a resource for support were the many opportunities to engage in culturally relevant community activities post-Katrina that developed social capital and social cohesion. Music played a unique role in the recovery process, for it was grounded in, and representative of, the city. Those musicians who were able to perform in and around New Orleans post-Katrina identified that they were engaging in community healing, and that the connection between audience and performer was a tangible and emotional experience that facilitated healing for both musicians and audiences.

Similarly, informant's reported that engagement in mentoring programs served to not only ensure the future of New Orleans music and culture, but also to build social capital and cohesion and ultimately, resilience. While the physical destruction of New Orleans was unsurpassed in terms of scale, scope, and cost, many wondered what would become of the future of New Orleans musical culture. The process of transmission of culture manifests itself through community connections and social support networks between families and neighbors to pass down music to the next generation. New Orleans lost many musicians to the diaspora following the mandatory evacuation of the city. Many have not returned and never will. This is a net loss for the city, and for the future of music, as an entire generation of musicians was displaced and was no longer engaging in the traditions of their native communities. Mentoring programs aimed at pairing professional musicians with young and aspiring musicians have proved to be invaluable in the continuities of the tradition of intergenerational transmission of culture, and have been supported and embraced by musicians, schools, the community, and musician-specific organizations dedicated to preserving the future of an "authentic" New Orleans.

This research is noteworthy for the development of an instrument to qualitatively measure resilience, the New Orleans Resilient Musician Scale (NORMS), derived from informant interviews. NORMS may be appropriate for other postdisaster at-risk populations, although further research is needed to parse out which factors are most representative of at-risk populations in general. An abridged version of NORMS can help to determine community resilience and the steps that can be taken to be consistent with other studies using the VGA approach (Chapman et al., 2012; Figley et al., 2013).

In addition to the myriad ways in which music was identified as building the capacity for community resilience, this research identifies how music may be a culturally relevant and appropriate method of alleviating stress and providing therapeutic benefits to those most in need. It is recommended that future research examine the potential of music as a therapeutic intervention for other at-risk populations, including the homeless and the incarcerated.

There are several limitations of this study that should be mentioned. The generalizability of the findings of this research is a limitation. While there are thousands of musicians in New Orleans (Rowell, 2007), only 10 were interviewed for this project and, therefore, cannot be considered to be representative of all musicians in New Orleans. Furthermore, sample recruitment was a peer-nominated process, which was limited to those musicians who received funding from the New Orleans Musicians Assistance Foundation and who answered the email request for nominations. Musicians who were not nominated may have been appropriate for inclusion, but for various reasons were either not selected or chose not to participate. Hence, another sampling strategy could have yielded different results. A further limitation could be the choice of COR and/or VGA for data collection and analysis, as alternate methods may have yielded different results. It is important to note that as no alternate scale of musicians resiliency exists with regard to disasters, further research would be needed before more meaningful conclusions can be made concerning musician risk, coping, and resiliency before, during, and after Hurricane Katrina.

New Orleans has a history of rebirth and renewal, stubbornly surviving slavery, tropical diseases, poverty, inequality, crime, and floods. These challenges have shaped people who have produced a culture that is optimistic, forgiving, adaptable, and community-based. This chapter has attempted to examine how culture and cultural outputs are not only valuable in their own right and as a source of employment in an informal economy, but also more importantly cultural outputs, such as music, are identified here as an integral factor in postdisaster community resilience, particularly through their ability to produce a specific language of cultural identity and continuity that provides shared solace, trust, community capital, social cohesion, and a sense of meaning to the inexplicable.

REFERENCES

Bonanno, L. F., Galea, S., Bucciarelli, A., & Vlahov, D. (2006). Psychological resilience after disaster: New York City in the aftermath of the September 11th terrorist attack. *Psychological Science*, *17*(3), 181–186.

Bonner, T. (2010). New Orleans and its Writers: Burdens of place. *Mississippi Quarterly*, *63*(½), 195–209.

Bourque, L. B., Siegel, J. M., Kano, M., & Wood, M. M. (2006). Weathering the storm. The impact of hurricanes on physical and mental health. *Annals of the American Academy of Political and Social Science*, *604*, 129–151.

Campbell, D. T., & Russo, M. J. (2001). *Social measurement*. Thousand Oaks, CA: Sage Publications.

Chapman, P. L., Cabrera, D., Varela-Mayer, C., Baker, M., Elnitsky, C., Figley, C., et al. (2012). Training, deployment preparation, and combat experiences of deployed health care personnel: Key findings from deployed U.S. Army combat medics assigned to line units. *Military Medicine, 177*(Issue 3), 270−277.

Coclanis, A. P., & Coclanis, P. A. (2005). Jazz funeral: A living tradition. *Southern Cultures, 11*(No. 2), 86−92.

Corley, C. (2010). Creative expression and resilience among holocaust survivors. *Journal of Human Behavior in the Social Environment, 20,* 542−552.

Coulson, S. (2010). Getting "Capital" in the music world: Musicians' learning experiences and working lives. *British Journal of Music Education, 27*(3), 255−270.

Dobson, M. C. (2011). Insecurity, professional sociability, and alcohol: Young freelance musicians' perspectives on work and life in the music profession. *Psychology of Music, 39*(2), 240−260.

Ekman, P. (1993). Facial expression and emotion. *American Psychologist, 48*(4), 384−392.

Elliott, J. R., & Pais, J. (2006). Race, class, and Hurricane Katrina: Social differences in human response to disaster. *Social Science and Research, 35,* 295−321.

Elrod, C. L., Hamblen, J. L., & Norris, F. H. (2006). Challenges in implementing disaster mental health programs: State program directors' perspectives. *The Annals of the American Academy of Political and Social Science, 604,* 152.

Figley, C., Cabrera, D., & Speciale, J. (2011). Combat stress injury markers predicting PTSD: Preliminary findings from in-depth interviews of combat medics between deployments. *European Psychiatry, 26*(1), 1810.

Figley, C., Morris, J., Corzine, E., Weatherly, C., Lattone, V., and Sujan, A. (2013). *Parish perspectives: Learning form terrebonne parish post-disasters*. Workshop presented at the International Critical Incidence Stress Foundation World Congress, Baltimore, MD, February 22, 2013.

Fredrickson, J., & Rooney, J. F. (1988). The free-lance musician as a type of non- person: An extension of the concept of non-personhood. *The Sociological Quarterly, 29*(2), 221−239.

Freedy, J. R., Saladin, M. E., Kilpatrick, D. G., Resnick, H. S., & Saunders, B. E. (1994). Understanding acute psychological distress following natural disaster. *Journal of Traumatic Stress, 7*(2), 257−273.

Gabe, T., Falk, G., McCarty, M., & Mason, V. W. (2005). Hurricane Katrina: Social-demographic characteristics of impacted areas. Washington, DC: Congressional Research Service, Library of Congress.

Hobfoll, S. E. (1989). Conservation of resources: A new attempt at conceptualizing stress. *American Psychologist, 44,* 513−524.

Hobfoll, S. E., & Lilly, R. S. (1993). Resource conservation as a strategy for community psychology. *Journal of Community Psychology, 21,* 128−148.

Jenkins, P., Laska, S., & Williamson, G. (2007). Connecting future evacuation to current recovery: Saving the lives of older people in the next catastrophe. *Generations, 31*(4), 49−52, Winter 2007/2008.

Jenson, J. M. (2005). Reflections on natural disasters and traumatic events (Editorial). *Social Work Research, 29*(4), 195−198.

Kaiser, C. F., Sattler, D. N., Bellack, D. R., & Dersin, J. (1996). A conservation of resources approach to a natural disaster: Sense of coherence and psychological distress. *Journal of Social Behavior and Personality, 11*(3), 459–476.

Kish, Z. (2009). My Fema People:" Hip-hop as disaster recovery in the Katrina Diaspora. *American Quarterly, 61*(3), 671–692.

Krauss, R. M., Chen, Y., & Chawla, P. (1996). Nonverbal behavior and nonverbal communication: What do conversational hand gestures tell us? In M. Zanna (Ed.), *Advances in experimental social psychology* (pp. 389–450). San Diego, CA: Academic Press.

Lee, E. O., Shen, C., & Tran, T. V. (2009). Coping with Hurricane Katrina: Psychological distress and resilience among African American evacuees. *Journal of Black Psychology, 35*(1), 5–23.

Le Menstrel, S., & Henry, J. (2010). "Sing us Back Home:" Music, place, and the production of locality in post-Katrina New Orleans. *Popular Music and Society, 33*(2), 179–202.

Linley, P. A., & Joseph, S. (2004). Positive change following trauma and adversity: A review. *Journal of Traumatic Stress, 17*, 11–21.

Nigg, J., Barnshaw, J., & Torres, M. R. (2006). Hurricane Katrina and the flooding of New Orleans: Emergent issues in sheltering and temporary housing. *The Annals of the American Academy of Political and Social Science, 604*(113), 113–288.

Norris, F. H., Perilla, J. L., Riad, J. K., Kaniasty, K., & Lavizzo, E. A. (1999). Stability and change in stress, resources, and psychological distress following natural disaster: Finding from a longitudinal study of Hurricane Andrew. *Anxiety, Stress & Coping: An International Journal, 12*, 363–396.

Oppenheimer, M. (2008). A physical science perspective on disaster: Through the prism of global warming. *Social Research, 75*(3), 659–668.

Pais, J. F., & Elliott, J. R. (2008). Places as recovery machines: Vulnerability and neighborhood change after major hurricanes. *Social Forces, 86*(4), 1415–1453.

Patterson, O., Weil, F., & Patel, K. (2010). The role of community in disaster response: Conceptual models. *Population Research Policy Review, 29*, 127–141.

Raeburn, B. R. (2007). *"They're Tryin' to Wash us Away": New Orleans musicians surviving Katrina*, . Journal of American History (94, pp. 812–819).

Regis, H. A. (2001). Blackness and the politics of memory in the New Orleans Second Line. *American Ethnologist, 28*(4), 752–777.

Roset-Lobet, J., Rosines-Cubells, D., & Salo-Orfila, J. M. (2000). Identification of risk factors for musicians in Catalonia (Spain). *Medical Problems of Performing Artists, 15*, 167–174.

Rowell, C. H. (2007). Reid wick. *Callaloo, 29*(4), 1322–1333.

Sakakeeny, M. (2006). Resounding silence in the streets of a musical city. *Space and Culture, 9*(1), 41–44.

Stige, B., Where music helps: Community music therapy in action and reflection, 2017, Routledge.

Smith, M. P. (1994). *Mardi Gras Indians*. Gretna, LA: Pelican.

Sweet Home New Orleans. (2010). *2010 State of the New Orleans Music Community Report*. <http://sweethomeneworleans.org-wp-content/uploads/2011/05/SHNO-2010-Music-Report.pdf>. Accessed 16.04.12.

Turley, A. C. (1995). The ecological and social determinants of the production of Dixieland Jazz in New Orleans. *International Review of the Aesthetics and Sociology of Music*, 26(1), 107–121.

Ursano, R. J., Fullerton, C. S., & Terhakopian, A. (2008). Disasters and health: Distress, disorders, and disaster behaviors in communities, neighborhoods, and nations. *Social Research*, 75(3), 1015–1028.

Voltmer, E., Schauer, I., Schroder, H., & Spahn, C. (2008). Musicians and physicians—A comparison of psychosocial strain patterns and resources. *Medical Problems of Performing Artists*, 23(4), 164–169.

Weil, F. (2010). *The rise of community engagement after Katrina. The New Orleans Index at Five*. Washington, DC: Brookings Institution and Greater New Orleans Community Data Center.

Williams, J. M., & Spruill, D. A. (2007). Surviving and thriving after trauma and loss. *Journal of Creativity in Mental Health*, 1(3), 57–70.

Winick, C. (1959). The use of drugs by Jazz Musicians. *Social Problems*, 7(3), 240–253. (winter, 1959–1960).

Zottarelli, L. K. (2008). Post-Hurricane Katrina employment recovery: The interaction of race and place. *Social Science Quarterly*, 89(3), 592–607.

CHAPTER 11

Resilience among vulnerable populations: The neglected role of culture*

Mark VanLandingham
Tulane University, New Orleans, LA, United States

CHAPTER OUTLINE

Introduction	257
The Resilience and Recovery Frameworks	258
Resilience	258
Recovery	259
Resilience and Recovery	259
Application of Current Frameworks to the Vietnamese-American Community in post-Katrina New Orleans	260
Recovery	260
Resilience	260
The Missing Piece: Culture	261
Application of an Expanded Framework to the Vietnamese-American Community in Post-Katrina New Orleans	262
Culture Confounders	262
Cultural Influences on Post-Katrina Recovery	263
Conclusions	264
References	264

INTRODUCTION

Disasters discriminate. Even among those disasters that affect wide swaths of geography and a broad range of communities, those who were disadvantaged before the event generally fare the worst afterwards. Hurricane Katrina flooded about 80% of New Orleans when the federal levees failed soon after the storm came across the city. Much of the heaviest flooding occurred in communities that

---

*A more lengthy treatment of culture, resilience, and disaster recovery can be found here: VanLandingham, M. (2017). *Weathering Katrina: Culture and recovery among Vietnamese-Americans*. New York, Russell Sage Foundation.

were located in some of the poorest—and lowest lying—sections of the city, like Central City and the Lower Ninth Ward. Rates of return among Whites far outpaced rates of return among Blacks, an inequitable feature of the recovery that can be explained by levels of damage and depth of flooding (Sastry & VanLandingham, 2009).

But predisaster features such as poverty and elevation do not fully explain patterns of postdisaster recovery. Flood waters reached many working-class, middle-class, and affluent sections of the city, too, such as Eastern New Orleans, Mid-City, and Lakeview. Driving through these areas provides a stark picture of differential recovery by neighborhood. Within Eastern New Orleans, for example, seemingly similar neighborhoods have experienced vastly different trajectories of recovery. The Vietnamese-American community centered in the Southeast corner of Eastern New Orleans had similar levels and ranges of pre-Katrina socioeconomic status as the African-American neighborhoods that surround it, and suffered similar levels of storm and flood-related damage. And yet on the major indicators of postdisaster recovery, the Vietnamese-American community is faring better than those that surround it.

Why?

THE RESILIENCE AND RECOVERY FRAMEWORKS
RESILIENCE

After a major disaster, why do some communities recover more quickly and completely than others? Obviously, communities that experience more impact from a catastrophe would be expected to take longer to recover. But what about communities that are affected to the same or to a similar degree? Why do some of these communities fare better than others? The major paradigm for explaining differences in postdisaster recovery among similarly affected communities is based on a concept called resilience.

Resilience is conceived by major disaster scholars to represent a set of attributes that enable recovery from disaster-related setbacks; it is the flip side of vulnerability (Cutter & Emrich, 2006; Cutter, Boruff, & Shirley, 2003). Fran Norris defines disaster-related community resilience as the "linking [of] a set of adaptive capacities to a positive trajectory of functioning and adaptation after a disturbance" (Norris, Stevens, Pfefferbaum, Wyche, & Pfefferbaum, 2008). The adaptive capacities they have in mind are economic development, social capital, information and communication, and community competence.

Strong *economic development* facilitates more resilience to potential disaster-related setbacks by providing adequate levels and distribution of wealth and jobs to the affected community. Dense and helpful *social capital* facilitates more resilience directly through connections with others who can be helpful in the rebuilding process and indirectly by providing links to other sources of help. Similarly, well-developed *information and communication* flows ease access to resources that can help. *Community competence* is less well defined, but refers to more general community-level features that facilitate collective efficacy. These four dimensions of resilience are more effective when they are robust, redundant, and can be deployed rapidly (Bruneau et al. 2003; Norris et al., 2008).

RECOVERY

Like resilience, the formal treatment of recovery in the disaster literature accords with intuition: recovery refers to the recapturing of a quality of life comparable to what affected individuals, families, and communities had before the disaster occurred. David Abramson and his colleagues propose five dimensions of postdisaster recovery: housing stability, economic stability, physical health, mental health, and social role adaptation. *Housing stability* helps ensure that members of the household—children in particular—regain some constancy in their day to day lives. *Economic stability* provides a steady income to the family, and reduces the need to be constantly on the lookout for income generating activities. *Health*, both physical and mental, underlies all of the other dimensions: without it, the ability to secure and maintain a stable home and job becomes all but impossible. Health is also an omnibus proxy measure for general well-being. *Social role adaptation* is a bit vague, but is intended to capture the degree of postdisaster engagement with other community members (Abramson et al., 2010).

RESILIENCE AND RECOVERY

The presumed links between resilience and recovery are direct and straightforward: better resilience leads to better recovery (Fig. 11.1). Again, this simple conceptualization corresponds well with our intuition: communities that had well-developed adaptive capacities in economic development, social capital, information and communication, and community competence should fare better in a postdisaster context than should communities that were less endowed with these capacities. That is, the more resilient communities should fare better on postdisaster housing stability, economic stability, physical health, mental health, and social role adaptation than should more vulnerable (less resilient) communities.

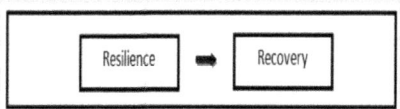

FIGURE 11.1

Resilience and Recovery

© Russell Sage Foundation. Reprinted with permission.

APPLICATION OF CURRENT FRAMEWORKS TO THE VIETNAMESE-AMERICAN COMMUNITY IN POST-KATRINA NEW ORLEANS

RECOVERY

On each of the standard measures of postdisaster recovery described above, the Vietnamese community is doing remarkably well. Regarding *housing stability*, using data that are representative for the city of New Orleans at the time of Katrina and data from a population register of the entire Vietnamese-American community in New Orleans taken just a few weeks before the hurricane, we find that the pace of return for Vietnamese Americans to New Orleans outpaced that of either Blacks or Whites (VanLandingham, 2017). Regarding *economic stability*, two years after Katrina, Vietnamese Americans had only about a third of the unemployment suffered by their non-Vietnamese neighbors living in communities with similar socioeconomic status that had suffered similar levels of flooding (VanLandingham, 2017). Regarding *physical and mental health*, standard health assessments based on the SF-36 show that while both dimensions plummeted for the Vietnamese (and everyone else) during the first year after Katrina, by the second year levels of physical and mental health had bounced back to their pre-Katrina levels for Vietnamese Americans (Vu & VanLandingham, 2012). Also, Vietnamese Americans showed much lower levels of Post-traumatic stress disorder (PTSD) than did other groups experiencing Katrina (Norris, VanLandingham, & Vu, 2009). Regarding *social role adaptation*, post-Katrina community mobilization within the Vietnamese-American community has been nothing short of remarkable (VanLandingham, 2017).

Was this remarkable recovery due to the fact that the Vietnamese community was more resilient?

RESILIENCE

Like many communities, the Vietnamese community in New Orleans exhibits features of both resilience and vulnerability. *Economic development* is moderate, both within the principal enclave in Eastern New Orleans and in the smaller enclaves elsewhere. Compared with immigrants from other countries in Asia, Vietnamese Americans do not fare as well on economic and occupational outcomes (Lee & Zhou, 2014). Compared with Whites and Blacks born in America, Vietnamese immigrants fare worse on measures of health insurance (Smedley, Stith, & Nelson, 2002). *Social capital* is dense and helpful within their ethnic group (Fu & VanLandingham, 2010), but outside of it, many Vietnamese Americans are quite isolated from social networks with members of other segments of American society. Similarly, *information and communication* flows well within the Vietnamese community but less well beyond the ethnic borders.

Community competence is difficult to assess, given the vagueness of the term. The Catholic Church provided the training ground for a few leaders who would emerge in Katrina's aftermath, but there is little evidence to suggest that community competence was significantly more developed in the Vietnamese community than in other communities prior to Katrina.

In sum, based on conventional views of community resilience, there really weren't any strong grounds for predicting that the Vietnamese community would recover quickly and well; and yet they did.

Might something be missing from these conventional views of resilience and recovery?

THE MISSING PIECE: CULTURE

Culture has been omitted from frameworks explaining postdisaster recovery—and from more general frameworks explaining a wide array of other outcomes—for good reason. Culture is both difficult to operationalize and politically fraught.

Many researchers still avoid discussions of culture due to politically charged debates in the 1960s and 1970s that invoked culture as a mechanism for perpetuating poverty across generations. Many scholars became discouraged from studying the connections between culture and outcomes because they did not wish to be seen as blaming the unfortunate for their own problems. While culture has experienced an academic resurgence of late, this legacy still discourages many social scientists—including disaster researchers—from engaging in a reckoning with cultural influences.

This academic squeamishness is reinforced by echos of these earlier debates in widely read takes on how culture affects success in the popular press, such as recent books by Amy Chua that lionize the success of some immigrant groups while implicitly disparaging others (Chua & Rubenfeld, 2014; Chua, 2011) without accounting for confounding influences of class and other sources of privilege. I think of these omitted variables as Culture Confounders (see Fig. 11.2). The omission of these Culture Confounders can make it appear as if culture is solely influencing outcomes (postdisaster recovery in this case), when in fact a host of other factors that cooccur with cultural factors are driving much of the differences across groups. Three examples of these Culture Confounders are specific patterns of migration characteristic among some groups that favor families with extensive economic and human capital ("selection"), advantages associated with living in closely knit ethnic enclaves ("barrio effects"), and social expectations of success among members of the broader society at destination ("stereotype promise") (VanLandingham, 2017).

But as implied in Fig. 11.2, I think culture—properly defined and properly delimited—can help explain differentials in postdisaster recovery, too. Most

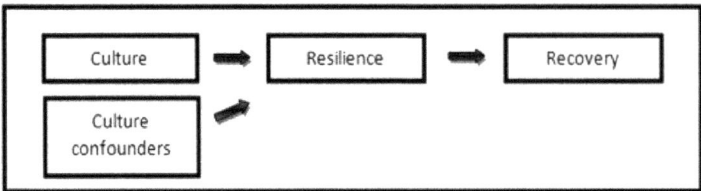

FIGURE 11.2

Culture and Its Confounders

© Russell Sage Foundation. Reprinted with permission.

recent definitions of culture contain the following elements: *the specific systems of beliefs, values, and meaning that members of a particular community use to weigh and consider their social world* (Alexander, Thompson, & Edles, 2012; Beldo, 2010). Survey researchers have focused on the beliefs and values part of the definition, because these features are easier to operationalize than is the "meaning" part. Unfortunately, research relying on beliefs and values as leverage for investigating the role of culture on a wide array of outcomes have not seen much success (Kim, 2002; Steinberg, Brown, & Dornbusch, 1996). Qualitative research focusing on the "meaning" part of the definition has been more fruitful, focusing on shared *narratives* about the cultural group's history; which in turn can lead to shared perspectives or *frames* through which one views the world and a shared sense of *symbolic boundaries* regarding who is included and excluded within the cultural group (Panter-Brick, 2015; Small, Harding, & Lamont, 2010; Clair, Daniel, & Lamont, 2016).

APPLICATION OF AN EXPANDED FRAMEWORK TO THE VIETNAMESE-AMERICAN COMMUNITY IN POST-KATRINA NEW ORLEANS

The rapid and robust recovery of the Vietnamese-American community after Hurricane Katrina was driven in part by features that appear to be cultural but in fact have little to do with it (i.e., by Culture Confounders). But culture, properly defined and delimited, played an important role, too.

CULTURE CONFOUNDERS

One set of pre-Katrina advantages that the Vietnamese community had over other communities that have done less well—and that has little to do with culture—is related to what demographers and other social scientists call "selection," which in this case refers to the selective migration to the United States of some of the best and brightest in South Vietnamese society. In the initial exodus, many high- and middle-level government and military officials and their families were evacuated to the United States and settled in places like New Orleans (Do, 2002; Lipman, 2014; Routledge, 1992; Stone & McGowan, 1980; Strand & Jones, 1985). This concentration of talent and experience helped in the establishment of ethnic communities that were well-functioning by the time refugees from subsequent waves arrived.

A second set of advantages that has little to do with culture is the benefits of living in closely knit ethnic enclaves. Migration scholars refer to these effects as "barrio effects" (Jargowsky, 2009). Earlier research on this same Vietnamese community in Eastern New Orleans by Min Zhou and Carl Bankston found that being surrounded by other Vietnamese immigrants helped to facilitate information flows within the community about perils and opportunities with regard to employment, education, and other topics (Zhou & Bankston, 1998).

A third set of advantages that has little if anything to do with culture is related to what social scientists call "stereotype promise." Vietnamese enjoy a positive reputation that has been nourished and reinforced by positive portrayals by official spokespersons and in the media since their arrival in 1975 (Dao, 2015; Lipman, 2014). These positive impressions contributed to a remarkable outpouring of post-Katrina support for the community by foundations, government, and businesses (VanLandingham, 2017).

CULTURAL INFLUENCES ON POST-KATRINA RECOVERY

Analysis of various types of qualitative data has also discerned a number of avenues through which cultural elements have facilitated community resilience and postdisaster recovery for the Vietnamese-American community. I think of these elements as "Cultural Abutments." One example is a shared *narrative of survival*. Central elements of this narrative include the exodus of many of the community's elders from North Vietnam to South Vietnam in 1954 to escape anticipated persecution from the new communist government; the subsequent exodus from South Vietnam to the United States after 1975, as their government and society were collapsing; and the subsequent reestablishment of a vigorous and successful community at these new destinations. This narrative of survival helped to put the disruption caused by Hurricane Katrina into perspective and also engendered a kind of collective self-confidence that the community would overcome these new challenges (VanLandingham, 2017).

A second example is a shared *frame of hierarchy*. Traditions of Confucianism and Catholicism reinforce a comfort with hierarchical social relations among Vietnamese Americans. Unlike many other Americans who are more egalitarian in perspective, the Vietnamese were well positioned to both produce and follow a chain of command after Katrina, which gave them some clear advantages navigating the chaos that ensued (VanLandingham, 2017).

CONCLUSIONS

The especially strong recovery of the Vietnamese-American community after Hurricane Katrina was facilitated by resilient features of their community which were in turn buttressed by several sources of community cohesion that are best described as culture. Culture is neglected in most current conceptualizations of predisaster resilience and postdisaster recovery because of challenges of operationalization and a legacy of political baggage. This is unfortunate, because an incorporation of cultural influences can help to explain why some communities fare better than others after a disaster, despite suffering similar levels of disaster-related damage and disruption and enjoying similar levels of predisaster affluence.

For the Vietnamese-American community, elements of culture that facilitated their recovery include shared narratives, such as one that focuses on a history of surviving previous crises. Their strong recovery was also facilitated by shared perspective or frames, such as one that casts social relations as hierarchical. Invoking culture as a panacea is misguided; other advantages were important, too. Noncultural features that facilitated their strong recovery and that are often confused with culture (culture confounders) include an initial wave of immigrants who were especially gifted and talented (selection), some natural advantages that come with being surrounded by coethnics (barrio effects), and some predispositions among the non-Vietnamese that made the Vietnamese seem especially worthy of assistance (stereotype promise).

Understanding what facilitates postdisaster recovery—and clarifying this distinction between influences that are cultural and influences that are not—is important for several reasons. First, such a clear understanding will help to avoid policy recommendations that are misguided and counter-productive. For example, one ill-advised policy recommendation that can arise from a misunderstanding of these mechanisms would advocate for those who are not Vietnamese to become more like them, as if a set of cultural attributes can be adopted at will. Second, a better understanding of what facilitated the recovery of the Vietnamese-American community will help us to better understand why some communities thrive—and others falter—after future disasters.

REFERENCES

Abramson, D. M., Stehling-Ariza, T., Park, Y. S., Walsh, L., & Culp, D. (2010). Measuring individual disaster recovery: a socioecological framework. *Disaster Medicine and Public Health Preparedness*, 4(Suppl. 1), 46–54.

Alexander, J. C., Thompson, K., & Edles, L. D. (2012). *A contemporary introduction to sociology culture and society in transition*. Boulder, CO: Paradigm Publishers.

Beldo, L. (2010). In H. J. Birx (Ed.), *Concept of culture. 21st century anthropology* (1, pp. 144–152). Thousand Oaks, CA: SAGE.

Bruneau, M. S., Chang, R., Eguchi, G., Lee, T., O'Rourke, A., Reinhorn, M., Shinozuka, K., Tierney, W. A., Wallace., & von Winterfeldt, D. (2003). A framework to quantitatively assess and enhance the seismic resilience of communities. *Earthquake Spectra, 19*, 733–752.

Chua, A. (2011). *Battle hymn of the tiger mother*. New York: Penguin Press.

Chua, A., & Rubenfeld, J. (2014). *The triple package*. New York: Penguin.

Clair, M., Daniel, C., & Lamont, M. (2016). Destigmatization and health: Cultural constructions and the long-term reduction of stigma. *Social Science & Medicine, 165*, 223–232.

Cutter, S. L., Boruff, B. J., & Shirley, W. L. (2003). Social vulnerability to environmental hazards. *Social Science Quarterly, 84*(2), 242–261.

Cutter, S. L., & Emrich, C. T. (2006). Moral hazard, social catastrophe: The changing face of vulnerability along the hurricane coasts. *Annals of the American Academy of Political and Social Science, 604*, 102–112.

Dao, V. (2015). *From the ground up: A qualitative analysis of gulf coast Vietnamese community-based organizations and community (Re)building in post-disaster Louisiana, Mississippi, and Alabama Ph.D*. Tulane University.

Do, P. V. (2002). Between two cultures: Struggles of Vietnamese American adolescents. *The Review of Vietnamese Studies, 2*(1), 1–18.

Fu, H., & VanLandingham, M. (2010). Mental and physical health consequences of repatriation for Vietnamese returnees. *Journal of Refugee Studies, 23*(2), 60–182.

Jargowsky, P. A. (2009). Immigrants and neighbourhoods of concentrated poverty: Assimilation or stagnation? *Journal of Ethnic and Migration Studies, 35*(7), 1129–1151.

Kim, R. Y. (2002). Ethnic differences in academic achievement between Vietnamese and Cambodian children: Cultural and structural explanations. *Sociological Quarterly, 43*(2), 213–235.

Lee, J., & Zhou, M. (2014). The success frame and achievement paradox: The costs and consequences for Asian Americans. *Race and Social Problems, 6*(1), 38–55.

Lipman, J. K. (2014). A refugee camp in America: Fort Chaffee and Vietnamese and Cuban refugees, 1975–1982. *Journal of American Ethnic History, Volume 33*(Number 2), 57–87.

Norris, F., VanLandingham, M., & Vu, L. (2009). PTSD in Vietnamese Americans following Hurricane Katrina: Prevalence, patterns, and predictors. *Journal of Traumatic Stress, 22*(2), 91–101.

Norris, F. H., Stevens, S. P., Pfefferbaum, B., Wyche, K. F., & Pfefferbaum, R. L. (2008). Community resilience as a metaphor, theory, set of capacities, and strategy for disaster readiness. *American Journal of Community Psychology, 41*, 127–150.

Panter-Brick, C. (2015). In L. C. Theron, et al. (Eds.), *Chapter 17: Culture and resilience: Next steps for theory and practice. Youth resilience and culture, cross-cultural advancements in positive psychology* (11, pp. 233–243). Dordrecht: Springer.

Routledge, P. J. (1992). *The Vietnamese experience in America*. Bloomington: Indiana University Press.

Sastry, N., & VanLandingham, M. (2009). One year later: Mental illness prevalence and disparities among New Orleans residents displaced by Hurricane Katrina. *American Journal of Public Health, 99*(Suppl 3), S725–S731.

Small, M. L., Harding, D. J., & Lamont, M. (2010). Introduction: Reconsidering culture and poverty. *Annals of the American Academy of Political and Social Science 629*, 6−27. (ArticleType: research-article/Issue Title: Reconsidering Culture and Poverty/Full publication date: May 2010/Copyright © 2010 American Academy of Political and Social Science).

Smedley, B., Stith, A., & Nelson, A. E. (2002). *Unequal treatment: Confronting racial and ethnic disparities in health care*. Washington, DC: Institute of Medicine Committee on Understanding and Eliminating Racial and Ethnic Disparities in Health Care.

Steinberg, L., Brown, B., & Dornbusch, S. (1996). *Beyond the classroom: Why school reform has failed and what parents need to do*. New York: Simon & Schuster.

Stone, S. C. S., & McGowan, J. E. (1980). *Wrapped in the wind's shawl: Refugees in Southeast Asia and the Western World*. San Rafael, CA: Presidio Press.

Strand, P. J., & Jones, W. J. (1985). *Indochinese refugees in America: Problems of adaptation and assimilation*. Durham, NC: Duke University Press.

VanLandingham, M. (2017). *Weathering Katrina: Culture and recovery among Vietnamese-Americans*. New York: Russell Sage Foundation.

Vu, L., & VanLandingham, M. (2012). Physical and mental health consequences of Katrina on Vietnamese immigrants in New Orleans: A pre and post-disaster assessment. *Journal of Immigrant and Minority Health, 14*(3), 386−394.

Zhou, M., & Bankston, L. (1998). *Growing up American: How Vietnamese children adapt to life in the United States*. New York, NY: Russell Sage Foundation.

CHAPTER 12

Faith-based organizations in Katrina: The United Methodist Church

Sarah Kreutziger[1], Ellen Blue[2] and Michael J. Zakour[3]

[1]Tulane University, New Orleans, LA, United States [2]Phillips Theological Seminary, Tulsa, OK, United States [3]West Virginia University, Morgantown, WV, United States

CHAPTER OUTLINE

Theoretical and Conceptual Foundations	268
The United Methodist Church in Katrina	269
Katrina-Related Outreach and Ministries	271
Cursillo	271
The United Methodist Church and Transformative Resilience	271
The Decision Process of Mergers and Closures	273
Outreach Ministry at First Grace	274
Outreach at Hagar's House	274
Outreach Through Luke's House	275
Mt. Zion UMC Outreach	276
United Methodist Women	276
Effects on Individual Katrina Survivors	277
A Recovery Imbued With Spirituality	277
Individuals and Resilient Recovery	278
Conclusion	280
References	280

During Katrina, I (Kreutziger) was forced to evacuate with my family, staying for almost a month at the Methodist Conference Center at Woodworth, Louisiana. My husband returned to New Orleans early to resume work at the hospital that had remained open throughout the disaster. This hospital had little air conditioning in the near-100°F heat for almost a month, yet many patients could not be discharged because they could not return home to damaged or destroyed residences. When I was finally able to return to my family home, our neighborhood was one of the few reoccupied areas near the devastated city. I and the other congregants of the Munholland United Methodist Church (UMC) needed to decide how they could best help other UMCs in New Orleans, and how they could respond to the needs of Katrina survivors. In this way, we became barefoot scholars learning about a resilient recovery.

Over a decade after Katrina, massive flooding again occurred in Southeast Louisiana. Such is life on the Gulf Coast after Hurricanes Katrina and Rita, Hurricane Gustav and Harvey, and the British Petroleum Oil Spill. As one of the Ninth Ward men interviewed by a local television station (WWL, February 7, 2017) said, "when the storms come, we clean up and keep going." Neighbors, added the reporter, were already helping each other pick up debris despite damage to their own properties. Just like the "Cajun Navy" that set out in pirogues and small vessels to pick up stranded victims from the massive flooding in Southeast Louisiana, post-Katrina residents take calamity in stride: "We keep calm and carry on."

This chapter examines the role of faith-based organizations in fostering resilience during the Katrina recovery and reconstruction periods. Other disasters have shown that volunteerism has the potential to aid in a resilient recovery and in community revitalization. Faith-based organizations are recognized by many Katrina survivors as highly effective in helping people recover and rebuild (Adams, 2013). Through the provision of volunteers, faith-based communities were leaders in a disaster-related utopian movement after Katrina. The United Methodist Church (UMC) provided thousands of volunteers, many of whom returned several times to aid in the recovery process. After the immediate response and recovery periods, the church also enacted outreach services to continue to promote resilience. A case study of the recovery services of the UMC in New Orleans after Katrina and Rita is presented to illustrate the role of faith-based organizations in fostering church, community, and survivor resilience. Stories of Katrina survivors aided by the United Methodist Committee on Relief (UMCOR) are recounted.

Over 2 million volunteers came to Southeast Louisiana to help with the recovery after Katrina, and the most prevalent were church-sponsored and faith-based programs. These included Catholic Charities and the Episcopal Diocese, the Baptists, Evangelicals, Mennonites, and the UMC. Other programs were based in Jewish synagogues, Buddhist centers, and Muslim congregations. Of the top ten private charities in terms of money raised, six were faith-based. These were the Salvation Army, Catholic Charities USA, the United Methodist Committee on Relief (UMCOR), International Aid, Feed the Children, and Habitat for Humanity (Adams, 2013).

THEORETICAL AND CONCEPTUAL FOUNDATIONS

The story of Katrina is the story of people helping each other to rebuild from scratch. It is the story of how volunteers came from all over the world to help in the recovery. The altruistic community that emerged after Katrina consisted of fellow survivors providing mutual aid, often before formal governmental response and recovery organizations were active (Barton, 2005). Survivors banded together to help one another, so that those most affected by the disaster were often the ones who offered their help first. The informal altruistic community tends to fade before the needs of all survivors are met, particularly poor and elderly survivors (Kaniasty & Norris, 2009). Often formal organizations will need to provide relief aid that neither the informal altruistic community nor official disaster relief organizations can offer. These formal organizations in disasters like Katrina prominently feature faith-based organizations. The response of faith-based, charitable, and nongovernmental grass-roots organizations was overwhelming in post-Katrina New Orleans (Adams, 2013). Of these organizations, the churches played the largest role in reconstruction.

Similar to the altruistic community, a series of utopian communities arose in the aftermath of Katrina, with both secular and faith-based auspices. Utopian communities are characterized by solidarity, generosity, and altruism (Solnit, 2009). From increased solidarity, civil society actions and collective efficacy empowerment can emerge. Disaster utopias can have an overtly religious nature. Faith-based organizations are typically able to mobilize large numbers of volunteers over an extended period of time. The utopian communities were often imbued with spiritual meaning and world-views. Volunteers view their aid with spiritual meaning, seeing their help as doing the work of the Supreme Being. The survivors who are aided recognize that government has not provided adequate aid for recovery, and they view the help and their recovery as a gift from their God.

Religious inspiration and revitalization movements are one aspect of the relief and recovery outreach of faith-based churches, as well as their efforts to reconstitute themselves after a severe disaster. These movements arise when one or more people envision a better world after disaster, and work as leaders for social and cultural change to bring about this more meaningful world (Wallace, 1956, 1956/2003). Religious inspiration is a form of both individual- and community-level resilience that helps to advocate for social justice after a disaster. The revitalization process is a vision of a new way of life by individuals and groups. Because so much has been destroyed by a disaster, it is often not possible to seek comfort in returning exactly to the way things were predisaster. Instead, a new world that is worth working for is envisioned that includes elements of the past combined with new realities.

THE UNITED METHODIST CHURCH IN KATRINA

One way that the UMC has combined an established mission with new realities is in the building of the disaster preparedness mission of the Louisiana UMC. Because of the lessons of Katrina, over the last 12 years the Louisiana UMC has taken a lead role in the denomination's long-standing mission of preparedness. Semiannually, opportunities open for training in Early Response Teams (ERT) whose purpose is to

- provide immediate assistance to those in greatest need,
- assure victims of the care and concern of the church, and
- give visibility to the United Methodist presence (Tate, 2010).

The ERT are trained by UMCOR, a division of the General Board of Global Ministries and the American Red Cross as a first line of volunteers who move quickly into crisis situations. They work under the direction of the Disaster Response Coordinators (DRCs) at every level of the church's hierarchical structure. Each local church has a DRC who answers to and communicates with the District Superintendent (an administrative clergyperson in each of six geographical areas of Louisiana) and the District DRC team in his/her district. The District DRCs, in turn, work with the Conference Executive Director and the Bishop of the Louisiana Conference. There are extensive guidelines for every level of response.

After initial rescue, evacuation, shelter, and other emergent resources are activated as needed, preparation for the long-term recovery begins. Long-term recovery has two major components: (1) rebuilding, repair, and new building; and (2) case management. During the long-term recovery process for Katrina, UMCOR set up four mission stations along the Gulf Coast, three in Louisiana and one in Mississippi. Each mission station had caseworkers, an estimator of damages, and a director. Federal Emergency Management Agency monies, directed through the Governor's office, paid salaries. Additionally, UMCOR created and directed Katrina Aid Today:

> A national case management consortium was funded through a $66 million grant from the Department of Homeland Security and Federal Emergency Management Agency (FEMA). That grant, funded by donations from foreign governments for post-Katrina rehabilitation, was the largest single government-managed grant ever given to the private sector for disaster response. At its peak, Katrina Aid Today had 138 offices in 34 states. Before closing on March 31, 2008, the program assisted more than 73,000 households with long-term recovery (Gilbert & Bloom, 2015).

The bulk of other resources came from other UMCs. From August 29, 2005, until the last grant was awarded by UMCOR in Mississippi in December 2013, hundreds of thousands of people were helped. In Louisiana alone, over 125,000 households were impacted. "United Methodists in that state multiplied about $35 million allocated by UMCOR with volunteer labor and other gifts, bringing the total to an estimated $130 million worth of in-kind donations" (Gilbert & Bloom, 2015). Local UMCs sent volunteers from all over the nation. Two years after the recovery began, 28,000 volunteers served with the church's Storm Recovery Center. By 2010, this Ministry has hosted 40,000 volunteers with over half returning one or more other times (Adams, 2013). Ginghamsburg UMC in Tipp City, Ohio, for example, sent over 85 teams. *Newsweek* magazine's survey ("Big Names in Katrina's Relief") ranked UMCOR's contribution sixth in the nation for the $70 million raised for Gulf Coast Recovery. Even the Council of Bishops raised two million dollars (Gilbert & Bloom, 2015).

The local UMC in New Orleans fostered the resilience of survivors by providing the resources and capabilities needed for recovery. As the Rev. Thomas

Hazelwood, then director of UMCOR, commented, as "local churches raised money, sent teams, and gathered supplies, every Sunday school class was talking about it, every worship service prayed about it [...] Because we were so focused as a church, I think it drove what UMCOR did." The Reverend Darryl Tate, former Conference Executive Director of Disaster Response, is clear about why the United Methodists engage in these endeavors:

> *The church has an obligation and a high calling to respond to the needs of human beings when they are facing some of the most challenging times in life. We know that in times of trouble people will turn to the church for both their physical and spiritual needs. While it is impossible to plan for specific situations which disasters may bring, we can prepare ourselves in a general way, so that when a disaster occurs we are able to act responsibly in the midst of chaos (Tate, 2010).*

KATRINA-RELATED OUTREACH AND MINISTRIES

After Katrina, a number of outreach ministries, described below, helped people and the community to recover more resiliently. Resilience here means to recover well and rapidly to pre-Katrina (or even higher) levels of biopsychosocial functioning (Zakour & Gillespie, 2013). It makes sense to include the spiritual aspect of wellness too, along with biopsychosocial functioning. For many Katrina survivors, recovery had an important spiritual component (Adams, 2013).

CURSILLO

One way that the UMC was revitalized after Katrina was through the Cursillo retreat experience. Cursillo, or *Cursillos de Cristiandad* (a short course in Christianity), is a laity-led, international faith-based movement that began in the Roman Catholic Church in Majorca, Spain, in 1944. This retreat experience aided in rebuilding a spiritual community during the Katrina recovery. Cursillo is a 3-day experience that looks at Christianity as a lifestyle from a New Testament perspective. Participants are encouraged to take the message into the community in "the Fourth Day" and gather regularly for accountability in reunion groups in various churches. Cursillistas, or those who have experienced the Weekend, are in leadership positions throughout the local New Orleans churches. Their network helped to rebuild the spiritual community in the disorienting days following Hurricanes Katrina and Rita.

THE UNITED METHODIST CHURCH AND TRANSFORMATIVE RESILIENCE

A resilient recovery can involve either restoring things to their former state before the disaster, or it can involve an opportunity to transform the community to function at higher and unprecedented levels compared to predisaster. The dilemma the UMC faced after Katrina was choosing between two competing goods: the restoration of the church with the comfort that could bring, and the opportunity for transformative change. In a devastated city where so many landmarks were gone that it was difficult for life-long residents to find their way around, returning New Orleanians desperately wanted to find something that had not changed. Many believed that something should be their church. Seeking Sanctuary, in the medieval sense of "safe haven," they thought the larger church should help them put things back exactly as they had been, even if there were only a few members left to be comforted by the continuity.

Disaster can provide a context to transform existing communities. Other United Methodists, particularly those in authority, saw the circumstances as an opportunity to live out the demands of the Gospel in ways that congregations had formerly insisted their aging buildings prevented them from doing. The chance and the imperative to reach out to their neighborhoods and their community—to change their way of being a church—were unprecedented.

Changing and not changing both had appeal, and doing either could seem like a ministry. Some of the churches that put their energy into simply restoring what had gone on before had some success at that, but over a decade after Katrina, none is truly thriving. The more interesting and ultimately more successful congregations came to understand that they would have to reach out in new ways. Simply adapting to new circumstances would not be enough; they would have to transform their missions and change their identities in fundamental ways.

Those changes have involved addressing problems that even before the storm appeared unresolvable. The scope of difficulties in an urban area like New Orleans was immense. New Orleans is a port city built on the slave trade with a large African-American population still feeling the effects of educational deprivation during slavery and Jim Crow segregation. Sin has consequences, and structural sin of this magnitude has consequences that linger for generations. In Katrina, these structural inequalities led to greater losses for African Americans and the poor, and to high levels of emotional distress (Adams, 2013).

Even in less complex locales, it is easy to think, "I'm only one person. What can I possibly accomplish?" or "We're just one congregation. How can we make any real difference?" One of the few positive effects of Katrina was destroying the belief that small efforts were not worth making. Overwhelming destruction and devastation made it apparent to almost everyone that it would require the small efforts of hundreds of thousands of individuals and groups joined together in order for the city to recover. Therefore, "joining together" was on the table in ways that it had not been before. This manifested both in congregational mergers, that would probably never have happened without the storm, and in the cooperation of stand-alone congregations to create and sustain new ministries and nonprofit agencies.

THE DECISION PROCESS OF MERGERS AND CLOSURES

The decision process for mergers and closures was important for fostering the resilience of the United Methodist Church's (UMCs) recovery. One story that combines a congregational merger and the creation of new nonprofits is that of First Grace UMC, a new multiracial congregation. Although First UMC and Grace UMC were located only about a mile apart, they had very little in common. It was not just a matter of race, though. First UMC had a White membership and Grace's members were African-American. Grace had flood insurance and a number of members who returned to New Orleans, but no one who came back had the wherewithal to see to an early repair of their heavily damaged structure. First UMC, on the other hand, had dwindled to only a couple of dozen members before the storm, but one of those who returned was able to secure a contractor and limit further damage very quickly.

Resilience is a major piece of Grace's ability to say yes to a merger that required the members to leave their own structure. It was already a merger of former congregations that had been forced to relocate—one when the creation of Storyville, a district where prostitution was legal, led the city to refuse admission of children to a church building located within the district, with police quite literally barring the way to the church one Sunday morning (Long, 2004).

The story of how both churches slowly came to decide that a merger would allow them to do more good than they could ever do apart and of how they even more slowly learned to thrive as one worshiping congregation is related in *In Case of Katrina* (Blue, 2016). In contrast to the very hierarchical governance of the Roman Catholic Church, the leaders of the UMC were democratic and participative. Grace's members were able to reconcile themselves to giving up their building and making a new home for themselves in a structure where they would have likely been unwelcome to join the year before. The primary enabling factor for this move was the process that the Louisiana conference leadership decided to use as they pondered how the churches in New Orleans should recover after the flood. Perhaps this process grew out of a disaster utopia of flexibility, egalitarianism, and flattened hierarchies (Solnit, 2009).

That process can be posed over against that of the Roman Catholic Church, whose hierarchy made the decisions about closures and mergers and found astonishing levels of resistance among its members as a result. Leaders of the Catholic Church were authoritarian in their approach; Catholic parishioners were finally arrested and taken to jail for occupying church buildings that the archdiocese decided to close without the input or agreement of the members.

The UMC had a nonhierarchical and feminist approach to church mergers and closures. Decision-making was not based on a vertical and steep chain of command. Then Bishop William Hutchinson, and Provost Don Cottrell of the Louisiana Conference, decided that insofar as it was possible, decisions about such matters would come from the ground up, made by parishioners who had returned and the pastoral teams appointed to seven groups of churches within the city. In an innately hierarchical institution such as the UMC, it is never possible to say that decisions were completely the result of collaboration. However, to a very large extent, it was the members of First and Grace UMCs who chose to move forward as a merged congregation.

They found the resilience required to make that journey largely because they were allowed to look for it and depend on it as they determined their futures; they were able to choose that future for themselves. Because the parishioners engaged in citizen participation and collective action to help make the decisions, their agency was strengthened, and this fostered resilience.

OUTREACH MINISTRY AT FIRST GRACE

For this consideration of resilience, the work of a group of young people at First Grace and the agencies they created are particularly appropriate topics. The Rev. Shawn Anglim and his wife, Dr. Anne Daniell, lived in Baton Rouge where Anglim was the UMC's campus minister at Louisiana State University (LSU). Both loved New Orleans and wanted to live there. Rev. Anglim had written a letter to the bishop expressing his hopes for the First UMC location as a ministry site, and after the storm, the bishop pulled the letter out and appointed Anglim to a pastoral team that served that neighborhood.

Several young people who had been involved in the campus ministry followed the pair to the city, and they and other young people who wanted to be part of the city's recovery began living in an intentional community in the church building, even before basic utilities had been restored. Some received money from AmeriCorps, but this government funding meant that they could spend only part of their time in church-related work. They were to learn the needs of their neighborhood and figure out how to address them. Soon, they developed the strategy of riding their bikes up and down the streets and getting to know the people who were resettling there. As schools began to reopen, they painted classrooms and tutored children. In this way, these young people became barefoot scholars and practitioners.

Early on, it seemed that the space under every bridge and overpass in the city was filled either with flooded-out cars or homeless people living in tents or simply with bedrolls. Everyone who returned needed almost everything, making it difficult to isolate which needs they should focus on, but the group decided to try to help with the issue of homelessness. Specifically, they chose to open a home for women and children to help them transition toward permanent housing.

OUTREACH AT HAGAR'S HOUSE

Angela Davis, who had been a seminary student at Princeton when Katrina made landfall, was one of those instrumental in purchasing a house and repurposing it for the project they call "Hagar's House." Hagar's House opened in late November 2007. Named for the servant of Sarah and the mother of Abraham's first child who was cast out into the wilderness with her child when Sarah finally bore a son of her own (Genesis 21), the facility serves women and children who have been subjected to domestic abuse. Its website calls it "a beautiful, clean, safe place to rest and call home." Residents form an "intentional community that is actively engaged in undoing the root causes of poverty." The holistic programming focuses "on physical health, emotional/spiritual health and social justice," healthy food, and a space for planting in the community garden.

As a part of the strategy to help residents move toward housing of their own, Hagar's House helps them get jobs that will allow them to pay rent and child care, and insists that the women participate in a program that lets them save 70% of their income. The intention is that they will have accumulated $3000 by the time they move. (see "First Grace Community Alliance," available at http://www.firstgracecommunityalliance.org/.)

Disaster recovery on the scale that followed Katrina requires enormous numbers of construction workers. This is one of the primary reasons that the percentage of Spanish-speaking residents in New Orleans increased approximately ten fold, as workers came from countries such as Mexico and Honduras. There was hope that the conference would increase its commitment to Hispanic ministry in the city, but not enough happened in that regard despite ongoing efforts.

Angela Davis, one of the people involved in creating ministries after the storm, went to law school at Loyola University and passed the bar. Her work at First Grace now includes running Project Ishmael, "a pro bono immigration legal resource for children in New Orleans." Ishmael was the son of Hagar who was cast out by Abraham with his mother after the birth of Isaac.

OUTREACH THROUGH LUKE'S HOUSE

Before Katrina, Rayne UMC had been in ministry with people who were homeless. The ministry was held each Tuesday, in the circle driveway in front of their church that fronts onto St. Charles Avenue, a street well served by public transportation. The ministry served sandwiches, cookies, and cold drinks to between 75 and 100 people. They also provided vouchers for a local shelter that charges $8 a night, and gave out other needed items. After the distribution, they held a chapel service for anyone who wanted to stay. They employed a seminary graduate part-time for this ministry.

Located on the high ground—the sliver by the river—Rayne did not flood. However, high winds knocked the large steeple off the roof and onto the circle drive and front yard. They were not able to get all the repairs completed and use their sanctuary again for 3 years.

The ministry with homeless individuals was suspended. The damage to the front meant that the program could not continue there. Both sides of the church are immediately up against the adjoining streets, and they could not move the project to the back because Rayne's early childhood program was able to reopen quickly and the back of the church is where children are dropped off and collected. Furthermore, the congregation adapted the church to house volunteers who came in to help rebuild. Yet, the needs of the people they had already been serving, as well as the increasing numbers of homeless individuals in the city, continued to trouble Pastor Callie Winn Crawford.

MT. ZION UMC OUTREACH

Mt. Zion UMC is located on Louisiana Avenue, barely over a mile away from Rayne. Its membership is primarily African American, as Rayne's is primarily White. While Crawford returned to the city early and provided a stabilizing force at Rayne, Mt. Zion's pastor chose not to return to Louisiana.

Long before Katrina, the sanctuary was in need of significant repairs, and Mt. Zion was not able to make them happen for years after the flood. In the meantime, they worshiped each Sunday in the fellowship hall. Charity Hospital, a huge medical facility that had served New Orleanians who lacked financial resources since 1939, did not reopen after the storm. The city's entire health care system was sketchy and fragile.

The dream of providing free health care to people in need took hold at Rayne and Mt. Zion. A health clinic was jointly established by the two congregations to provide for unmet health care needs in the community. The clinic took place on Tuesdays of each week, and volunteer doctors, nurses, and eventually mental health professionals provided care for people who had no access to health care after Katrina. This new ministry at Luke's House helped compensate for the loss of health care professionals after Katrina, particularly for the severe shortage of mental health professionals.

UNITED METHODIST WOMEN

The United Methodist Women (UMW) provided resources to support volunteers, organizations, and other disaster recovery groups in the local community. Local UMCs are required by the UMC *Book of Discipline* to have a chapter of the UMW, although many do not. Women working outside of the home, community demands for volunteers from greatly decreased number of women available, and a change from the "do something" ethic that inspired earlier generations for social reform has been severely diminished by differing current values and priorities (Kreutziger, 1997). Nevertheless, in the aftermath of the storm, the women who could stepped up to help. Many joined the work teams, but the majority reappraised the call for "radical hospitality" (Schnase, 2007) and moved into traditional roles of responsibility. Women (and men) cooked for work teams, bought groceries and other supplies, cleaned up before and after the helping visitors left the church residences, entertained the volunteers, and scheduled the team visits.

At Munholland UMC in Metairie, LA, members supported Katrina volunteers by turning Sunday school classrooms into dormitories, installing a washer and dryer next to the education building, and providing one meal a day for those who stayed there. They stored snacks for the volunteers to take to work sites and provided emergency resources as needed. The pastor, Rev. Gene Finnell, often talked about the challenges of running a church continually, as the congregation did for 2 years.

The UMW raised funds for nonprofits engaged in disaster relief, and helped to provide the resources needed by the community for a resilient recovery. Each year, the UMW hosts a White Elephant Sale to raise money for missions. The year-long effort collects "white elephant" items (gently used) to sell at greatly reduced prices on the last Saturday of September. The women sell jams, jellies, soups and other baked goods, provide a hot lunch, sponsor a jewelry room and used book room, sell plants, and haul in donated furniture for resale. Whatever is not sold goes to another church as contribution to their subsequent Garage Sale. The event usually brings in between $17,000 and $18,000 that goes entirely to various nonprofit groups in Louisiana, India, and Haiti.

When Hurricane Katrina made landfall on August 29, 2005, most of the inventory for the White Elephant Sale had been collected and the majority of the preparation completed for that year. Rather than cancel the sale, UMW President Phyllis Zansler offered the donated items to the community for free. Many came and picked up toys, food, linen, kitchen supplies, electronic devices, and decorative pieces without cost. When the UMW finished, it had made six thousand dollars from paying customers and donations that went to local ministries.

United Methodist Men (UMM) stepped up as well during the Katrina recovery. The UMM performed outreach to help survivors in the local community. They became the community "husbands" by going into the neighborhoods to do carpentry and other minor repairs for widows and those without the means to hire others to do so at a time when workmen were in very short supply. They continue that mission today, over 12 years after Katrina.

EFFECTS ON INDIVIDUAL KATRINA SURVIVORS
A RECOVERY IMBUED WITH SPIRITUALITY

Many of the people served by the UMC and other faith-based charities reported that their recovery was imbued with a new spirituality and commitment to faith. Just as the volunteers saw their services as an act of faith, the people served developed a shared meaning of Katrina in spiritual terms. Phrases such as "before Katrina" and "after Katrina" came to define people's life narrative, and many people framed the recovery period as "after the flood" (Adams, 2013). The years of struggle toward recovery following Katrina were seen by many as a "test of faith" (Adams, 2013). A renewed commitment to faith came from the recovery experience, and many saw New Orleans and Southeast Louisiana as a center for spirituality after Katrina.

INDIVIDUALS AND RESILIENT RECOVERY

In the following section, we describe the effects of the UMC on individual survivors. The UMC's fostering of resilience is emphasized in these cases. At a recent installation luncheon for spouses of physicians employed in New Orleans, a widowed person in attendance stated "If it weren't for the Methodists, I wouldn't have survived Katrina" (J. Lothscheutz, personal communication, May 3, 2017). She went on to say that her Catholic church had not been there for them after her house flooded, but the Methodists were. A group from Michigan had cleaned out her house, helped her move, and had done major repair work later when they returned. She remains loyal to her Catholic faith, but expresses deep gratitude years later for the spiritual descendants of John Wesley, a man who saw the world as his parish.

Her statement echoed other stories from friends who, likewise, expressed appreciation for United Methodist outreach. A retired tax accountant before Katrina lived uptown in a house that was destroyed in Katrina. Mr. H is currently disabled because of two previous strokes and was unable to take care of himself, much less repair his home after the Storm. When he and his wife signed up for help, a team of Methodists from other states "mucked" out the house and discarded all their ruined furniture over several weeks while living in one of the UMC's buildings that housed the multiple work crews arriving daily.

The couple sold what was left of their home and moved to Florida with no intent of retuning to Louisiana. While there, the infrequent churchgoers decided to visit a UMC because of the kindness, they say, of their Louisiana rescuers. Since the Florida church emphasized similar community outreach, Mrs. H engaged in numerous missions to build, serve food, or reach out to others in need. Three years after the floods that followed Hurricane Katrina, they returned to New Orleans to be closer to Mr. H's daughter from a previous marriage. When they returned, Mr. H decided to attend a UMC. They joined Munholland UMC soon after visiting, passing the church they grew up in to get there.

Since their return, the couple are active in an ecumenical bible study, attend Sunday school and worship services regularly, and Mrs. H continues her work in community outreach despite having the additional responsibility of caring for a spouse who uses a wheelchair. Mr. H has been hospitalized several times with malignant cancer and other health issues, yet they remain strong in the face of every adversity and say openly that the prayers and support of the church community sustains them and allows them to persevere. In a personal email, Mrs. H recently described her husband as a "tuff 'ol bird with at least 22 lives." The daughter they moved back to be near, graduated first in her law school class this year.

More poignant is the story of a colleague who lost more than his home during Katrina. J and his young wife evacuated before Katrina to Florida to a family-owned condominium. During their stay, his wife developed severe stomach pains that forced her into the hospital several times, the last time literally as a refuge from Hurricane Wilma that hit Florida not long after New Orleans' devastation. J stayed near the hospital while tests revealed a deadly cancer diagnosis. He made the decision to return to New Orleans to live temporarily with a friend until he could find more permanent lodging while restoring their home. He and the friend's husband, a physician at Ochsner Medical Foundation—the only New Orleans hospital that stayed open during Katrina—spent eight hours on the phone arranging an angel flight back to the city for her admission to Ochsner's Main Campus. In order to be admitted, J's wife had to arrive by midnight or her room would be given to others in the very overcrowded medical establishment. (The hospital could not discharge many treated patients because they had no place to go.).

J's friends met the helicopter flight at almost midnight on Wednesday, November 16, and accompanied her by ambulance to Ochsner while J prepared to drive back to the city from Florida. When he arrived the afternoon of Friday, November 18, his friends, who are active United Methodists, arranged for a team from Missouri to meet him and assess the damage to his home. A nurse was part of the team to make sure that repair of the house would include provisions for an ill person. A retired United Methodist pastor was asked to go to the hospital to offer spiritual and moral support to the besieged couple. The pastor visited daily while J's wife remained in the hospital and followed her during subsequent admissions. Other members of the friends' church visited regularly as well.

In the meantime, the work teams helped clean the house until J returned to prepare for the short-term homecoming of his wife. After a four-day Thanksgiving break, his wife was readmitted to the hospital and died on Wednesday, December 14, 2005. The memorial service was held at a nearby UMC on the following Sunday, followed by a repast at the home of their friends. Had the UMC congregation not offered their church for this service, it is likely there would have been no memorial in the devastated city. J joined a Sunday school class, attended Cursillo and enjoyed the support of many friends until a job offer moved him to another State.

His personal experience gave new impetus to his academic research and a depth of understanding that few others have. Over the years since Hurricane Katrina, J has shared his experiences in his teaching and multiple publications and presentations. He has offered testimony in his church and counsel to others going through difficult times. He remains strong in his faith, and believes that his Christian world-view sustained him after his assumptive world was shattered by the Katrina disaster. He has experienced posttraumatic growth: a realignment of life-priorities and adoption of new life goals. From the disaster utopia during the Katrina recovery, J has experienced renewed purposefulness, compassion for others, belonging to a greater whole, and a sense of doing work that matters (Solnit, 2009).

Another survivor, who transported his critically ill mother by boat to the care of other relatives, only to see her die a couple of weeks later, came back to rebuild a church slated to close. He and his wife continue to be leaders in that church. Another person, despairing over losing his home twice, first to bad weather and then to a fire, admitted he, "dropped to my knees exhausted [saying] 'God, I know I'm a strong military guy but I'm done. I am out of hope" (Gilbert & Bloom, 2015). A United Methodist Committee on Relief (UMCOR) van drove up just then with 11 volunteers to pray with him, hang sheetrock, and subsequently lift his spirit.

One of the public service announcements playing repeatedly by a local television station (WWL, Fall of 2005) focused on a woman surrounded by destruction, but with a big smile on her face. She excitedly declared that when she looked around at her destroyed home, all she could say was, "oh God, I have nothing." Then after reflection, she realized, "I have God, so I have everything." Her statement summarizes it all for many of us who survived Katrina and other challenges as we regrouped and rebuilt. Religious faith provides sanctuary, a community to help us in difficult times, the challenge to offer compassion when it has been received, and a meaning system that gives eternal hope: perfect ingredients for building resilience despite great loss.

Similar stories abound. Rev. Tom Hazelwood, former head of UMCOR, "is still amazed that within the first year following Katrina, every US congregation seemed to have 'a laser focus' on responding. Local churches raised money, sent teams and gathered supplies. 'Because we were so focused as a church, it drove what UMCOR did'" (Gilbert & Bloom, 2015).

CONCLUSION

The long-term viability of the disaster utopia, and of transformative resilience, is an issue more than a dozen years after Katrina. The establishment of volunteerism, their organization and direction, is one way to make Katrina's disaster utopias last. Volunteerism is one important component of transformative resilience that translates into rapid recovery and preparedness for the next disaster. Volunteers from the UMC and other denominations are part of a latent disaster community present in the United States and elsewhere. Members of the disaster community in Katrina believed in civil society and the "beloved community" first spoken about by Martin Luther King Jr.

REFERENCES

Adams, V. (2013). *Markets of sorrow, labors of faith. New Orleans in the wake of Katrina.* Durham, NC: Duke University Press.

Barton, A. H. (2005). Disaster and collective stress. In R. W. Perry, & E. L. Quarantelli (Eds.), *What is a disaster? New answers to old questions* (pp. 125–152). New York: International Research Committee on Disasters.

Blue, E. (2016). *In case of Katrina: Reinventing church in post-Katrina New Orleans*. Eugene, OR: Cascade Books.

Gilbert, K.L., & Bloom, L. (2015, August 21). Mission reborn from Katrina's destruction. *United Methodist News Service (UMNS)*. Retrieved from ⟨http://www.umc.org/news⟩.

Kaniasty, K., & Norris, F. H. (2009). Distinctions that matter: Received social support, perceived social support, and social embeddedness after disasters. In Y. Neria, S. Galea, & F. H. Norris (Eds.), *Mental health and disasters* (pp. 175–200). Cambridge: Cambridge University Press.

Kreutziger, S. K. (1997). Wesley's legacy of social holiness. In R. E. Richey, D. M. Cambell, & W. B. Lawrence (Eds.), *Ecclesiology, mission, and identity* (Vol. 1, pp. 137–147). Nashville, TN: Abingdon Press.

Long, A. P. (2004). *The Great Southern Babylon: Sex, race, and respectability in New Orleans, 1865–1920* (pp. 136–138). Baton Rouge: Louisiana State University Press.

Schnase, R. C. (2007). *Five practices of fruitful congregations*. Nashville, TN: Abingdon Press.

Solnit, R. (2009). *A paradise built in hell. The extraordinary communities that arise in disaster*. New York: Penguin Books.

Tate, D.A. (2010, October 28). *Disaster response Louisiana annual comprehensive plan* (p. vii). Retrieved from www.La-UMC/DisasterResponse

Wallace, A. F. C. (1956). *Tornado in Worcester. An exploratory study of individual and community behavior in an extreme situation* (Committee on Disaster Studies, Publication 392). Washington, DC: National Academy of Sciences-National Research Council.

Wallace, A. F. C. (1956/2003). Mazeway resynthesis: A biocultural theory of religious inspiration. In R. S. Grumet (Ed.), *Revitalizations & mazeways: Essays on culture change* (Vol. 1, pp. 164–177). Lincoln, NE: University of Nebraska Press.

Zakour, M. J., & Gillespie, D. F. (2013). *Community disaster vulnerability: Theory, research, and practice*. New York: Springer.

CHAPTER 13

Collective efficacy, social capital and resilience: An inquiry into the relationship between social infrastructure and resilience after Hurricane Katrina

Paul Kadetz

Drew University, Madison, NJ, United States

CHAPTER OUTLINE

Introduction	284
The Context of This Analysis	285
Identifying and Fostering Resilience: The Essential Lens of an Assets-Based Approach	286
Do Needs-Based Approaches to Change Create Need?	286
The Relationship Between Neoliberalism and the Creation of Need	287
"Acts of Faith" or Tyranny: From Neoliberalism to Normative Community Development	287
Assets-Based Approaches: The Difference Between Listening and Telling	290
A Tale of Two Cities: Two Studies of Resilience in Post-Hurricane Katrina New Orleans	291
They're Called "Evacuees": Semantics or Representations of Structural Violence?	291
Disaster Capitalism and Forced Disaster Migration: The Merger of Neoliberalism and Structural Violence	292
Positive Deviance and the Language of Resilience	295
The Relationship of Structural Violence to Resilience	297
The Impact of Inequality: Neoliberalization, Individual Competition, and the Erosion of Resilience	298
Conclusion: Understanding Resilience in the Complex System of a Community	300
References	301

INTRODUCTION

Not all communities are equally resilient. Not all communities have the ability to positively adapt to, withstand and recover from risk, adversity, and socioenvironmental shocks. Norris, Stevens, Pfefferbaum, Wyche, and Pfefferbaum (2008, p. 127) define community resilience as a process in which a network of adaptive capacities is linked "to adaptation after a disturbance or adversity." Thus, community resilience moves beyond coping and survival to active engagement and agency (Brown & Kulig, 1996). This chapter attempts to identify why some communities may prove more resilient than others in their ability to tolerate, and even emerge stronger from, marked adverse social and environmental changes. The answer, at least in part, may lay in a group's collective efficacy and social capital.

A group's success in "shaping their social and economic lives lies partly in a shared sense of efficacy to bring their collective influence to bear on matters over which they can have some command" (Bandura, 2000, p. 78). Collective efficacy can be understood as "the capacity of a group to regulate its members according to desired principles—to realize collective, as opposed to forced, goals" (Ledogar & Fleming, 2008, p. 28). The concept of collective efficacy stems from Bandura's original conceptualization of self-efficacy: "beliefs in one's capacity to organize and execute the courses of action required to produce given attainments" (Bandura, 1977, p. 3). One's perceived self-efficacy influences "whether people think erratically or strategically, optimistically or pessimistically; what courses of action they choose to pursue; the goals they set for themselves and their commitment to them; how much effort they put forth in given endeavors; the outcomes they expect their efforts to produce; how long they persevere in the face of obstacles; their resilience to adversity; how much stress and depression they experience in coping with taxing environmental demands; and the accomplishments they realize" (Bandura, 2000, p. 75). Thus, at the collective level, "people's shared beliefs in their collective efficacy influence the types of futures they seek to achieve through collective action, how well they use their resources, how much effort they put into their group endeavor, their staying power when collective efforts fail to produce quick results or meet forcible opposition, and their vulnerability to the discouragement that can beset people taking on tough social problems" (Bandura, 2000, p. 76). In general, Bandura (2000, p. 78) identifies, "The higher the perceived collective efficacy, the higher the groups' motivational investment in their undertakings, the stronger their staying power in the face of impediments and setbacks, and the greater their performance accomplishments." Thus, collective efficacy can be understood to be a central component of collective resilience. However, a shortcoming of this conceptualization is the exclusion of any mention of the structural biases faced by certain groups, leaving the false impression that all social groups function on a level playing field, with equal access to resources, which, as we will discuss, is most definitely not accurate.

According to Bhandari and Yasunobu (2009, p. 480), social capital can be defined as "a collective asset in the form of shared norms, values, beliefs, trust, networks, social relations, and institutions that facilitate cooperation and collective action for mutual benefits." This definition captures much of the multidimensional aspects of social capital. Social capital—and its elements of social trust, norms of reciprocity and particularly social cohesion—is foundational for a community's capacity to effectively respond to social change and catastrophic events, such as disasters and humanitarian crises.

Although much of the literature regarding resilience focuses on the single dimension of social capital, this chapter will illustrate the multidimensionality of resilience that requires the inclusion of both social capital and collective efficacy.

THE CONTEXT OF THIS ANALYSIS

Disasters and humanitarian crises provide unique situations through which to examine collective efficacy, social capital, and community resilience. Although portrayed as a great "leveler" of social stratification, disasters and humanitarian crises are not equitable in their impact across a population and will challenge existing levels of resilience. Vulnerability, resulting from poor social capital and collective efficacy, is inevitably worsened by disasters; reducing the potential for a resilient recovery of such communities.

The flooding of the city of New Orleans during Hurricane Katrina resulted in the permanent displacement of (primarily) lower income African-American residents. Many of those displaced were unable to return to New Orleans due to post-disaster changes to housing, education, and health-care sectors, which effectively eradicated former sources of the city's public welfare mechanisms that served to protect lower income residents. The case example of the social impact of Hurricane Katrina on two different minority groups in the city of New Orleans illustrates how community resilience is dynamic, specific to given contexts and, thereby, needs to be examined from a complex systems framework.

Cote and Healy (2001, p. 41) identify that "norms of reciprocity and networks help ensure compliance with collectively desirable behaviour. [However], in the absence of trust and networks ensuring compliance, individuals tend not to cooperate." In addition to the structural violence that disproportionately impeded the recovery and return of low-income African Americans, historical structural violence and horizontal inequality directed toward generations of low-income African-Americans in New Orleans fostered a paucity of collective efficacy and social capital, resulting in scarce social cohesion and a poverty of trust. This served to erode many low-income African-American communities before and (even more so) after Hurricane Katrina, and effectively hampered a significant number of African-American "evacuees" from returning to their homes and to their city. In sharp contrast, a small community of former Vietnamese refugees who demonstrated high social cohesion, trust and equity before the disaster, returned to successfully rebuild their community, whilst defying city ordinances attempting to reallocate their community land into a nonresidential green space. This case example will illustrate the importance and centrality of collective efficacy and social capital in fostering resilient communities.

This qualitative research of 155 in-depth semistructured interviews with a purposive sample of pertinent stakeholders, conducted over a period of 2 years, identifies how the complex relationships between social cohesion, trust, and equality can impact social capital and collective efficacy. This research also examines the ability of communities to withstand the deleterious impacts of structural violence and economic neoliberalization and to ultimately improve community resilience and well-being in the wake of disasters and humanitarian crises.

IDENTIFYING AND FOSTERING RESILIENCE: THE ESSENTIAL LENS OF AN ASSETS-BASED APPROACH

This section examines how assets-based approaches may offer a more appropriate theoretical framework for group interventions, as well as for a theoretical approach to inquiry in the social sciences in general, than the more commonly employed needs-based approaches. Assets-based approaches build upon the existing collective efficacy and social capital of a particular group. The determination of social need according to a normative (i.e., Eurocentric and modernist) standard of progress is a common framework for the social sciences, and one may argue, a central weakness of many of the social sciences that assume the universality of a particular paradigm. Unsurprisingly, this framework may also contribute to the ultimate failure of interventions that are based upon such a paradigm. The paradigm embedded in community and state "development" can be problematized for its reified representation of economic and social progress—constructed by powerful high-income nations—as an irrefutable, universal and natural law. Furthermore, the imposition of this construction on groups that do not share this teleology, nor this understanding of "progress," inevitably results in an unequal relationship of dependency.

DO NEEDS-BASED APPROACHES TO CHANGE CREATE NEED?

Community interventions are not inherently beneficent. Interventions directed by sources external to a given group can yield positive and negative results, as well as intended and unintended outcomes affecting community well-being. The normative needs-based approach to community change and development assumes that expertise, that lies outside of a community, is necessary to determine community needs and to distinguish how those needs can be met. In a needs-based approach, populations are assessed according to predetermined criteria of the basic requirements for human and community development, regardless of context. Thus, what is identified to be needed, or missing in a given group, is determined according to a value system and criteria of basic needs that is often foreign to the group assessed. Furthermore, by determining "what is missing or wrong," a process is set in motion that presumes that the knowledge of wrong and right is solely the domain of the external expert, who holds the only solution and/or intervention to rectify what s/he determines to be wrong. Right and wrong, which normatively follow a modernist (and increasingly global and neoliberal) paradigm, often remains external to the values of many communities. Thus, the formation, dissemination and practice of this expertise are built on the premise that solutions to community resilience exist solely outside of the community.

THE RELATIONSHIP BETWEEN NEOLIBERALISM AND THE CREATION OF NEED

Searching for external needs within a group may then lead to the creation of these needs in that group. Capitalism requires a perpetual creation of market needs for consumers to fill, and we can argue that needs-based, or more accurately "needs-created," community interventions fosters a dependency on the market, particularly in the form of technology, regardless if that technology is appropriate for a given context. Neoliberalism can be defined as an extreme form of capitalism, which embraces the free market; unfettered by state oversight and regulations. Such market *self-regulation* allows corporations, particularly transnational corporations, to act as autonomous legal agents that can, in effect, take legal action against sovereign states that are thwarting the free market and corporate profitability; as witnessed in the sanctions resulting from the arbitration process of the World Trade Organization.

In practice, however, the impacts of neoliberalism reach well beyond market profitability and extend into the foundations of the social contract and the responsibilities of the state. For example, neoliberalism was embedded in the contingencies of the structural adjustment programs of World Bank and International Monetary Fund (IMF) loans to low-income countries from the 1980s through the 1990s, and is currently present in the harsh "austerity" programs of IMF / European Union loans to European countries, such as Greece, Ireland, Spain, and Portugal. Neoliberal ideology has become embedded in international development and the subsequent restructuring of economies that includes the privatization of public-owned enterprises and the removal of state responsibilities for social welfare. As Klein (2007) argues in *The Shock Doctrine*, elites can, and often do, take advantage of the chaos of a physical, economic, or humanitarian disaster and quietly engage neoliberal policies that restructure social welfare and state ownership of production to the benefit of the elites and to the exclusion of the vulnerable.

"ACTS OF FAITH" OR TYRANNY: FROM NEOLIBERALISM TO NORMATIVE COMMUNITY DEVELOPMENT

To better assess the development of resilience, we turn to the critics of normative development; the so-called postdevelopment theorists. According to postdevelopment theorists (see, e.g., Escobar, 1995; Ferguson, 1994; Sachs, 1992), development interventions in low-income countries cannot escape from a paradigm which is ethnocentric, teleological, and modernist. This paradigm is perpetuated by development experts. Parpart (1995, p. 225) states: "These experts became, and continue to be, essential to the development enterprise, as development policies and programs are largely predicated on the assumption that development problems can be reduced to technical (i.e. 'solvable') problems which involve the transfer of Western technical expertise to the developing world". Thus, although technologies, such as biomedicine, are commonly portrayed in development (and colonial) discourses as inherently beneficent to recipients and somehow *apart from* the influences of politics and economics; technologies, like biomedicine, are a sociocultural output and, thereby, very much *a part of* the sociocultural forces of politics and economics.

Alternative development, sustainable development, basic-needs approaches, and any other modification to the original development paradigm do not overcome many of the shortcomings identified by these critiques. Regardless of their intentions, these development approaches share certain limiting features including:

- their construction by experts who are external to the locus of implementation;
- their dependence on closed-simple systems models that can, purportedly, be easily translated to any context and thereby universalized;
- their foundation on a Western Enlightenment form of liberal humanism and the complete dismissal of local knowledge, local assets, and other forms of social capital.

In other words, all of these approaches are limited by their ethnocentrism and inability to even consider, much less include, a recipient group's perspectives, knowledge, or strengths that may differ or exist independently from those of the interventionist. The postdevelopment theorists conclude that regardless of any modification, the idea of development itself is the wrong answer to the question of how to intervene with deprivation and disparity. Thus, the problem is the paradigm.

The area of the normative development paradigm that is most challenged by the reality of social capital and collective efficacy is identified in the representations of the agentless agent, or the agent whose agency must be defined and delivered by the external expert, similar to Spivak's (1988) warning against Western intellectual attempts to render a voice to the voiceless subaltern. Furthermore, unlike Foucault's depiction of a herd of human rats in a Kafkaesque social maze (see, e.g., Foucault, 1977, 1980; Foucault, Graham, Gordon, & Miller, 1991), Scott (1985, p. xvi) locates agency in resistance, which he defines as "the ordinary weapons of relatively powerless groups: foot dragging, dissimulation, desertion, false compliance, pilfering, feigned ignorance, slander, arson, sabotage [...] forms of class struggle [that] have certain features in common. They require little or no coordination or planning; they make use of implicit understanding and informal networks; they often represent a form of individual self-help; they typically avoid any direct, symbolic confrontation with authority." Yet, as will be discussed for the Vietnamese community of East New Orleans, as well as identified in research I conducted in the rural Philippines and in indigenous Guatemala, people can, and do, simply refuse to comply with authority (Kadetz, 2014).

Such contradictory representations of agency particularly problematize the community participation discourse. Although community participation has been identified as a potential means to overcome ethnocentric restrictions of development and other interventions meant to address community needs (Nelson & Wright, 1995), the concepts of "participation," "empowerment," and "capacity building" all share a connotation that something must be *done to* a group, with group members acting as participants, for their own betterment; all of which has been primarily determined by outside experts. For Cleaver (2001, p. 37), "empowerment" is an especially problematic concept for "it is often unclear exactly who is to be empowered: the individual? the community? the poor?" Furthermore, the question of how people in such categories might actually exercise agency is often side-stepped. In many policy documents, an apolitical individualization of this concept is identified, whereby the [conception of the liberal] individual is expected to take opportunities offered by health and development projects to "better" themselves and the development of their community. Thus, "participation" may ultimately be far from "empowering."

Cooke and Kothari (2001) cast participation as a form of "tyranny". Nelson and Wright (1995, p. 1) propose that: "'participation,' if it is to be more than a palliative," must involve "shifts in power". Cleaver (2001, p. 36) concludes: "There is little evidence of the long-term effectiveness of participation in materially improving the conditions of the most vulnerable people or as a strategy for social change. Participation has therefore become an act of faith in development; something we believe in and rarely question.". Deshler and Sock (1985) offer a more nuanced assessment of participation. They distinguish "Genuine Participation," which can include either citizen control or cooperation with delegated power or partnership agreements between citizens and agencies, from "Pseudo-Participation," which can include placation, consultation, or information without power sharing, and potentially with manipulation.

However, the definition of community capacity as "the potential of a community to build on its strengths in order to achieve its goals" is a move toward the inclusion of the concepts of collective efficacy and social capital, and thereby does offer an alternative to the normative community participation discourse (Jackson, Cleverly, & Poland, 2003, p. 345). This concept of community capacity begins to address the importance of leveraging the assets that may already lie within a group, rather than identifying what is externally determined to be needed for the group.

ASSETS-BASED APPROACHES: THE DIFFERENCE BETWEEN LISTENING AND TELLING

In contrast to needs-based approaches, assets-based approaches assume that all communities inherently possess strengths and assets that can be leveraged for the development of the entire community. In assets-based approaches, expertise is located within communities. Hence, communities can build upon their existing assets to best decide for themselves the direction of their development, or, in the case of a crisis, their redevelopment. Assets-based approaches, therefore, employ the community's own value system and assess what answers already lie within the community.

Positive deviance is an assets-based approach that can be particularly useful in identifying and fostering resilience in interventions. In a positive deviance approach, members of a group are distinguished according to who is thriving along a given criterion, as determined by the group, in which the majority of the group is not thriving (Berggren & Wray, 2002). The member(s) determined to be thriving is/are then assisted to disseminate their given successful practice to the group (Berggren & Wray, 2002). Hence, in a positive deviance approach, both the solution and expertise already lie within the group. The external "expert," then, is present solely to help identify and *facilitate* the process of disseminating internally based solutions to enhance social capital and collective efficacy, rather than to *direct* the process according to their own externally based solutions (Berggren & Wray, 2002). Thus, rather than providing a means to "do" an intervention "to" a community, a positive deviance approach enables a community to do for itself. Community resilience is thus rendered an organic internal, rather than imposed external, process. Appropriateness for the context is self-determined by the community. Furthermore, dependency on external knowledge, technology, or resources is thereby thwarted or minimized, whilst self-sufficient sustainability is maximized. Hence, in this context, community participation may be more accurately depicted as the participation of an external expert/facilitator with the community, on the community's own project, rather than the reverse. Collective efficacy and social capital are thereby derived from within a group, rather than provided or doled-out by funders and technocrats. Thus, an assets-based approach offers the potential to both identify and enhance collective efficacy and social capital within a group, which is particularly needed for the resilience to withstand and recover from disasters and humanitarian crises. Hence, the value of an assets-based approach is its centrality in the development of collective efficacy and social capital; the development of which, it may be argued, are central to any type of community development.

Understanding the relationships between needs-based approaches, neoliberalism, and exclusion is imperative in order to understand the relationships between collective efficacy, social capital, and resilience, as exemplified in the case of two post-Hurricane Katrina populations.

A TALE OF TWO CITIES: TWO STUDIES OF RESILIENCE IN POST-HURRICANE KATRINA NEW ORLEANS

The flooding of the majority of the city of New Orleans on August 29, 2005, due to breaches to flood protection structures in and around the greater metropolitan area during Hurricane Katrina, resulted in the permanent displacement of primarily lower economic, female, African-American, residents of New Orleans (Sastry & Gregory, 2014). Many residents who were evacuated before and after the hurricane were unable to return to New Orleans due to the postdisaster rapid privatization in housing, education, and health-care sectors, which effectively eradicated the former sources of the cities' public welfare mechanisms protecting lower income residents. In addition to these marked welfare changes, entire neighborhoods were being erased from the map when city planners, armed with the rhetoric of public safety and protection, redesignated these (predominantly lower income and African-American) areas as nonresidential "green spaces." Subsequently, many of the residents of these areas did not return to rebuild their homes. However, in an Eastern corner of the city, in a neighborhood known as New Orleans East, a predominantly Vietnamese refugee community, whose neighborhood was designated as a green space, returned to rebuild their community to their own specifications and against the wishes of city officials. Similar to the bureaucratic changes to the designation of place and home were the changes in the designation of personhood and identity.

THEY'RE CALLED "EVACUEES": SEMANTICS OR REPRESENTATIONS OF STRUCTURAL VIOLENCE?

The official designation of "evacuee" for the residents of New Orleans who left their homes before, during, and after Hurricane Katrina was employed by the US Government and in particular in the rhetoric of the Federal Emergency Management Agency. Initial American media reports of "refugees," "disaster refugees," and "internally displaced persons" were quickly replaced by the hegemonic "evacuee." Yet in interviews, particularly with African-American informants, the term evacuee was strongly rejected as a political designation and the terms refugee, migrant, or homeless were emphatically demanded to be used in its place. "I am *not* an 'evacuee,'" stated one African-American 68-year-old female living in Treme'. "I did not volunteer to leave and I was not given the opportunity to decide if, or when, I was coming back ... is that an evacuee?"

According to international law, the use of the word "refugee" to describe those fleeing from environmental pressures and disasters is not accurate (Brown, 2008). The United Nations' 1951 Convention and 1967 Protocol (UNHCR, p. 16) state: "a refugee is a person who owing to a well-founded fear of being persecuted for reasons of race, religion, nationality, membership of a particular social group, or political opinion, is outside the country of his nationality, and is unable to or, owing to such fear, is unwilling to avail himself of the protection of that country." Furthermore, the designation of a refugee is technically contingent on the crossing of an internationally recognized border, whereas "someone displaced within their own country is considered an 'internally displaced person'" (Brown, 2008, p. 14).

The International Organization for Migration (IOM) (2007, pp. 1−2) proposes the designation of "environmental migrant." "Environmental migrants are persons or groups of persons, who, for compelling reasons of sudden or progressive changes in the environment that adversely affect their lives or living conditions, are obliged to leave their habitual homes, or chose to do so, either temporarily or permanently, and who move either within their country or abroad". However, for the purposes of this chapter, the term "forced disaster migrant" may be most accurate, particularly due to the acknowledgment of the structural violence that is embedded in this forced migration.

The designation of refugee has often been resisted by governments due to fears that the state then will need to offer the same protections and interventions to disaster victims as reserved for political refugees (Brown, 2008). (Though, at the time of this writing, in 2017, it is quite debatable exactly which protections and interventions are afforded to political refugees in Europe and the United States.) However, the less urgent term "evacuee" suggests a temporary displacement, and thereby a lesser need for long-term state intervention and assistance. Hence, it is assumed that the evacuee does not require state intervention and the responsibility for return and recovery is then ultimately assigned to the agency and desires of the individual and community, and thereby is contingent on the social capital and collective efficacy of communities.

As discussed above, neoliberal policies disengage the state from the responsibility of social welfare, whereby all state responsibility is ultimately shifted to community and individual shoulders. Thereby, the purported independence and responsibility of the evacuees to return to their homes without requiring state intervention is effectively legitimized by the neoliberal state in its language. Yet, in the case of the low-income, predominantly African-American, residents who were evacuated from New Orleans, rapid privatization of imperative sectors of city welfare during their absence often obstructed their ability to return. One 55-year-old diabetic African-American woman who was evacuated to Houston stated: "How can I return? How? They closed Charity Hospital and tore down my kids' school. What is there for me to go back to? I want to go home, of course, but its better here."

DISASTER CAPITALISM AND FORCED DISASTER MIGRATION: THE MERGER OF NEOLIBERALISM AND STRUCTURAL VIOLENCE

Klein (2007) identifies that disasters effectively create a *tabula rasa* upon which local governments and venture capitalists can take advantage of the ensuing confusion to rapidly privatize, neoliberalize, and potentially permanently displace lower socioeconomic residents. This is an extreme form of structural violence in which the vulnerable may be permanently excluded and forced out from their communities, often rendered permanently unable to return home. The anthropologist Paul Farmer adopted Sociologist Johan Galtung's (1969) concept of structural violence. According to Farmer (2004, p. 307), structural violence is "violence exerted systematically by everyone who belongs to a certain social order [...] the concept of structural violence is intended to inform the study of the social machinery of oppression." Hence, social inequality fueled by bias can become embedded in social institutions (or structures) to effectively marginalize groups and exclude them from social benefits. When the inability for a given group to meet basic needs is institutionalized, this is known as structural violence.

Thereby, these processes and forces conspire to constrain individual agency and the agency of a given group. Both have been argued to be central to community resilience. And resilience is grounded in the concept of agency or "the capacity for meaningful, intentional action" (Brown & Kulig, 1996, p. 41). According to Bandura (2000) collective efficacy is a group level property, rather than merely the sum of the efficacy of individual members. Many outcomes people seek "are achievable only through interdependent efforts. Hence, they have to work together to secure what they cannot accomplish on their own" (Bandura, 2000, p. 75). Collective agency includes "People's shared beliefs in their collective power to produce desired results" (Bandura, 2000). The constraint of individual and collective agency is illustrated by the plight of many African-Americans in the United States, whose ancestors were kidnaped and sold as slaves in order to fuel the early American industrial economy. In addition, their present social and economic conditions in many inner-city slums, their frequent targeting by police violence and extrajudicial killings, as well as their inordinate generations of lost and incarcerated young males, reflect the structural violence of a bias that has become so deeply embedded across social structures and institutions in the United States, as if to seem predestined to those at the receiving end of this violence. Thereby, African-Americans are often rendered permanent outsiders in their own country. Farmer argues that neoliberal policy and the resulting inequitable distribution of wealth are primary reasons for the perpetuation of poverty and inequality. Hence, in explaining the possible lack of collective efficacy in many African-American communities in New Orleans, Bandura (2000, p. 78) identifies that those who are "disaffected from the political system, believing it ignores their interests have little faith that they can influence governmental functioning through collective initiatives." Clearly, these factors may be among the many that facilitated the permanent displacement of low-income African-American's after hurricane Katrina.

Klein's (2007) theory of disaster capitalism is also corroborated by the research conducted for this chapter. Privatization was rapidly employed in post-Katrina New Orleans in the three fundamental welfare sectors of housing, education, and health care. The New Orleans City Council unanimously voted in 2007 to allow the US Department of Housing and Urban Development to destroy 4500 low-income public housing units (of the total 5100 pre-Katrina units) and to replace these with "mixed-income" housing (Flaherty, 2010). However, according to informants, a majority of these units were actually middle-income housing with far fewer units reserved for low-income residents. Furthermore, Stivers (2007, p. 48) identifies how "public housing, even projects with minimal or no storm damage, has been closed." Thereby, the city council eradicated the possibility of public housing for a majority of low-income forced migrant families.

The New Orleans Public School system was dramatically reorganized post-Katrina and presently more than half of New Orleans school-age children attend the 70 privatized, performance-driven, charter schools (Childress, DeSimon, & Rupp, 2010; Horne, 2011). "The OPSB, the only elected school board in the city, [currently] controls just 6 traditional schools" (Dixson, Buras and Jeffers 2015, p. 296). Informants were particularly incensed over this post-Katrina development. One 38-year-old parent, who defined herself as an "activist seeking the destruction of charter schools" argued how there is a complete lack of accountability or parental input in the charter school model. "Who can I go to there? No one wants to hear the parents. No one wants to take responsibilities for what our children are being taught." And the focus of the curriculum of charter schools has been criticized by many for being performance driven and ultimately racist and sexist in its favoring of white males. Dixson et al., (2015, p. 288) argue "it is difficult to ignore the manner in which White supremacist ideology has been normalized in the reform as it has historically in US public education." They point to "the disproportionate enrollment and admissions practices" whereby a majority of the students in Flower ranked and performing schools are overwhelmingly African American, while a majority of White students attend the top ranked schools and identify how "black teachers and administrators have been displaced and replaced by a younger and whiter teaching force. Indeed, a majority of the decision-makers and policy makers are White and transplanted from cities like Chicago, New York, and Boston. They have no historical memory of the culture of New Orleans nor do they know firsthand what public education means to the community writ large" (Dixson et al., 2015, p. 297).

Finally, in the health-care sector, Charity Hospital was the sole public hospital in New Orleans and served the needs of low-income residents of the area. Though initially flooded, several former hospital staff informants reported the hospital was deemed fit to reopen. One surgeon from Charity Hospital reported: "I was there after the military clean-up. It was perfectly operable and capable of being reopened after the clean-up. It was the choice of the Board of Louisiana State University, which owned the building. They refused to allow the hospital to reopen." This choice left area residents without a public hospital. This gap not only created insurmountable issues for low-income patients in need of in-patient and out-patient care for physical diseases, but also for much needed mental health interventions (Rhodes, Chan, & Paxson, 2010). This is corroborated by Ott (2012, p. vii): "Charity Hospital was closed as a result of disaster capitalism. LSU, backed by Louisiana state officials, took advantage of the mass internal displacement of New Orleans' populace in the aftermath of Hurricane Katrina in an attempt to abandon Charity Hospital's iconic but neglected facility and to supplant its original safety net mission serving the poor and uninsured for its neoliberal transformation to favor LSU's academic medical enterprise."

According to Pyles and Cross (2008, p. 385) "African-Americans continue to be disproportionately affected in health, financial, employment, and housing indicators in post-Katrina New Orleans." Thereby, low-income forced disaster migrants who depended on public assistance were often effectively prohibited from returning to New Orleans. However, postdisaster recovery in New Orleans did not only result in individual forced disaster migrants, but the forced migration of entire communities via the radical rezoning of residential areas. Through a process known as green-spacing, city planners rezoned vulnerable low-lying residential areas (usually home to both poorer and African-American residents) to nonresidential park areas. Although the plans ultimately failed, many communities, particularly African-American communities, were discouraged from rebuilding their homes and communities. However, one community returned and quickly began to rebuild, aggressively challenging the plans of city council members.

POSITIVE DEVIANCE AND THE LANGUAGE OF RESILIENCE

In 1975, a group of Catholic North Vietnamese refugees who were being held in camps in the United States were invited by the Archbishop of New Orleans to form a community. As a result, a new parish, New Orleans East, was formed in 1980 of approximately 6000 Vietnamese residents. The activities of the Vietnamese residents of New Orleans East centered around its central Church, Mary Queen of Vietnam. Less than 5 months after their evacuation, the majority of Vietnamese residents returned to New Orleans East to rebuild their community. This heretofore quiet and compliant community of former refugees was converted into activists refusing their green-spaced designation and almost immediately upon return, took the rebuilding of their community into their own hands. More importantly the rebuilding of their community was specific to their needs and desires; a development that could only be effectively executed from within the community. No other community in New Orleans appeared to go to such lengths to not only return, but rebuild itself on its own terms, as opposed to passively "participating" with city council mandates.

It is important to note that many of the African-American communities studied were of a similar socioeconomic status to the Vietnamese community. Thus, we can ask why the Vietnamese community was able to return and rebuild whilst many communities of predominantly low-income African-American residents were not. In comparing the rate of African-American and Vietnamese evacuees returning to New Orleans, Li, Airriess, and Chen (2010, p. 113) identify that the Vietnamese evacuees "returned earlier and in greater numbers" compared to other members of their own community and compared to the African-Americans in their study sample. "By early May, seven months after evacuating, approximately 80 percent of Vietnamese households had returned in those neighborhoods closest to the parish church and by late June, return rates had increased to 90 percent for the entire study area" (Li et al., 2010). This is in contrast to a return rate in late June of 70%−80% in all other neighborhoods assessed. These differences may, at least in part, lie in factors that facilitated not only the development of higher social capital in the Vietnamese community, whilst thwarting social capital in nearby low-income African-American communities, but moreover may be accounted for by differences in collective efficacy. For collective efficacy takes into account the gross inequalities that exist from one community to another, the geographic isolation of racial and ethnic minority groups, and the "concentrated disadvantage" that characterizes many minority communities (Ledogar & Fleming, 2008, p. 28).

From an assets-based positive deviance assessment, we can identify several assets of the Vietnamese community in New Orleans East, that may differentiate it from other affected communities and that may have supported its resilience in postdisaster recovery. First, this is a highly cohesive community of three generations of refugee families who came to the United States together from North Vietnam. Chan, To, and Chan (2006, p. 298) define social cohesion as a "state of affairs concerning both the vertical and the horizontal interactions among members of society as characterized by a set of attitudes and norms that includes trust, a sense of belonging and the willingness to participate and help." Informant interviews and observation identified these characteristics in this community. One 25-year-old informant reported "We are very close. Many of our families came from the same areas of Vietnam. We are all involved in the church and we all look out for one another."

The organization of the community around one shared venue for social interaction, the church, also appeared to play a significant role in social cohesion. Cohesiveness was fostered by the insularity of a community whose central engagement with one church helped to reinforce community identity. Many have begun to recognize the potential of faith, religion, and spirituality to foster collective efficacy and social capital (see, e.g., Merino, 2014; Ledogar & Fleming, 2008). Li et al. (2010, p. 113−116) note "one of the key reasons for the early return of Vietnamese [to New Orleans] is the supportive role of the parish church that functioned as the logistical center of the rebuilding process [and] the deep family-based social networks that evolved among the refugee population that arrived more than thirty years ago." They further identify "especially in the context of disasters, religious institutions are a foci of social networks because institutional structures can make a difference to levels of participation and thereby, influence the formation of social capital and collective efficacy. Thus, ethnic-based religious institutions have always been a source of ontological trust, especially compared to larger scale institutions such as governments" (Li et al., 2010, p. 106). This insight may offer an important distinction between this community and neighboring communities that might have been more reliant on local, state, and national government institutions for support, that historically have not fulfilled their promises.

The New Orleans East community was socially stratified according to church hierarchy. The priest of the local church, Father Vien, and church staff, stood at the top of this hierarchy. Interviews identified that tenacious leadership was abundantly displayed by Father Vien and his staff who worked tirelessly to advance through the morass of city, state, and federal bureaucracy to secure the mass of permits and funds required for rebuilding their community. "We would not have succeeded without Father Vien," reports one 23-year-old church staffer. "He would not give up and he would not let us bend to the whims of the City Council. He kept us strong in a very difficult time." This hierarchy is further identified in the administrative structure of the community, which is divided into units for which church designated community leaders are held responsible for members of their specific community unit. As a function of this highly organized and hierarchical community structure, communication and coordination was so remarkable that there was only one reported incident of death during the flooding and, quite exceptionally, every community member was traceable after the flooding and subsequent evacuations; a marked issue for most other families in the Gulf Coast region, much less for entire communities. Aldrich (2010, p. 7), details how "when 500 signatures were needed to prompt Entergy — the local utility — to restore electrical power to the neighborhood, more than 1000 residents" of New Orleans East signed the petition within a day.

This highly organized community structure may have also been instrumental in supporting rapid community mobilization. Thus, the resilience of this community reflected in its positive deviant status to return and rebuild after the disaster, when other communities did not, may, at least in part, be an outcome of collective efficacy, social cohesion, and social capital. The social capital of New Orleans East is evidenced in its leadership, organization, resource mobilization, and management; similar to Labonte's (1993) conception of community participation. The development of trust, fostered by groups moving together through these difficult and changing geographies, appears to have been essential to the social cohesion fostered in this community.

Yet, the social trust that has been essential to the cohesiveness of this community network both before and after the disaster was found to be seriously eroded in many other postdisaster communities, including in similarly ethnically homogeneous communities.

THE RELATIONSHIP OF STRUCTURAL VIOLENCE TO RESILIENCE

Pyles and Cross (2008) maintain that in the case of many low-income African-American communities, trust may have eroded because the actual flooding of New Orleans was not due to the hurricane itself, but to the breaching of the levees protecting the low-lying city that were built by the Army Corps of Engineers and subsequently deemed to have been insufficiently maintained. "Natural disasters are often followed by a period of community togetherness; residents and volunteers may bond together to rebuild. [Whereas], technological disasters, with the ability to place blame upon individuals and agencies, often foster distrust and discontent among residents [and] weaken relationships in communities" (Pyles & Cross, 2008, pp. 384−385).

Yet, the Vietnamese community held the same knowledge of causality as many African-American communities; but whatever amount of trust that may have eroded toward people outside of their community, particularly, as observed, toward city government officials, did not weaken the interpersonal relationships within the Vietnamese community nor lessen the community's resilience. However, the fact that the flooding and ensuing death and destruction of an overwhelming proportion of poor African-Americans was due to human technical error, particularly that of a governmental entity, may have severed whatever fragile trust that remained of an historically perpetually marginalized population, whose ongoing inequitable treatment by state social structures served to build a history of mistrust. Hence, we may question if social trust, which appears to be foundational to social cohesion and social capital, is as affected by the specific cause of a disruptive event, as much as by other corroborating social factors that may have eroded trust and rendered communities more vulnerable long before the given event. These other factors include years of bias, racism, and inequality embedded and promulgated through social structures and the bureaucratic agents of the state. And unlike this Vietnamese community, when African-American communities attempt to organize and apply their collective agency, they have, more often than not, been undermined or criminalized, as is evident in the current demonization of the Black Lives Matter movement (see, e.g., Elmasry & el-Nawawy, 2016; Obasogie & Newman, 2016). Thus, the historical and ongoing structural violence directed toward low-income African-American communities in the United States may not only have thwarted the development of trust in these communities, but may ultimately have effectively undermined the capacity for social cohesion and the development of social capital and collective efficacy.

However, structural violence alone may not adequately explain the exclusion experienced by African-American communities in New Orleans. Political economy, specifically the role of income and resource inequality, may similarly impact social cohesion, collective efficacy, and social capital.

THE IMPACT OF INEQUALITY: NEOLIBERALIZATION, INDIVIDUAL COMPETITION, AND THE EROSION OF RESILIENCE

Abundant research has linked income equality to social well-being. Kawachi, Kennedy, Lochner, and Prothrow-Stith (1997, p. 1497) identify how a "growing gap between the rich and the poor affects the social organization of communities and that the resulting damage to the social fabric may have profound implications for the public's health." Wilkinson and Pickett (2006) found that more egalitarian societies are healthier societies with greater longevity. But, they also identify that socioeconomic factors other than income, such as social stratification, may be linked to inequality. However, one can argue that income inequality serves to perpetuate social stratification. Roca and Helbing (2011, p. 11370) assert that well-being in modern societies relies on social cohesion. And Coburn (2000, p. 135) proposes "a particular affinity between neoliberal (market-oriented) political doctrines, income inequality, and lowered social cohesion". Neoliberalism, it is argued, produces both higher income inequality and lowered social cohesion. Considering that neoliberalism champions competition and the cult of the individual and individual responsibility, it is apparent why neoliberalism would not inherently foster social cohesion.

Furthermore, as discussed, the neoliberal policies resulting in the rapid and dramatic restructuring of economies, often referred to as "economic shock" by some economists, can result in massive unemployment and wreak havoc on social cohesion, social networks, social trust, and ultimately on social capital and collective efficacy (Klein, 2007). Yet, in an important study of the rapid "shock" privatization of the former Soviet Union, Stuckler, King, and McKee (2009) identify social engagement (a proxy for social cohesion) as a particularly important factor in the ability of communities and individuals to withstand economic disasters. They identified that those countries with the highest amounts of membership and involvement in civil organizations seemed to reflect the most resilience through this economic crisis. They conclude "rapid mass privatization as an economic transition strategy was a crucial determinant of differences in adult mortality trends in postcommunist countries; [however], the effect of privatization was reduced if social capital was high" (Stuckler et al., 2009). Thus, social cohesion and social capital can act as buffers in disasters and crises and serve to build community capacity and resilience.

Interestingly, financial wealth, in and of itself, guarantees neither social capital nor greater resilience in disasters and humanitarian emergencies. In an examination of the 2004 Tsunami in Sri Lanka, Munasinghe (2007, p. 10) problematizes a normative approach to wealth and vulnerability: "the poor were relatively more resilient than the wealthy [because] the poor are better able to fall back on traditional, informal mutual-help networks, whereas the rich who are more dependent on mechanical devices, abundant electricity, water, food and the services of domestic aides, simply cannot cope with their absence."

Thus, the complexity of the relationships of social capital and collective efficacy to disasters and humanitarian crises cannot be understood via simple reductive formulae. However, even if we can direct our community interventions toward the development of collective efficacy and social capital, will communities be successfully resilient to then withstand the shocks of economic, political, humanitarian, and environmental disasters and crises? Unlike most low-income communities in New Orleans, the Vietnamese community was able to leverage collective agency and social resilience for decentralized grassroots community interventions and were not, thereby, stuck waiting for the federal government and concomitant private partnerships, that would have resulted in, yet, another layer of structural violence that was experienced by other resource-poor communities.

CONCLUSION: UNDERSTANDING RESILIENCE IN THE COMPLEX SYSTEM OF A COMMUNITY

Community resiliency refers to the capacity of community members to "engage in projects of coordinated action within the context of their community despite events and structures that constrain such projects" (Brown & Kulig, 1996, p. 43). But how can communities build their capacity for such agency? According to Brown and Kulig (1996, p. 30), capacity building can be achieved through "the transformation of social structures;" a dynamic exemplified by the Vietnamese community of New Orleans East. Specifically, Norris et al. (2008, p. 127) identify that such transformations can be achieved through collective resilience, in which "communities must reduce risk and resource inequities, engage local people in mitigation, create organizational linkages, boost and protect social supports, and plan for not having a plan, which requires flexibility, decision-making skills, and trusted sources of information that function in the face of unknowns. [In general,] Communities with more trust, civic engagement, and stronger networks can better bounce back after a crisis than fragmented, isolated ones" (Aldrich, 2010, p. 5).

The community of New Orleans East reflects how "organized communities can better mobilize and overcome barriers to collective action" (Aldrich, 2010, p. 7). Furthermore, this community illustrates how shared beliefs of a collective may bring a community together more than shared beliefs of individualism and individual responsibility. As Aldrich (2010, p. 8) notes: "Private citizens with a long-term stake in the community will be the most motivated to rebuild and possess the greatest capacity to do so while isolated individuals will be less likely to do either."

Although the majority of the literature on resilience emphasizes social capital, and though resilience may be better fostered via grassroots development that supports the assets that already exist within a community, social capital must be understood as one factor among myriad factors that impact community social systems and the community's resilience to withstand disasters and crises. For example, Pyles and Cross (2008, p. 388) note: "The potential hazard when analyzing social capital data is the belief that increasing social capital is a panacea for community problems. Without confronting power structures and changing policies, the practices of building community and strengthening assets may fall short of remedying inequities."

Rolfe (2006, p. 19) states: "The relationships between the dimensions of ecological capital, resilience and community well-being are interactive (as opposed to causal, linear or circular) and mutually reinforcing." Hence, rather than examine a community as a simple-closed system that is cut off from other social factors, this research suggests that we need to examine communities as complex, open, and multidimensional systems affected by the interplay of social variables both internal and external to the community. We then may begin to understand how the purported elements of social capital and collective efficacy react to one another in dynamic ways that lead to the emergence of other factors.

Furthermore, trust, social cohesion, norms of reciprocity, and civic engagement will all be dynamically affected by other social factors, such as structural violence and inequality, as well as by social changes imposed by higher level political economic forces, such as neoliberalization.

Social networks must be understood in their multidimensionality, as illustrated by the complexity of relationships of the East New Orleans community of Vietnamese refugees, which Airriess, Li, Leong, Chen, and Keith (2008, p. 1344) describe as possessing "a deep historical memory of refugee experiences, a shared faith, [and] a profound attachment to place [that] have contributed to the strong community identity, and this coupled with the harnessing of church-centered social capital across multiple scales was critical to their post-Katrina recovery." All of these characteristics support community resilience by enhancing the capacity of a distinct community or cultural system "to absorb disturbance, reorganizing while undergoing change to retain key elements of structure and identity that preserve its distinctness" (Fleming & Ledogar, 2008, p. 3).

Finally, this research argues that resilience cannot be perceived merely as an afterthought, but as something to foster in an ongoing manner, irrespective of periods of social shocks. Resilience need not be reactive and situational but can also be preventative, ongoing, and proactive to effectively reduce vulnerability. As Brown and Kulig (1996, p. 42) identify "To be resilient an individual or community is not merely returning" to a status quo, "but is able to grow and mitigate against such events in the future."

Thus, we can conclude that employing *both* an assets-based and complex systems approach to resilience may help to reveal the myriad relevant factors and their interplay that contribute to community resilience, as has been identified here in studying these postdisaster communities. Factors that foster community cohesiveness such as shared long-term networks and shared community identity, central organizations to which the community adheres, and established trust have been identified in this research as central to postdisaster resilience and recovery. However, all of these factors can be thwarted through social processes, such as structural violence, barriers to organization, and political economic neoliberalization that may interfere with community identity, agency, cohesion, and trust, as we identified in many of the low-income African-American residents who were unable to return to their communities and homes. In this manner, community resilience to disasters and humanitarian crises can ultimately be undeveloped and undermined, irrespective of social capital and collective efficacy.

REFERENCES

Airriess, C. A., Li, W., Leong, K. J., Chen, A. C. C., & Keith, V. M. (2008). Church-based social capital, networks and geographical scale: Katrina evacuation, relocation, and recovery in a New Orleans Vietnamese American community. *Geoforum*, *39*(3), 1333—1346.

Aldrich, D.P., "Fixing recovery: Social capital in post-crisis resilience" (2010). Department of Political Science Faculty Publications. Paper 3. Retrieved from: ⟨http: //docs.lib.purdue.edu/pspubs/3⟩.

Bandura, A. (1977). Self-efficacy: Toward a unifying theory of behavioral change. *Psychological Review*, *84*(2), 191.

Bandura, A. (2000). Exercise of human agency through collective efficacy. *Current Directions in Psychological Science*, *9*(3), 75−78.

Berggren, W. L., & Wray, J. (2002). Positive deviant behaviour and nutrition education. *Food and Nutrition Bulletin*, *23*(4), 7−8.

Bhandari, H., & Yasunobu, K. (2009). What is social capital? A comprehensive review of the concept. *Asian Journal of Social Science*, *37*(3), 480−510.

Brown, D. D., & Kulig, J. C. (1996). The concepts of resiliency: Theoretical lessons from community research. *Health and Canadian Society*, *4*(1), 29−49.

Brown, O. (2008). *Migration and climate change*. Geneva: International Organization for Migration.

Chan, J., To, H. P., & Chan, E. (2006). Reconsidering social cohesion: Developing a definition and analytical framework for empirical research. *Social Indicators Research*, *75*(2), 273−302.

Childress, S., DeSimon, J., & Rupp, N. G. (2010). *Public education in New Orleans: Pursuing systemic change through entrepreneurship*. Boston: Harvard Business School Publishing.

Cleaver, F. (2001). Institutions, agency and the limitations of participatory approaches to development. In B. Cooke, & U. Kothari (Eds.), *Participation: The new tyranny*. London: Zed Books.

Coburn, D. (2000). Income inequality, social cohesion and the health status of populations: The role of neo-liberalism. *Social Science & Medicine*, *51*(1), 135−146.

Cooke, B., & Kothari, U. (Eds.), (2001). *Participation: The new tyranny*. London: Zed Books.

Cote, S., & Healy, T. (2001). *The well being of nations, the role of human and social capital, organisation for economic cooperation and development* (pp. 1−118). Paris: OECD.

Deshler, D., & Sock, D. (1985). Community development participation: A concept review of the international literature. In *International League for Social Commitment in Adult Education*. Ljungskile, Sweden.

Dixson, A. D., Buras, K. L., & Jeffers, E. K. (2015). The color of reform race, education reform, and charter schools in post-Katrina New Orleans. *Qualitative Inquiry*, *21*(3), 288−299.

Elmasry, M. H., & el-Nawawy, M. (2017). Do black lives matter? A content analysis of New York Times and St. Louis post-dispatch coverage of Michael Brown protests. *Journalism Practice*, *11*(7), 857−875.

Escobar, A. (1995). *Encountering development: The making and unmaking of the third world*. Princeton, NJ: Princeton University Press.

Farmer, P. (2004). An anthropology of structural violence. *Current Anthropology*, *45*(3), 305−325.

Ferguson, J. (1994). *The anti-politics machine: "Development," depoliticization, and bureaucratic power in Lesotho*. Minneapolis, MN: University of Minnesota Press.

Flaherty, J. (2010). *Floodlines: Community and resistance from Katrina to the Jena Six.* Chicago: Haymarket Books.

Fleming, J., & Ledogar, R. J. (2008). Resilience, an evolving concept: A review of literature relevant to aboriginal research. *Pimatisiwin, 6*(2), 7.

Foucault, M. (1977). *Discipline and punish: The birth of the prison.* Middlesex: Penguin.

Foucault, M. (1980). Two lectures. In C. Gordon (Ed.), *Power/knowledge: Selected interviews and other writings 1972–1977.* New York: Pantheon Books.

Foucault, M., Graham, B., Gordon, C., & Miller, P. (1991). *The Foucault effect: Studies in governmentality: With two lectures by and an interview with Michel Foucault.* Chicago: University of Chicago Press.

Galtung, J. (1969). Violence, peace, and peace research. *Journal of Peace Research, 6*(3), 167–191.

Horne, J. (2011). New schools in New Orleans. *Education Next, 11*(2). Retrieved from: http://educationnext.org/files/ednext_20112_Horne.pdf.

International Organization for Migration (IOM). (2007). *Discussion note: Migration and the environment, ninety-fourth session.* MC/INF/288, 1–2. Retrieved from: ⟨https://www.iom.int/jahia/webdav/shared/shared/mainsite/about_iom/en/council/94/MC_INF_288.pdf⟩.

Jackson, S., Cleverly, S., Poland, B., et al. (2003). Working with Toronto neighbourhoods toward developing indicators of community capacity. *Health Promotion International, 18*(4), 339–350.

Kadetz, P. (2014). Risk and resistance: Creating maternal risk through the imposed biomedical construction of risk in the rural Philippines. In D. Lavell-Harvard, & K. Anderson (Eds.), *Indigenous mothering as global resistance, reclaiming and recovery.* Bradford, Ontario: Demeter Press.

Kawachi, I., Kennedy, B. P., Lochner, K., & Prothrow-Stith, D. (1997). Social capital, income inequality, and mortality. *American Journal of Public Health, 87*(9), 1491–1498.

Klein, N. (2007). *The shock doctrine: The rise of disaster capitalism.* New York: Henry Holt.

Labonte, R. (1993). Community development and partnerships. *Revue canadienne de sante publique, 84*(4), 237.

Ledogar, R. J., & Fleming, J. (2008). Social capital and resilience: A review of concepts and selected literature relevant to aboriginal youth resilience research. *Pimatisiwin, 6*(2), 25.

Li, W., Airriess, C. A., Chen, A., et al. (2010). Katrina and migration: Evacuation and return by African-Americans and Vietnamese Americans in an eastern New Orleans suburb. *The Professional Geographer, 62*(1), 103–118.

Merino, S. M. (2014). Social support and the religious dimensions of close ties. *Journal for the Scientific Study of Religion, 53*(3), 595–612.

Munasinghe, M. (2007). The importance of social capital: Comparing the impacts of the 2004 Asian Tsunami on Sri Lanka, and Hurricane Katrina 2005 on New Orleans. *Ecological Economics, 64*(1), 9–11.

Nelson, N., & Wright, S. (1995). *Power and participatory development. Theory and practice.* London: ITDG Publishing.

Norris, F. H., Stevens, S. P., Pfefferbaum, B., Wyche, K. F., & Pfefferbaum, R. L. (2008). Community resilience as a metaphor, theory, set of capacities, and strategy for disaster readiness. *American Journal of Community Psychology, 41*(1–2), 127–150.

Obasogie, O. K., & Newman, Z. (2016). Black lives matter and respectability politics in local news accounts of officer-involved civilian deaths: An early empirical assessment. *Wisconsin Law Review*, 541.

Ott, K. B. (2012). *The closure of New Orleans' Charity Hospital after Hurricane Katrina: A case of disaster capitalism*. Unpublished Dissertation. University of New Orleans.

Parpart, J. (1995). Deconstructing the development 'expert'. In M. Marchand, & J. Parpart (Eds.), *Feminism/postmodernism/development*. London: Routledge.

Pyles, L., & Cross, T. (2008). Community revitalization in post-Katrina New Orleans: A critical analysis of social capital in an African-American neighborhood. *Journal of Community Practice*, *16*(4), 383–401.

Rhodes, J., Chan, C., Paxson, C., et al. (2010). The impact of Hurricane Katrina on the mental and physical health of low-income parents in New Orleans. *The American Journal of Orthopsychiatry*, *80*(2), 237–247.

Roca, C. P., & Helbing, D. (2011). Emergence of social cohesion in a model society of greedy, mobile individuals. *Proceedings of the National Academy of Sciences*, *108*(28), 11370–11374.

Rolfe, R. E. (2006). *Social cohesion and community resilience: A multi-disciplinary review of literature for rural health research* (pp. 1–27). Halifax: Department of international development studies faculty of graduate studies and research Saint Mary's University. Retrieved from: https://pdfs.semanticscholar.org/4942/90ec68a89855d7a77b0125-b54e080755ad46.pdf.

Sachs, W. (Ed.), (1992). *The development dictionary: A guide to knowledge as power*. London, UK: Zed Books.

Sastry, N., & Gregory, J. (2014). The location of displaced New Orleans residents in the year after Hurricane Katrina. *Demography*, *51*(3), 753–775.

Scott, J. C. (1985). *Weapons of the weak: Everyday forms of peasant resistance*. New Haven: Yale University Press.

Spivak, G. C. (1988). Can the subaltern speak? In C. Nelson, & L. Grossberg (Eds.), *Marxism and the interpretation of culture* (pp. 271–313). Urbana, IL: University of Illinois Press.

Stivers, C. (2007). "So poor and so black": Hurricane Katrina, public administration, and the issue of race. *Public Administration Review*, *67*(s1), 48–56.

Stuckler, D., King, L., & McKee, M. (2009). Mass privatisation and the post-communist mortality crisis: A cross-national analysis. *Lancet*, *373*, 399–407.

UNHCR. Convention and protocol relating to the status of refugees. UNHCR, Geneva. Retrieved from: ⟨http://unhcr.org.ua/files/Convention-EN.pdf⟩.

Wilkinson, R. G., & Pickett, K. E. (2006). Income inequality and population health: A review and explanation of the evidence. *Social Science & Medicine*, *62*(7), 1768–1784.

CHAPTER 14

Dynamics of early recovery in two historically low-income New Orleans' neighborhoods: Tremé and Central City

Nancy B. Mock[1], Paul Kadetz[2], Adam Papendieck[1] and Jeffrey Coates[3]

[1]*Tulane University, New Orleans, LA, United States* [2]*Drew University, Madison, NJ, United States* [3]*National Conference on Citizenship, Washington, DC, United States*

CHAPTER OUTLINE

Background 305
    Focusing Recovery Efforts: Theoretical Underpinnings 306
    Assessing Information and Data at Multiple Levels to Aid Recovery 307
RALLY and Neighborhood Action Research 308
    The Neighborhood Context 310
    Neighborhood Dynamics and Change in Tremé and Central City 312
Neighborhood Residents' Perceptions and Social Infrastructure 317
Significant Findings of RALLY 321
Conclusion 323
References 325
Further Reading 328

BACKGROUND

Postdisaster community level analysis is essential for long-term community viability and resilience. This chapter analyzes the early recovery efforts following Hurricane Katrina in two low-income New Orleans neighborhoods; Tremé and Central City. This analysis is based on a novel project that created a unique partnership and collaboration between academics and communities. The results demonstrate the importance of using neighborhood-level mixed-methods research and multilevel analysis to better inform the recovery process. Lessons derived from this research can inform similar recovery efforts in future recovery contexts.

FOCUSING RECOVERY EFFORTS: THEORETICAL UNDERPINNINGS

The recovery of New Orleans following Hurricane Katrina has been complicated by preexisting long-term socioeconomic and demographic decline (Kates, Colten, Laska, & Leatherman, 2006). The population of New Orleans proper declined from approximately 630,000 in 1960 to an estimated 437,186 in July 2005 (Frey, Singer, & Park, 2007) and 389,617 in 2015 (U.S. Census Bureau, 2016). During the past four decades, New Orleans has become increasingly poor and exhibited the second highest prevalence of extreme poverty among major U.S. cities before Hurricane Katrina (U.S. Census Bureau, 2001). Although the overall poverty rate of New Orleans of 27% was statistically consistent from 2000 to 2013, extreme poverty dropped from 39% in 2000 to 30% in 2013 (Plyer, Shrinath, & Mack, 2015). However, the reduction in extreme poverty may be more of a reflection of the reduction of 33,000 poor residents in New Orleans between 2000 and 2013 (Plyer et al., 2015).

Although recovery is one of the most significant, time consuming, and expensive elements of disaster management (Chang-Richards & Wilkinson, 2016; Kates et al., 2006; Stehr, 2006), the theory and practice of disaster recovery is relatively underdeveloped (Mileti, 1999; National Research Council, 2006; Smith & Wenger, 2007). Recovery has been identified as "the least understood aspect of emergency management" (Smith & Wenger, 2007, p. 234). Indeed, there is neither a standard nor a widely accepted definition of recovery in place.

Early research defined recovery in terms of reconstructing the built environment by the distinct phases of restoration (often synonymous with early recovery), reconstruction, and betterment reconstruction. However, more recent definitions recognize the complexity, nonlinearity, and social nature of the process. In contrast, Smith and Wenger (2007, p. 237), define recovery as "the differential process of restoring, rebuilding, and reshaping the physical, social, economic, and natural environment through pre-event planning and post-event actions."

Recovery is a highly complex process that can lead to an acceleration of either positive or negative trends in affected areas (Farazmand, 2014; National Research Council, 2006) or, in some cases, a dramatic turnaround in sustainable community development (Kates et al., 2006). This turnaround is often a result of great infusions of disaster or recovery assistance, regional trends, or direct effects of the disaster itself.

In general, disasters exacerbate preexisting social inequities, and recovery, thereby, is not an equitable process (Beaudoin, 2007). Those populations with limited capabilities to avoid the damaging effects of a disaster (i.e., lacking social, physical, natural, economic, and political resources before a disaster) are far more vulnerable to disasters. These populations tend to be at a disadvantage for resilient disaster recovery because of their lack of access to adequate resources. Those with less access to resources prior to the disaster will typically have even less access to resources postdisaster. This inequity affects the process of recovery, both in terms of time and quality of reconstruction. Thus, the most vulnerable populations are the least likely to recover in a resilient and sustainable manner. Hence, to not employ a targeted needs assessments at the community level, and thereby, to potentially compromise appropriate community interventions, is ultimately to perpetuate inequity. A stronger approach requires analysis of both the recovery at various levels of analysis, including individual, household, street, community, municipality, region, and state, as well as of the various stakeholders involved in the recovery process (Smith & Wenger, 2007). Without a greater variety and quality of economic, social capital, communication, and community competence variables networked and made accessible to communities after a disaster, the more likely predisaster vulnerability (primarily) will determine a less resilient recovery.

One of the great challenges to a resilient recovery in New Orleans is determining how to address social and environmental vulnerability, while balancing equity and cultural preservation. Achieving equity in resource distribution during the recovery process is a particular challenge in a strong federalist system such as the United States, in which national, state, and local recovery agendas may be in conflict. Much of the early community recovery activity in New Orleans was carried out by private citizens and nonprofit organizations, owing to a highly contentious and protracted planning process among local and state authorities. These private and nonprofit organizations and citizens formed a variety of networks and alliances to plan, mobilize resources, and coordinate efforts. Thus, while public attention was focused on planning zones and on New Orleans as a whole, private efforts were directed toward smaller areas, usually neighborhoods. (Neighborhood, a term used interchangeably in the literature with communities, is not easily defined. According to Wayland & Crowder (2002) the World Health Organization's definition of community includes all persons and organizations within a "reasonably circumscribed geographic area" in which there is a sense of interdependence and belonging. Yet, communities and neighborhoods are often self-defined and, thereby, defy any universal definition).

ASSESSING INFORMATION AND DATA AT MULTIPLE LEVELS TO AID RECOVERY

Community-based and grassroots organizations are often organized at the neighborhood level and, therefore, directly benefit from neighborhood data for their activities. Analysis of neighborhoods provides insight of the microcommunities within a designated area and can improve chances of successful intervention by guiding the effective placement of community centers and service delivery points (Pickett & Pearl, 2001).

Although analysis of individual factors ignores social, cultural, and environmental influences on the individual, analysis solely at an aggregated ecological level can hide individual variability within a given geographic area. Furthermore, some individual-level variables (such as socioeconomic status and ethnicity) may only be understood relatively; within the context of a given group (Diez-Roux,

2000). Hence, for both theoretical and practical reasons, assessment and monitoring of the recovery process at multiple levels of analysis is recommended (Smith & Wenger, 2007).

Multilevel analyses help to prevent incorrect inferences, such as when sole individual variables are considered at the exclusion of any contextual or nested data. In essence, a multilevel model of a neighborhood is derived from all possible variables for that neighborhood (Subramanian, Duncan, & Jones, 2001). In public health, for example, the neighborhood provides an appropriate level by which to understand group influences on health, including accessibility of services, quality of infrastructure social support, and community attitudes toward health (Pickett & Pearl, 2001). Multilevel models have been used to understand these complex relationships between individual and group-level factors that affect health (Diez-Roux, 2000, 2001).

From a practical perspective, successful recovery must address local needs and capacities (Olshansky, 2005) and must include collaboration with the community in the recovery process (Berke & Campanella, 2006; Esnard, 2003; Olshansky, 2005). Furthermore, adapting government programs to accommodate local needs and capacities is key (Olshansky, 2005). In the context of New Orleans, the neighborhood level of analysis is particularly significant. New Orleans has the highest nativity rate of any metropolitan area in the United States (Campanella, 2006) and narrative accounts from evacuees suggest that many low-income New Orleanians have never traveled beyond their neighborhood, much less outside of New Orleans. As a result, neighborhood-specific cultures have evolved in New Orleans that may result in more disparate needs across New Orleans' neighborhoods (Scribner, Cohen, Kaplan, & Allen, 1999).

However, in practice, the evidence base, particularly of systematic information to guide recovery efforts, is often limited, in part because information systems can be profoundly affected during disasters (Comfort, 2006). In New Orleans, few resources were invested in systematic population monitoring and the few that were invested could only be analyzed at the planning zone level. New Orleans is divided into 13 zones; though there are 73 neighborhoods in New Orleans, several of which are quite substantial in geographic scope and population. As a result of these planning zones, population planning and monitoring data were collected at highly aggregate levels until the late summer of 2006 (Stone, Lekht, Burris, & Williams, 2007).

RALLY AND NEIGHBORHOOD ACTION RESEARCH

In the wake of Hurricane Katrina, New Orleans' universities and scholars were challenged to contribute to the complex recovery process. Nancy Mock, former head of a graduate program in complex emergency and disaster management at Tulane University School of Public Health and Tropical Medicine, enlisted interested students to assist in New Orleans recovery efforts. From this effort, Mock established the Recovery Action Learning Laboratory (RALLY) project, a nonprofit organization staffed by two faculty and more than thirty Tulane students. The mission of the RALLY project was participatory action research involving "practitioners as both subjects and co-researchers" (Argyris and Schön, 1989, p. 613). The research was in support of recovery efforts (via collecting and applying information for community action), and it also aimed to foster "laboratories of learning" in which academia and communities could experiment in evidence-based recovery and revitalization.

The RALLY project quickly established itself within the New Orleans neighborhoods of Tremé and Central City by forming functional partnerships with community organizations. The first RALLY activity was a community-based service delivery program in Tremé. All research activities were carried out by, for, and with community organizations.

The project adopted a multimethod research approach, with inductive and deductive methods to identify and quantify important determinants of community recovery. An initial problem was to identify and define "community" in each neighborhood. Diverse stakeholder groups were active or had interests in these two neighborhoods, including businesses that operated in or provided services to neighborhood residents, neighborhood evacuees dispersed across the US, temporary residents, and returning residents.

Client surveys, household probability surveys of neighborhoods, routine program monitoring and analysis, focus groups, ethnographic surveys, asset mapping, and geospatial analysis and representation were used to gather data. For example, client entry surveys were administered by food pantry staff due to illiteracy. These surveys were used to determine food preferences, community priority needs, interests regarding the development of a community center, and situational awareness of problems within a given community. As food pantries were being developed into community centers, client entry surveys would provide the data necessary to best adapt a community center to a community's needs. Client surveys were also used as a quality improvement tool to monitor food pantries and their evolution into community centers.

Household probability surveys were used to estimate the total population, in terms of sociodemographic characteristics, residential intentions, access to resources/services, and perceived barriers to family and community recovery.

Windshield surveys (conducted by driving around a community and assessing the community through observation) were initially used to collect data on the rate of return of recovery activity in a neighborhood. In Tremé, students performed monthly windshield censuses of the neighborhood, as well as a rapid census four-months after Hurricane Katrina.

Focus group assessments were used to identify initial perceived barriers to recovery, sources of community problems, priorities for neighborhood interventions, the fine-tuning of program strategies and to discuss the findings of the surveys with community residents for the purpose of developing recommendations for the neighborhood's recovery. In general, community forums were employed to explore the interpretation of quantitative data and community-based identification of priority interventions. These assessments also provided investigators with opportunities for active interactions with community leader key informants. Service statistics were collected to understand both service use patterns and client preferences for services.

Household probability surveys were implemented in each of the neighborhoods during the summer of 2006. In the fall of 2007, a census of one of the target areas within Central City, known as the Hoffman Triangle, was conducted. In general, similar instruments and methods were used in both neighborhoods. However, sample designs differed according to differences in neighborhood population, geographic size, and information needs of residents in the two neighborhoods. For example, although household surveys were sampled using both systematic and cluster sampling methods, the larger neighborhood of Central City was surveyed with a modified cluster design, whereas the relatively smaller neighborhood of Tremé employed a systematic sampling of households.

The sampling frame consisted of all residences in each neighborhood. Residency status was determined to the greatest extent possible, but doing so often proved difficult with survey techniques, because as many as 80% of residences were not easily identified as occupied. Therefore, a method was devised to estimate occupancy using proxy reports and visual cues. An analysis of the sociodemographic characteristics of refusals to participate in surveys (e.g.; approximately 20%) indicates that African-Americans were significantly less likely to respond than other groups.

Data was analyzed with SPSS 15.0 (Statistical Package for the Social Sciences) and Epi Info. Secondary data analysis comparing RALLY neighborhood data with United States census data from 2000 was used to determine changes in neighborhoods.

THE NEIGHBORHOOD CONTEXT

The two neighborhoods studied were chosen for their geographic importance to New Orleans recovery. This relevance included (1) their vulnerability and proximity to the Mississippi River, which could facilitate early recovery activities (Fig. 14.1); (2) their historical and cultural significance; (3) the amount of significant flooding to the neighborhoods (Figs. 14.2 and 14.3), and the variable levels of damage within the neighborhoods (see Figs. 14.4 and 14.5); and (4) the varying degrees of gentrification in the neighborhoods prior to Hurricane Katrina.

The New Orleans neighborhood of Tremé was a French-speaking, Creole community known as America's oldest black neighborhood (Crutcher, 2010). Tremé was home to Congo Square where, since the early 1700s, slaves and free people of color gathered to socialize, barter, and celebrate African traditions (Johnson, 1991). It evolved into one of the first multiracial districts in the early 19th century(Johnson, 1991). It is remarkable for an era, in which America was

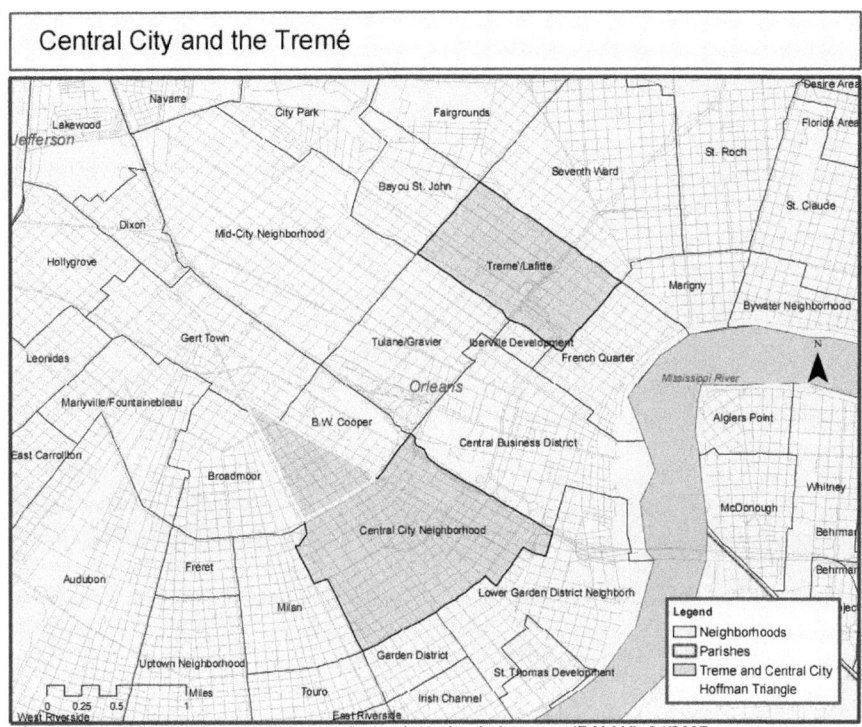

Map produced by New Orleans Recovery Action Learning Laboratory (RALLY), 04/2007

FIGURE 14.1

Central City and the Tremé.

still immersed in slavery, for freed slaves to acquire, purchase, or own property. Among other noteworthy firsts, Tremé is known as the home of the first African-American newspaper, the first African-American Roman Catholic Parish, the first literary salons (at a time when teaching blacks to read in the United States was illegal), and the first anthology of African-American poetry, as well as home to a cornucopia of famous jazz musicians (Crutcher, 2010). However, despite this rich history, Tremé is also characterized by high poverty (56.9% in 2000 and 41.2% in 2015), which has prevailed since the 1960s (Greater New Orleans Community Data Center, 2005; The Data Center, 2016), when it housed two of the city's largest public housing developments (Iberville and Lafitte) and the Interstate-10 Expressway, which government officials carved directly through this historic black neighborhood (Kadetz, 2007). In 2005, Tremé had a total of 865 public housing units and in 2013 there were 445 units (Greater New Orleans Community Data Center, 2005; Lovett, 2013, p. 29).

Central City has been the home to German, Irish, Italian, and Jewish immigrants since the early 1830s, when developers "reclaimed" it from mosquito-

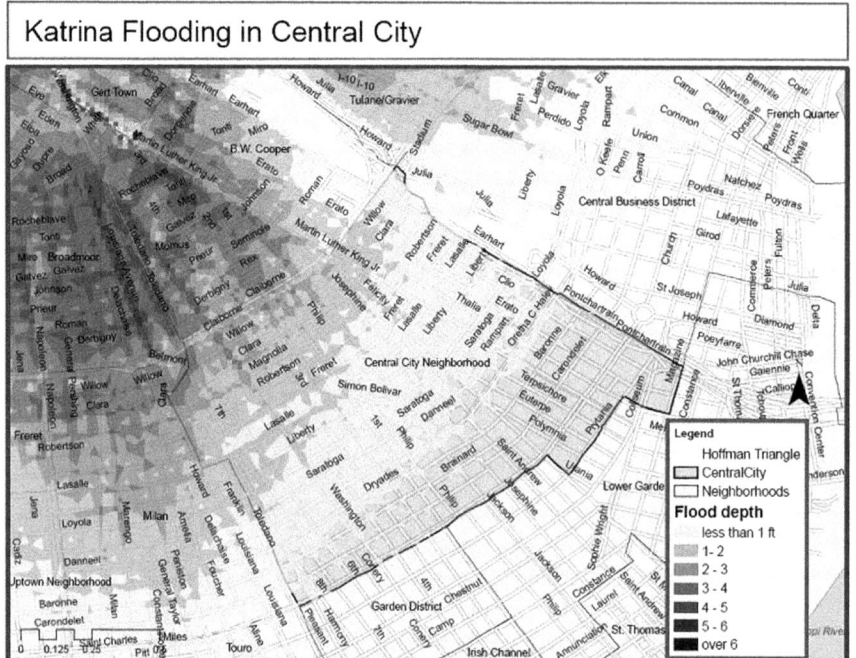

FIGURE 14.2
Katrina Flooding in Central City.
Map Source: RALLY. Data Source: FEMA (flooding), GNOCDC (neighborhoods).

infested swampland 3 to 10 ft below sea level. In the latter half of the 20th century, Central City received an influx of African-Americans. By the 2000 census, this neighborhood housed two major public housing complexes (Peete and Guste, totaling 3000 units) and suffered a poverty rate of 49.8% (75% among children under age 5 years). It is one of the largest neighborhoods in New Orleans and had one of the highest crime rates prior to Hurricane Katrina (City of New Orleans, 2006, Kadetz, 2007). As of 2013, the total number of subsidized public housing units was reduced to 661, a fivefold reduction (Lovett, 2013).

NEIGHBORHOOD DYNAMICS AND CHANGE IN TREMÉ AND CENTRAL CITY

Prior to Hurricane Katrina, Central City and Tremé were among the most disadvantaged neighborhoods in New Orleans. They were socially and demographically quite similar with the exception that Central City was a larger neighborhood in terms of both land area and population (Fig. 14.1, Table 14.1). Both

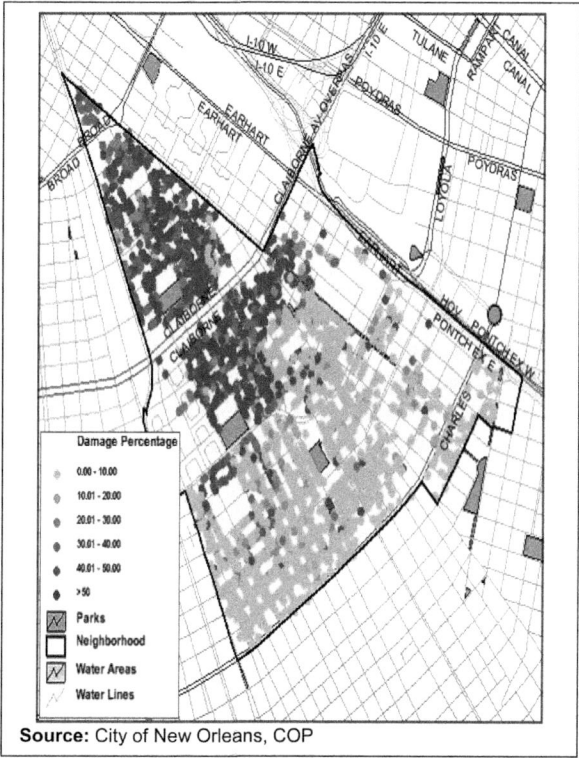

FIGURE 14.3

Storm Damage in Central City.

Map Source: RALLY. Data Source: *City of New Orleans, COP.*

neighborhoods also include high-income residences, particularly at neighborhood boundaries. However, as noted above, poverty was significant in both neighborhoods. According to RALLY data, residents of both neighborhoods tended to be members of female-headed households with children living below the poverty line. Approximately one-half of the households in each neighborhood owned a car or some form of personal transportation and were active in the labor force at the time of the survey. Approximately 60% of householders in each neighborhood had completed high school. Both neighborhoods had been experiencing population declines prior to Hurricane Katrina, with Tremé's population declining 5% between 1990 and 2000 and Central City's declining by 15% within the same time frame (Campanella, 2002).

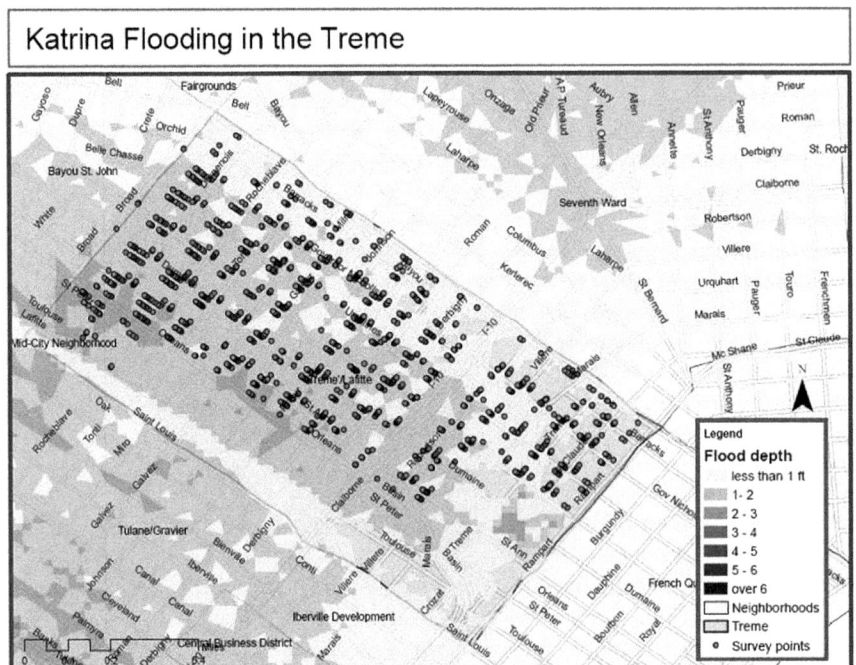

FIGURE 14.4

Katrina Flooding in the Tremé.

Map Source: RALLY. Data Source: *FEMA (flooding), GNOCDC (neighborhoods)*.

In Tremé, where earlier time-series data were available, the return of residents to the wealthier and higher elevation areas bordering the French Quarter and on the Esplanade Ridge was relatively rapid. By December 2005, Tremé had a minimum population of approximately 870 people (RALLY, 2006) (Figs. 14.6 and 14.7). The profile of returned residents was predominantly male, older, and small household size. Most returned residents were homeowners with long histories in the neighborhood. These early returnees expressed a sense of safety and relief from the rampant crime that plagued the neighborhood prior to Hurricane Katrina (RALLY, 2006).

By the summer of 2006, the household composition of New Orleans was more similar to pre-Katrina levels in terms of household headship, size, race, and socioeconomic status, although homeowners were still more heavily represented than tenants. In Tremé, there was still a high degree of family fragmentation; more than 20% of households were missing members who lived with them before the Hurricane and a similar percentage had new members living in the household. School and employment were the major reasons cited for family fragmentation.

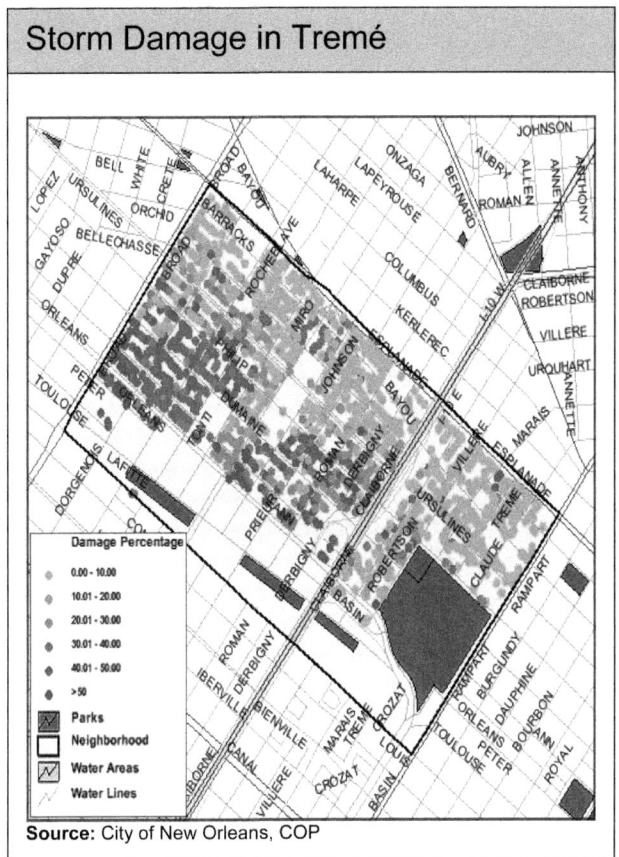

FIGURE 14.5

Storm Damage in Tremé.

Map Source: RALLY. Data Source: *City of New Orleans, COP.*

Most of those residing in Tremé in the summer of 2006 were long-term residents (RALLY, 2006).

Tremé's location, architecture, and cultural appeal made it a particular target for post-Katrina gentrification. According to the Multiple Listing Service, the cost of real estate increased by nearly 10% from pre-Katrina levels through September 2007. An active battle over public housing persists to the present (Warner, 2007). Hence, although the early recovery of the neighborhood suggests a return to the familiar, Tremé's future and that of its residents remains unclear.

Central City also contains a smaller population today than pre-Katrina, but resident characteristics have changed little since Hurricane Katrina. Racial and

Table 14.1 Demographic Data from US Census 2000 for Tremé, Central City, and Orleans Parish

	Tremé	Central City	Orleans Parish
Basic Demographics			
Population	8853	19,072	484,674
Total households	3429	8147	188,251
Total family households	4016	2064	112,977
Black	92.4%	87.1%	66.6%
White	4.9%	9.9%	26.6%
Hispanic	1.5%	1.6%	3.1%
Other	1.2%	1.4%	3.7%
Households			
Households with children under 18	42.0%	32.5%	35.3%
Female householder with children under 18	32.2%	24.0%	17.7%
Children living with mother only	61.7%	58.3%	39.2%
Income and Poverty			
Average household income	$19,564	$23,237	$43,176
People living in poverty	56.9%	49.8%	27.9%
Children under 5 in poverty	75.2%	88.1%	43.0%
Total families below poverty level	1138	1936	26,988
Female householder (no husband present) with own children under 18	74.5%	68.2%	65.2%
Households with no vehicle available	55.6%	56.5%	27.3%

U.S. Census Bureau (2000).

economic status is comparable today to pre-Katrina rates. As in Tremé, homeowners returned with greater frequency than did tenants in Central City. But unlike Tremé, Latino migrant workers (primarily from Honduras) began to arrive as part of the recovery effort, living in densely inhabited residences clustered within the neighborhood of Central City (Fig. 14.8).

Examining the dynamics in the Hoffman Triangle area of Central City illustrates the importance of a subcommunity level of analysis (Figs. 14.7 and 14.8). The 2007 US census within the Hoffman Triangle supported observations that the population was slow to return to this highly affected section of Central City. Only one-third of residences were occupied in 2007 compared with 78.8% in 2000. Homeowners returned with a higher frequency in early recovery. However, by the fall 2007, only 30% of the surveyed households were living in homes they owned. This percentage was more similar to the profile of the community before the hurricane. Furthermore, though the Latino migrants avoided participation in the US

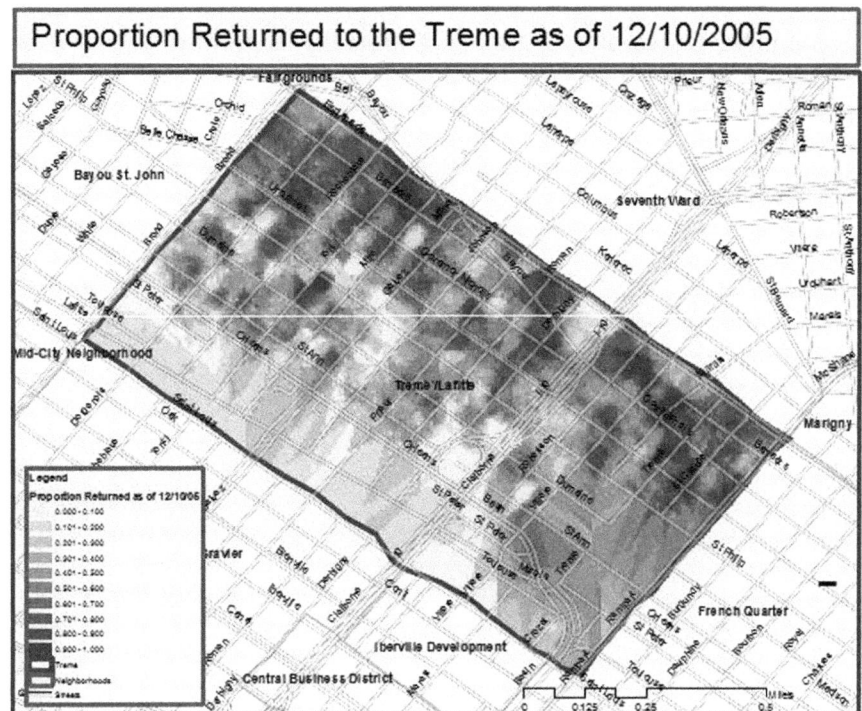

FIGURE 14.6

Proportion Returned to Tremé.

Map Source: RALLY. Data Source: *GNOCDC (neighborhoods), RALLY (2006) Treme Survey in 2005 (occupancy)*.

census, they were more amenable to participating in the RALLY survey, which identified their needs as a subcommunity within the neighborhood of Central City.

NEIGHBORHOOD RESIDENTS' PERCEPTIONS AND SOCIAL INFRASTRUCTURE

Concerning the reduction of vulnerability and the development of community resilience, the early reconstitution of these neighborhoods was not encouraging. Both neighborhoods by 2006 were characterized by significant threats to public safety. Residents in both neighborhoods cited crime and safety as a primary concern. Residents perceived neighborhoods to be less safe than they had been before Hurricane Katrina. Simple interventions such as street lighting were cited as a

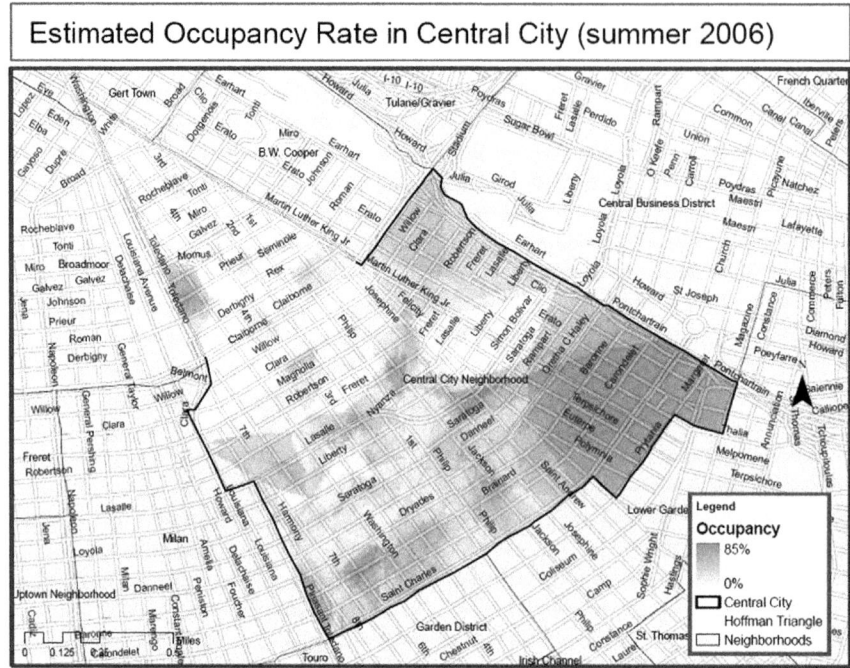

Map produced by New Orleans Recovery Action Learning Laboratory (RALLY), 10/2006
Sources: GNOCDC (neighborhoods), RALLY summer 06 Central City Survey

FIGURE 14.7

Central City Estimated Occupancy Rate.
Map Source: RALLY. Data Source: *GNOCDC (neighborhoods), RALLY (2006) summer 06 Central City survey*.

high priority. Residents in both neighborhoods felt that lighting and policing techniques were serious problems. While poor lighting and blighted and abandoned housing were historical issues in these neighborhoods, both issues were greatly exacerbated by Hurricane Katrina and resulted in higher levels of crime. Similarly, education and programs to keep youth "off the streets," which community members also felt were important, were even more limited since Hurricane Katrina (RALLY, 2006).

In both neighborhoods, resident's reports reflected a profound loss of community and connectivity. Loss of community and the lack of information about the recovery process exacerbated anxiety about the future and created social service access issues. Residents cited information and communication about service availability and recovery activities as important, particularly with respect to health services. Residents of a postdisaster community require information in order to be able to negotiate transitional health care and other social services. Low-income residents in both neighborhoods relied heavily on the public Charity Hospital system, whose main hospital facility never reopened. Residents frequently

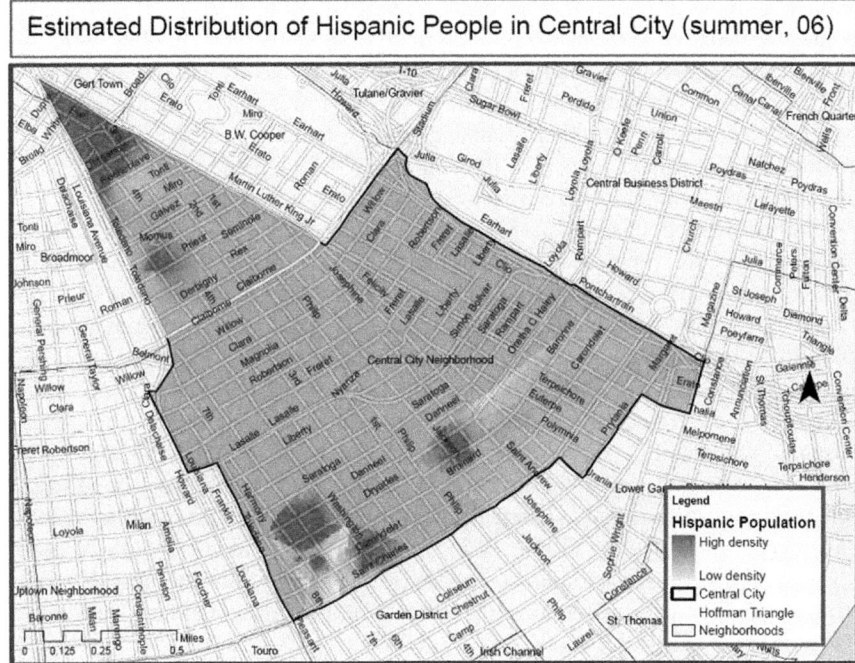

FIGURE 14.8

Estimated Distribution of Hispanic People in Central City.

Map Source: RALLY. Data Source: *GNOCDC (neighborhoods), RALLY (2006) summer 06 Central City survey.*

requested more proactive attempts to get services and programs to the community (Rudowitz, Rowland, & Shartzer, 2006).

Low-income, female-headed households were particularly vulnerable. These households more often reported insecurity and personal safety as a major constraint, than did households headed by males. Women more often occupied residences that were unsafe and whose basic services were often inadequate. Women also more frequently reported problems finding employment, and in general, their income was lower after Katrina. Day care facilities were slow to reopen and were only at 39% of pre-Katrina capacity nearly 3 years later (Liu & Plyer, 2008). At the city level, the markedly limited social and transportation services reinforced the sense that the early recovery environment was particularly hostile to low-income residents. Only one-half of pre-Katrina public transportation routes were in service as of November 2007, with buses operating at 19% of pre-Katrina levels.

The reopening of public schools began in earnest in fall 2006. The poor performance of the pre-Katrina school system in New Orleans did not foster confidence among residents that adequate schools would be open for their children. As late as January 2007, public schools had waiting lists of children (Johnson, 2007). In the Hoffman Triangle area of Central City, nearly all residents (84%) cited the reopening of their school as one of the most important steps toward recovery. However, despite excellent planning and community support, the recovery authority did not support reopening the Hoffman Triangle area school, which was a major defeat to community-driven recovery efforts (RALLY, 2006).

Another marked change after Katrina was the racial composition of each neighborhood. As reported above, RALLY project staff found that Latino residents may have been significantly underrepresented in census surveys due to large groupings of Latino migrant workers in residences and sometimes in make-shift shelters. The population of Latinos in Central City was not at all uniformly distributed, but rather clustered across the neighborhood (Fig. 14.8).

The lack of resources needed for a resilient recovery serves to increase household vulnerability to future disasters. In general, households in Central City and Tremé reported being far more vulnerable after, than before the disaster. Twice as many respondents reported a decrease in income compared to respondents reporting an increase (Table 14.2). In both neighborhoods, household heads indicated a high degree of interest in home ownership within the neighborhood, but report that help in negotiating the process and securing funding were major constraints to home ownership.

Findings from both communities illustrate the importance of accessible distributed social service points and outreach; a consideration that was not adopted by public services. The use of the only functioning food pantry in Tremé and its surrounding areas dropped dramatically as a function of distance from the pantry (Fig. 14.9). Use declined by approximately 25% with every two blocks from the

Table 14.2 Post-Katrina Changes in Income for Tremé and Central City.

		Tremé	Central City
Increased	%	16.5	24.1
	n	19	52
	CI	(10.7, 24.5)	(18.6, 30.5)
Decreased	%	40.4	33.3
	n	45	72
	CI	(31.8, 49.6)	(26.0, 41.6)
Stayed the same	%	43.1	39.4
	n	50	85
	CI	(34.4, 52.2)	(32.2, 47.0)
Don't know	%		2.8
	n		6
Confidence Interval 95%	CI		(1.3, 5.7)

Greater New Orleans Data Committee (2008).

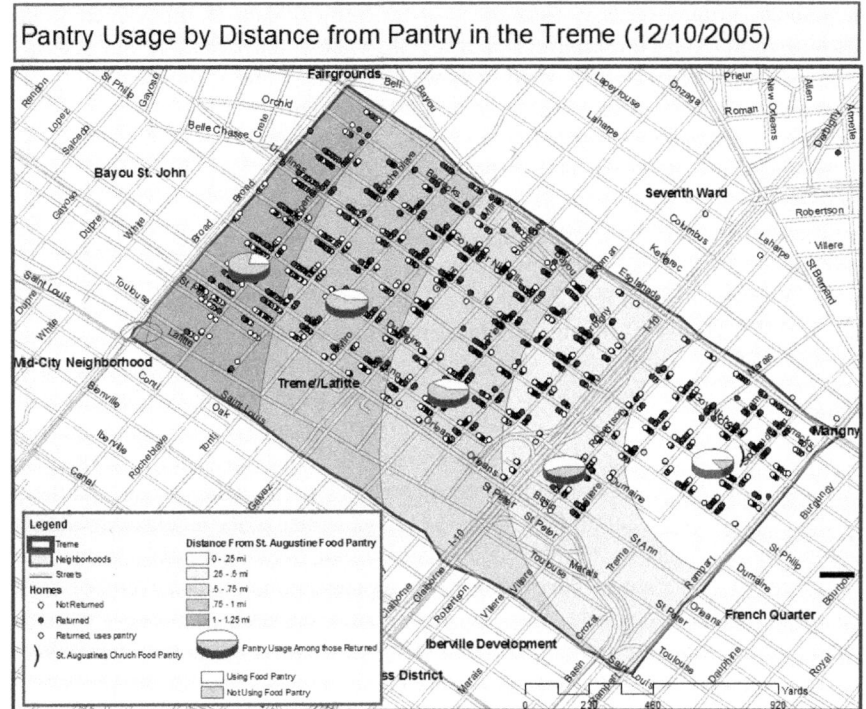

FIGURE 14.9

Food pantry service usage in Treme decreases with distance from service delivery point.
Map Source: RALLY. Data Source: *GNOCDC (neighborhoods), RALLY Treme Survey 2005 (pantry usage).*

pantry. A later RALLY survey of the Hoffman Triangle area of Central City, indicated the need to locate safe haven youth programs within the neighborhood (RALLY, 2006). These findings were in direct conflict with New Orleans redevelopment planning initiatives that focused on larger planning zones as the unit of planning for schools, community centers, and social services.

Myriad community-led initiatives, such as the development and advocacy for community-based schools, the establishment and revitalization of neighborhood organizations, the emergence of faith-based organizations as community centers, and numerous other efforts have been present in the neighborhoods, but public recovery funds were slow to reach the neighborhoods and slow to be distributed.

SIGNIFICANT FINDINGS OF RALLY

The RALLY learning laboratory experience resulted in several lessons. First, participatory action research can be an important catalyst to the recovery process.

Stakeholder groups eagerly used the RALLY project data as an advocacy and planning tool. When the progress of recovery wanes, understanding the determinants of slow recovery can spark renewed optimism about the possibility for isolating and addressing recovery constraints. Furthermore, community members can successfully assume leadership for conducting and analyzing action research.

Specifically, RALLY data has been used extensively to guide neighborhood targeting and intervention designs. For example, the Central City Safety Initiative, an alliance of multiple organizations concerned with safety and development, has used the RALLY data to determine the need for community centers and safe havens. Organizations in the Central City Safety Initiative which have used RALLY data include: Baptist Community Missions, Youth Empowerment, New Orleans Development Foundation, New Orleans Police and Justice Foundation, and Crime Stoppers. Data on the importance of street lighting was used by the New Orleans Police Department to bring lighting back into the Central City neighborhood in a more timely manner than where this data was not available (RALLY, 2006). The assessment of the migrant Latino worker population, primarily in the Hoffman Triangle section of Central City (Fig. 14.8), generated data essential to design appropriate interventions for this community within a community. Community-prioritized youth activities have been a major concern in the recovery of these neighborhoods. RALLY data was used to mobilize resources that resulted in the opening of a community swimming pool in Central City. Also, the data documented the critical importance of physical proximity of services to be detrimental to service use, as witnessed above in the example of food pantry utilization as a function of accessibility (Fig. 14.9). These findings resulted in a change in strategy from one of fewer large community centers to the concept of more numerous and smaller neighborhood-based services, as well as in the need to extend community centers beyond school-based centers. Hence, addressing community needs at a zone level will most likely not provide adequate service to residents because of poor access.

Second, the recovery process for major catastrophes, such as Katrina, is long term (Kates et al., 2006). The willingness of a community to participate in action research is correlated to recovery success (Olshansky, 2005). Survey response was nearly universal during the first three months after Katrina, a point in time when survey interaction was often described by informants as a "therapeutic experience" (RALLY, 2006). Thereafter, community enumerators and those who had built trust within the community were most able to effectively conduct the research (RALLY, 2006).

Furthermore, early recovery research following a catastrophic disaster requires face-to-face interviews, often in the absence of usable sample frames and channels to reach households. This work can prove time-intensive. A particularly challenging aspect of survey research is determining whether buildings and residences are inhabited. The RALLY project developed occupancy indicators in collaboration with other research groups in New Orleans. Contacting people is another concern. People can be living in unconventional spaces such as commercial buildings or even vehicles, family members may be displaced, and the nature of recovery employment itself can all make residents difficult to reach. Another challenge for accurate demography is the existence of temporary populations, such as migrant workers and displaced residents from other parts of the city. Displaced residents present other challenges due to the lack of conventional approaches to contact them, as well as the uncertainty of their intentions to return; though, lists provided by churches and neighborhoods have proved useful.

A final finding from the RALLY project concerns the importance of recovery clustering of settlement patterns and the implications for sample design in neighborhood research. Fig. 14.8 demonstrates the clustering of Latino residents in Central City. This clustering is a typical repopulation pattern in disaster areas. Because of this pattern, traditional approaches to survey design that use cluster sampling methods may be inappropriate. Simple random sampling or systematic sampling of households could provide more accurate estimates of household traits.

In general, post-Katrina recovery proceeded very slowly and gains may have been modest due to the lack of microlevel data in all New Orleans neighborhoods. In the example of the Central City initiative, initial funding was based on RALLY data and intention was microfocused, thereby facilitating solutions at a manageable level for postdisaster recovery. However, output levels, such as the community pool utilization in Central City, has been collected, but secondary data in terms of impacts was not easily accessible to RALLY. The RALLY surveys occurred early in the recovery process and further follow-up would be valuable.

The problem with less finely granular data is that it may mask important details needed to regenerate the diverse neighborhoods that comprise many urban settings. These neighborhoods correspond most closely to the sociological construct of community. In many cases they reflect the cultural assets of urban systems. Community surveys uncovered interesting clustering of new immigrant populations and microlevel dynamics of property recovery (Figs. 14.6–14.8). These factors were important for identifying areas that have recovery momentum. Microlevel data is also important for continuous monitoring of neighborhood-specific concerns. For example, a high clustering in neighborhoods was demonstrated in RALLY's microlevel data, which macrolevel data would not have provided. Microlevel data is especially beneficial in terms of being able to identify where in neighborhoods facilities and infrastructure belong. Context specific information and data will reveal where best to locate these interventions. Hence, microlevel data will provide the means by which to utilize macrolevel data. Understanding aspects of the resilience of recovery, and the dynamics of community recovery, is essential to understand the recovery process in general. Clearly, it would not be possible to fully understand these dynamics within a macrolevel analysis alone.

CONCLUSION

Although the Disaster Management Act of 2000 necessitates extensive risk assessments and planning, there is no requirement to consider vulnerable populations (Juntunen, 2006). To create an accurate portrayal of postdisaster neighborhood needs, a comprehensive evaluation of neighborhood contexts and effects must be performed. The importance of identifying disparity across and within neighborhoods rests on a methodology that can distinguish a neighborhood (Kawachi & Subramanian, 2007). We conclude that if recovery is not assessed at the appropriate levels of analyses, important recovery needs will be missed.

This study illustrates the importance of data at the community and neighborhood levels for early, appropriate, and resilient disaster recovery. Microlevel data are imperative to facilitate decision making affecting communities, particularly in recovery efforts. For example, data from the Hoffman Triangle section of Central City resulted in multiagency intervention specific to the needs of the multiethnic and multiracial populations of this neighborhood.

According to Adger, Hughes, and Folke (2005, p. 1036), "Social and ecological vulnerability to disasters is influenced by the buildup or erosion of resilience both before and after disasters occur." The development of general resilience before disasters is able to counter susceptibility to disasters, and reduce the overall level of vulnerability of a community. A resilient disaster recovery also helps reduce the vulnerability of a community in future disasters. Resilience of the population and the geographic area is a major goal of postdisaster recovery. Investing in the resilience of postdisaster neighborhoods will be an outcome of detailed multilevel neighborhood analyses during recovery. Resiliency can be bolstered by creating detailed vulnerability analyses of neighborhoods prior to a disaster using participatory action research (Godchalk, 2003).

In general, this research also illustrates the benefits of participatory action research for resiliency in disaster recovery. The barefoot scholarship or insider research conducted by resident-informants, students and faculty from New Orleans, proved essential to fostering trust and collaboration. Only through the participation of community members, especially in terms of participatory action research, can communities gain the resources needed for resilience in recovery and prevention and hold the elite accountable for the structures perpetuating inequality. This trust and collaborative approach to neighborhood recovery led to a unique working model of academic-community parternship for recovery, putting data collection, analysis, and dissemination tools at the service of local communities.

Undertaking neighborhood assessments can be a daunting activity. However, projects such as RALLY illustrate the value of incorporating academia to aid in resilient recovery. Local universities are particularly well positioned to undertake microlevel participatory action research in support of recovery efforts, and this work can in turn become part of the university's own recovery. We have illustrated how universities can and should bring their expertise to local action research in support of recovery. The partnership created through RALLY merits consideration for replication in other New Orleans neighborhoods and more broadly as a model for future disaster recovery scenarios.

In this instance, data collection was conducted according to the dimensions of analysis deemed necessary for an understanding of microlevel needs. The Greater

New Orleans Community Data Center, RALLY, and similar organizations actively collected and analyzed information concerning New Orleans as a whole; across multiple levels of analyses. Thus, we argue for a multilevel approach to recovery on the logical grounds established in this chapter. As this study demonstrates, it is imperative to use appropriate levels of analyses to accurately determine population needs, inequities, and disparities. Including neighborhood assessments improves policy and the use of funds allocated for recovery, particularly in the most imperative components of social infrastructure such as community safety, housing, public schools, health care, and vocational training (Turner, 2006).

Finally, in framing the complex processes of recovery in terms of political ecology, we can better understand the parts comprising the whole, and how the parts interact with one another, as well as with the whole. The larger the lens for analyzing the micro environment, the more compromised the ability to understand the multidimensionality of the recovery for a given region. Hence, by starting with bits of information about a neighborhood in recovery, a better understanding of the neighborhood, as well as of the region, will be established. To not account for resiliency during recovery and reconstruction, as well as the levels of vulnerability that develop during the recovery process, is to perpetuate injustice, poverty, and dependency (Turner, 2006), in addition to the many other inequities in the socioeconomic structure of a city (Yarnal, 2007).

REFERENCES

Adger, W., Hughes, T., Folke, C., et al. (2005). Social-ecological resilience to coastal disasters. *Science, 12*(309), 1036–1039.

Argyris, C., & Schön, D. A. (1989). Participatory action research and action science compared: A commentary. *American Behavioral Scientist, 32*(5), 612–623.

Beaudoin, C. (2007). News, social capital and health in the context of Katrina. *Journal of Health Care for the Poor and Underserved, 18*, 418–430.

Berke, P., & Campanella, T. (2006). Planning for postdisaster resiliency. *Annals of the American Academy of Political and Social Science, 604*, 192–207.

Campanella, R. (2002). Time and place in New Orleans, past geographies in the present day. Gretna: Pelican.

Campanella, R. (2006). Geographies of New Orleans: Urban fabrics before the storm. Lafayette: Center for Louisiana Studies.

Chang-Richards, Y., & Wilkinson, S. (2016). The insurance industry and integrated project management frameworks in post-disaster reconstruction: Recovery after the 2010 and 2011 Christchurch earthquakes. In P. Daly, & R. M. Feener (Eds.), *Rebuilding Asia following natural disasters: Approaches to reconstruction in the Asia-Pacific region.* (pp. 339–366). Cambridge: Cambridge University Press.

City of New Orleans. 2006. *Crime statistics.* Available from http://secure.cityofno.com/portal.aspx?portal=50&tabid=12.

Comfort, L. (2006). Cities at risk: Hurricane Katrina and the drowning of New Orleans. *Urban Affairs Review, 41*(4), 501–516.

Crutcher, M. E., Jr (2010). *Tremé: Race and place in a New Orleans neighborhood* (Vol. 5). Athens: University of Georgia Press.

Diez-Roux, A. V. (2000). Multilevel analysis in public health research. *Annual Review Public Health, 21*, 171–192.

Diez-Roux, A. V. (2001). Investigating neighborhood and area effects on health. *American Journal of Public Health, 91*(11), 1783–1789.

Esnard, A. M. (2003). Beyond semantics and the immediate postdisaster period: Community quality of life as an overarching theme for sustaining collective action. *Natural Hazards Review, 4*(3), 159–165.

Farazmand, A. (2014). Crisis and emergency management: Theory and practice. Boca Raton: CRC Press.

Frey, W., Singer, A., & Park, D. (2007). Resettling New Orleans: The first full picture from the census. Washington, DC: Brookings Institution, Metropolitan Policy Program. Available from http://www.brookings.edu/~/media/Files/rc/reports/2007/07katrinafreysinger/20070912_katrinafreysinger.pdf.

Godchalk, D. (2003). Urban hazard mitigation: Creating resilient cities. *Natural Hazards Review, 4*(3), 136–143.

Greater New Orleans Community Data Center. (2005). Tremé/Lafitte neighborhood snapshot. New Orleans: GNOCDC. Available from http://www.gnocdc.org.

Johnson, J. (1991). New Orleans's Congo Square: An urban setting for early Afro-American culture formation.. *Louisiana History: The Journal of the Louisiana Historical Association, 32*(2), 117–157.

Johnson, R. (2007). New Orleans students still waiting for schools. *People's Weekly World Newspaper*. August 2007. Available from http://www.pww.org/article/articleview/10538/1/358/.

Juntunen, L. (2006). Addressing social vulnerability to hazards. *Disaster Safety Review, 4*(2), 3–10.

Kadetz, P. (2007). Tremé' table project. *APHA Action Newsletter, March, 2007*, 2.

Kates, R., Colten, C., Laska, S., & Leatherman, S. (2006). Reconstruction of New Orleans after Hurricane Katrina: A research perspective. *PNAS, 103*(40), 14653–14660.

Kawachi, I., & Subramanian, S. V. (2007). Neighborhood influences on health. *Journal of Epidemiology and Community Health, 61*, 3–4.

Liu, A., & Plyer, A. (2008). State of policy and progress. The New Orleans index. New Orleans: Greater New Orleans Community Data Center. Available from www.gnocdc.org.

Lovett, J.A. (2013). *Tragedy or triumph in post-Katrina New Orleans? Reflections on possession, dispossession, demographic change and affordable housing*. Available from: http://urbanlawjournal.com/files/2013/05/TragedyorTriumphPost-Katrina.pdf.

Mileti, D. (1999). Disasters by design: A reassessment of natural hazards in the United States. Natural hazards and disasters: Reducing loss and building sustainability in a hazardous world: A series. Washington, DC: Joseph Henry Press.

National Research Council, Committee on Disaster Research and the Social Sciences. (2006). Facing hazards and disasters: Understanding human dimensions. Washington, DC: National Academy of Science Press.

Olshansky, R. 2005. How do communities recover from disaster? A review of current knowledge and an agenda for future research. In: *Address presented at association of collegiate schools of planning conference*. Kansas City. October 27, 2005.

Pickett, K., & Pearl, M. (2001). Multilevel analyses of neighborhood socioeconomic context and health outcomes: A critical review. *Journal of Epidemiology and Community Health*, *1*(55), 111−122.

Plyer, A., Shrinath, N., & Mack, V. (2015). The New Orleans index at ten. Measuring greater New Orleans' progress toward prosperity. New Orleans: The Data Center. Available from https://s3.amazonaws.com/gnocdc/reports/TheDataCenter_TheNewOrleansIndexatTen.pdf.

RALLY Project. (2006). Final report on the summer 2006 survey for the department of Justice's Weed and Seed Project. New Orleans: RALLY. Available from http://rally-foundation.org/projects/rally/rallydocs/RALLY%20Final%20Report%202006%20Survey%20for%20DoJ%20Weed%20and%20Seed.pdf.

Rudowitz, R., Rowland, D., & Shartzer, A. (2006). Health care in New Orleans before and after Hurricane Katrina. *Health Affairs*, *25*(5), 393−406.

Scribner, R., Cohen, D., Kaplan, S., & Allen, S. (1999). Alcohol availability and homicide in New Orleans: Conceptual considerations for small area analysis of the effect of alcohol outlet density. *Journal of Studies on Alcohol.*, *60*(3), 310−316.

Smith, G., & Wenger, D. (2007). Sustainable disaster recovery: Operationalizing an existing agenda. In Rodriguez, Quarantelli, & Dynes (Eds.), *Handbook of disaster research*. New York: Springer.

Stehr, S. (2006). The political economy of urban disaster assistance. *Urban Affairs Review*, *41*(4), 492−500.

Stone, G., Lekht, A., Burris, N., & Williams, C. (2007). Data collection and communications in the public health response to a disaster: Rapid population estimate surveys and the daily dashboard in post-Katrina New Orleans. *Journal of Public Health Management & Practice*, *13*(5), 453−460.

Subramanian, S., Duncan, C., & Jones, K. (2001). Multilevel perspectives on modeling census data. *Environment and Planning*, *33*, 399−417.

The Data Center. (2016). *Treme'/Lafitte statistical area*. Available from: http://www.datacenterresearch.org/data-resources/neighborhood-data/district-4/treme-lafitte/.

Turner, M. A. (2006). Building opportunity and equity into the new New Orleans: A framework for policy and action after Katrina. Washington, DC: Urban Institute. Available from http://www.urban.org/UploadedPDF/900930_building_opportunity.pdf.

U.S. Census 2000. Available from https://www.census.gov/census2000/states/us.html.

U.S. Census Bureau. 2001. *American community survey. Supplementary survey*. Washington, DC: U.S. Census Bureau. Available from http://www.census.gov/acs/www/Products/Profiles/Single/2001/SS01/Narrative/160/NP16000US2255000.htm.

U.S. Census Bureau. 2016. *American community survey. Supplementary survey*. Washington, DC: U.S. Census Bureau. Available from https://www.census.gov/quickfacts/table/PST045215/2255000.

Warner C. 2007. Unanimous. *The Times-Picayune*. 12/21/2007. Available from http://blog.nola.com/times-picayune/2007/12/unanimous.html.

Wayland, C., & Crowder, J. (2002). Disparate views of community in primary health care: Understanding of perceptions influence success. *Medical Anthropology Quarterly*, *16*(2), 230−247.

Yarnal, B. (2007). Vulnerability and all that jazz: Addressing vulnerability in New Orleans after Hurricane Katrina. *Technology in Society*, *29*, 249−255.

FURTHER READING

Rushford, N., & Thomas, K. (2015). Disaster risk, vulnerability and resilience: An emergent socio-ecological perspective. In N. Rushford, & K. Thomas (Eds.), *Disaster and development: An occupational perspective.* (pp. 21–32). Edinburgh: Elsevier.

Part IV

Conclusion and Lessons Learned

CHAPTER 15

The Katrina catastrophe and science: Does experiencing a catastrophe at *"ground zero"* have impacts on the professional performance/identity of social scientist survivors?

Shirley Laska[1,2]

[1]*University of New Orleans, New Orleans, LA, United States*
[2]*Lowlander Center, Gray, LA, United States*

CHAPTER OUTLINE

Ground Zero Manifested	331
The *Ground Zero* Impact: Reasoning for the Study	332
Study Methods	333
Findings	335
Trajectory of Research	335
Assessment of Own Research	336
Research Challenges and Benefits of Being Ground Zero Survivor Researchers	337
Witnessing	337
If/How Ground Zero Experience Changed Respondents Professionally	338
Implications for Researchers and Discipline	340
References	342

GROUND ZERO MANIFESTED

As Hurricane Katrina slammed into the Gulf coast in late August 2005, many colleges and universities in Southeast Louisiana and Southern Mississippi were in its direct path. Numerous social scientists lived in the area and taught in these institutions. Some of their Katrina experiences were harrowing; others caused them long-term suffering. For some, physical survival was a real challenge. In order to survive, one faculty member and his wife literally swam for their lives from an upstairs window to the house next door. Eventually rescued by volunteers in a boat already filled with 15 other survivors, they were deposited on high ground at Lake Pontchartrain only to wait another two days for removal. In the interim, eight hundred survivors slept on the airport runway as helicopters constantly flew overhead and they were "guarded", as if they were "prisoners or refugees" (Belkhir, 2015, p. 122). As the tidal surge overtook their neighborhood another faculty member was trapped in the upper floor of his home with his wife and five young children, including a small baby, until they were rescued and taken to a public "shelter of last resort."

Other impacted professors evacuated safely before the storm, but their homes were destroyed or damaged by floodwaters and mold to the point of requiring demolition or needing total interior reconstruction. Others lost their university offices and research materials to flood water and mold. Evacuated faculty members were "sheltered" away from their homes and universities, the duration *for all* being measured in months. The shelters, while likely the homes of friends and relatives as opposed to public shelters, were still crowded and stressful. And when those who lost their homes returned they were housed in tiny, contaminated FEMA (Federal Emergency Management Agency) trailers for months on end, some in faculty FEMA "villages" as repairs were repeatedly delayed.

One social-science program was completely canceled after Katrina and, thus, the faculty members employed by that university were forced to leave and to seek positions in other colleges and universities. Some social scientists made personal decisions to leave their positions in the damaged *ground zero* universities and moved away from the area even though their universities recovered. For some, the trauma of the event was so great that they were not even able to confront their damaged homes or retrieve their belongings. Of those who left their positions in *ground zero*, some stopped teaching altogether, others acquired new academic positions.

THE *GROUND ZERO* IMPACT: REASONING FOR THE STUDY

As we have past the 12th anniversary of the Hurricane, conversations have occurred about the social and institutional changes along the Gulf coast since the catastrophe and the nature and magnitude of those changes. Similar questions are asked about individuals who went through Katrina: If and how they have changed? One of my own colleagues who experienced the most dramatic and traumatizing rescue that I was aware of for an academic, actually had a stroke during the 2012 Hurricane Isaac disaster event. He survived and recovered; but it is obvious that the stress of having gone through Katrina changed him, including challenging his capacity to experience another, albeit much less severe, event. We came to call this "the layering effect" in subsequent research that we conducted after yet another "layer," the BP (British Petroleum) oil catastrophe (Laska et al., 2015).

Given the impacts observed on Katrina survivors in general, there is reason to assume that being directly impacted by a catastrophe could affect the life-world and world views, as well as the professional performance / self-image, of academic survivors. This research asks the question: "Does experiencing a catastrophe affect the professional understanding and performance, and even the sense of professional self of scientists?" This question could be asked in the context of any major disaster when the researcher of human dynamics and impacts of the event are members of the "studied." Public health scientists who studied the health impacts of oil exposure in the BP oil catastrophe would be an example from another catastrophe. Did Katrina change the *ground zero* social scientists affected by Hurricane Katrina?

This work emerged from the invitation by the Southern Sociological Society ("Southerns") to be their Distinguished Lecturer for 2011–12. My goal was to give a lecture on a topic that was relevant to the research experiences of the social scientist audience. Lectures were given at the annual meetings of the Southerns in New Orleans and then at the University of Alabama at Birmingham, University of North Carolina at Charlotte, North Carolina State (Raleigh) and East Carolina University (Greenville). I wanted the lecture to be a "talk among ourselves"—about our professional selves and our discipline.

STUDY METHODS

The defined population for the study was social science professors and ABD (all but dissertation) doctoral students, with an emphasis on sociologists who lived/worked within the core area affected by Hurricanes Katrina and Rita. To cast a broader geographic net, some researchers a distance from the eye of the impact were interviewed if they had conducted research on the storm that required that they come into the *ground zero* area frequently.

A sample of 15 respondents was achieved by means of a snowball sampling process and by conversations with university department heads for their recommendations. Because the research was "sandwiched" between applied, disaster-related, projects that I was doing, it was not possible for me to interview a larger sample and, therefore, I prefer to label this research "exploratory." To the extent that I received consensus from the respondents on quite a few questions that I asked, I feel comfortable that the observed patterns would likely have also been manifested in a larger sample.

The sample had the following characteristics: The career stages of the group at the time of the event are eight senior faculty, three midcareer, two junior faculty, and two doctoral candidates. The universities in which they were located spanned the coast from Mobile, AL, to Lafayette, LA (Fig. 15.1).

The disciplines represented are: sociologists ($n = 12$), an anthropologist, a geographer, and a political scientist (1). Their specialties ranged from

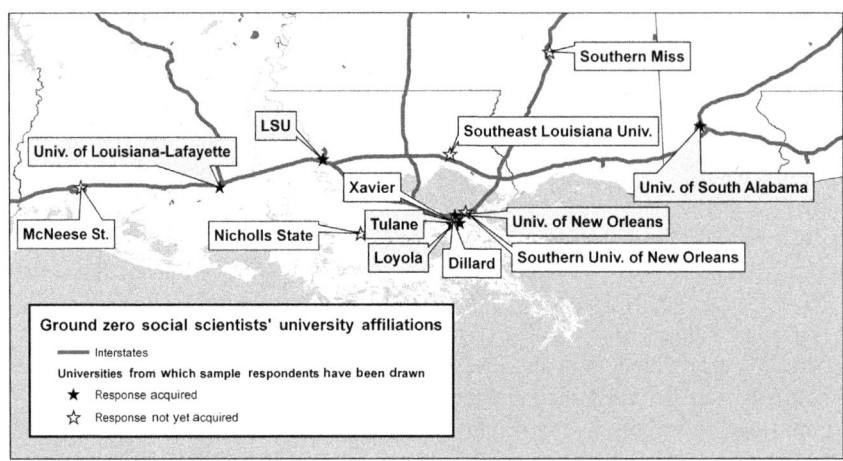

FIGURE 15.1

Ground Zero Social Scientists' University Affiliations

environmental sociology ($n = 5$) to hazards/disasters ($n = 4$), urban political economy ($n = 2$), race/gender/social movements ($n = 1$), political institutions ($n = 1$), demography ($n = 1$), and an ethno-videographer ($n = 1$).

The interviews were conducted via phone 6.5 years after the event. I typed the answers as I conducted the interviews and shared the transcripts with the respondents so that any errors in my transcribing their responses could be corrected. I returned the transcripts for review to all but one respondent, and of those 14, all but 1, (i.e., 13), returned the reviewed transcriptions back to me. There were very few corrections made by the respondents, but of course the corrections that did occur are very important in terms of achieving accurate data.

The interview guide consisted of five themes: (1) the respondents' accounts of their career research trajectories and if and how they had changed post-catastrophe, (2) their own assessment of the quality of their research and if/how being survivors at *ground zero* affected their research, (3) their research challenges, (4) their experience of "witnessing" the event as researchers, and (5) if/how *ground zero* experiences changed them professionally. (Due to space constraints of this chapter's format, one theme of the study—assessment by *ground zero* researchers of out-of-the area researchers and their work—is not included. If the reader is interested in these findings, please contact the author.)

The respondents were comfortable responding to my invitation to be interviewed and to the broad questions I asked. They replied with very little prompting. We will not know how much each respondent had thought about the topics before the interview, but when asked, answers to the questions flowed very quickly and thoughtfully some 6.5 years after their post-Katrina professional life began to be implemented. No one refused my invitation to be interviewed.

FINDINGS

The themes of the impacts emerged very easily from my framing of the questions and what the respondents wanted to discuss. My questions were not a probe of psychological challenges of the respondents, nor of personal recovery challenges. When the latter were important to the respondent with regards to their research and they felt comfortable describing them, then such themes emerged in their responses.

TRAJECTORY OF RESEARCH
Research topic selection
Nine of the fifteen respondents had already conducted disaster research; six had not. The entre' into the research was different for those who had done earlier disaster research, and those who had not. Those who had studied relevant topics, refined their research questions based on their observations and experiences. Those who had not, began reading disaster and environmental sociology literature. Some, who fit neither category, emphasized disasters; if they were environmental specialists they added disaster research to their literature review. Others entered into new specialties because they were invited to join research teams that were out of their pre-Katrina specialties or research methodologies.

Those who were environmental or disaster specialists described how they saw themselves as flexible in being able to select and implement a research topic. One indicated that s/he was always thinking of him/herself as able to do "windshield field work" when an environmental event occurred and this instance was no different. Those who described themselves as applied disaster specialists found the pathway to conducting applied research to be "organic", because the research is "refined from disaster to disaster." The applied social scientist is continually in observation mode for the refinement and application of the theories that they study. The difference this time was that it was a catastrophe and they were survivors. Another respondent felt that what they were observing was so familiar to them, (i.e., they had studied it frequently in the past), that the "(f)irst thing I decided, [was] there was no need to collect any data. By June, 2006, I was educating, communicating."

Fate of previous topic(s)
Four respondents continued on the same research themes that they had been studying before Katrina. Nine continued within the same research "thread" but now with a disaster theme. For two respondents, their pre-Katrina research thread was present, but they established a significantly new research agenda.

Influence on future research
There was almost universal agreement by the respondents that having been at *ground zero* would influence their future research beyond the time of the

interview. Some made reference to the impact on topics selected and others to the methods used. One noted "Definitely because *I am grounded in place*." [author's emphasis] One who studied urban traumas felt that the increase in such events would make this topic more important as his/her career progressed. After Katrina, another switched to a different hazard, "wild fires," but reported that they would still study the same dimensions—inequality and political economy of place. S/he added another interesting comment: "The event (Katrina) enabled me to practice a graduate education specialty that I had never applied—environmental." Serendipitous outcome. The respondent seemed pleased to be able to do so. One respondent who focused on methods believed that s/he would take the methods (s)he applied to their Katrina research back to their pre-Katrina topic of Third World dynamics.

ASSESSMENT OF OWN RESEARCH

Quality

Given the research and personal conditions of the respondents I did not know what to expect when I queried them about their assessment of the quality of their Katrina research. Overwhelmingly, they concluded that the research they had undertaken on the catastrophe was very good. "Best research I ever did, even with the tremendous methodological challenges." "More complete in my understanding and I have more to contribute." "We are not detached from the scene [. . ..] We monitor changes [. . ..] We have first-hand experience and information." Finally, "*ground zero* experiences helped me in the field methodologically [. . ..] Enabled me to do better research." Another expressed it: "Because it was your experience too, it gave me the ability to listen in a different way."

Respondents continued to speak in similar positive tones about how their being *ground zero* survivors benefitted their research. "Being at *ground zero* helps us to understand much better." Another commented that location and experiences enabled them "to see what out-of-area colleagues can't see from afar." So many research dimensions are different for the *ground zero* social-science observer: "We are not detached. [stated positively] We monitor. We have first-hand experience. We have access." Finally, one respondent noted that s/he had much more respect for the whole process [of what was happening] "because you do not have the luxury to step out whenever you want." And finally, one respondent commented that s/he learned about social structure dynamics by virtue of how the challenges of Hurricane Katrina manifested themselves *for them*. The respondents' problems "were really a public issue." This response is supported by Haney and Elliott (2013) in which they argue: "By experiencing the event themselves, they [survivor researchers] gain unique access to the collective experiences of those involved" (p. 14).

Bias

Only a few respondents felt their research might be biased. More noted that they themselves might be *called* biased as in being a "cheerleader" for the region. One respondent who was concerned that their research was biased explained: "I think it is just a reaction (by me)." Another acknowledged, "(s)ometimes coming in from the outside allows you to see things more clearly." This response would be in line with the practices of traditional methods that admonish the researcher to not get too close to one's subjects.

RESEARCH CHALLENGES AND BENEFITS OF BEING GROUND ZERO SURVIVOR RESEARCHERS

Four themes emerged from posing this third question: (1) effect of personal and institutional challenges on being able to practice their profession, (2) reaching the residents (i.e., selecting a representative sample), (3) biasing the research with their own experiences, and (4) retaining quality research in such difficult circumstances. The last concern had a characteristic that was not likely to be present in nondisaster research: public scrutiny. "Given that we were in the public eye, we had to be even more careful with our research," one respondent notes. Concurring, another said: "I am harder on myself than ever before the storm. I think the stakes are higher [....] Because we lived in an area so misconstrued in the media, we must be very careful and more rigorous. We must represent it right." Another stated: "Continuing interest by those outside of the area helped me sustain my research commitment." And while a burden, that accountability was not rejected as inappropriate or un-useful. "I want people to hold me accountable," one said. Location and experience strongly motivated the respondents to do research and to make it the best that they could while feeling a heavy burden: "I felt I had a ton of bricks on me, while at the same time I was stimulated by the opportunity."

WITNESSING

Ground zero social scientists had unusual opportunities to witness the catastrophe. Once they returned, they were able to move freely about the community as soon as debris was removed from the streets. Why collect such stories represented by images for this research? Because I believed they speak to the powerful "data" that the respondents were "capturing." These images can help us understand the impact of the *ground zero* experience on respondents. I asked the respondents to offer "an image/event *connected to their research* that remains in your memory, a memory that is haunting."

The respondents selected images of visual experiences, observing trauma, relational (social) experiences, and complexity of research experiences observed. The following are the commentaries they gave to the witnessing they described:

1. *Visual or physical*: "brown projecting death and destruction"; "gray stain"; "flood line so deep"; "destroyed contents of respondents house piled on curb"; "cars covered with gray slime and bathtub-like lines."
2. *Observing trauma*: "dead body"; "nothing where communities used to be"; "giant mountain of cots, bottles and trash ... a huge mountain of debris left by this mobilization of evacuees"; "absence of people"; "suicide threats in phone interviews"; "complete visual sense of helplessness [...] crying, faces of disbelief"; "was like the end of the city [...] by virtue of the destruction, everything stood still and hopeless."
3. *Relational*: "neighbors left behind"; "sharing with survivors of restored homes and what was salvaged", (which was described as similar to respondent's recovered belongings); blackboard message in abandoned church school that included "a thank you and a condemnation [...] which remained for years as an homage, a kind of holy shrine"; "elation when we received notice that our friends had been found, mixed with deep disappointment that our companions who had not experienced the storm were annoyed by our emotional response."

The respondents were asked to provide photographs of the images they described. Fig. 15.2 is one of the photos submitted with the respondent's narrative provided:

Respondents struggled with what they saw. The respondent who submitted the photo described the "wall of cars as being as much an emotional impact as an intellectual one." Another said: "I knew it was going to happen [...] existential realization." Another explained how s/he kept a distance from what (s)he was observing: "You psychologically retreat to the concepts such as 'corrosive community,' 'collective trauma.'" "They let me fool myself that I have an understanding of what I am seeing." Robert Lifton (1997) calls this experience psychic numbing. Finally, one respondent wished for future soothing of what they had seen: "Time is amnesia."

4. *Complexity of research observations*: A respondent said: "I learned about the murder of the son of one of my study respondents while visiting their home to celebrate its restoration...highs and lows of what people go through as part of the experience of a disaster..."

IF/HOW GROUND ZERO EXPERIENCE CHANGED RESPONDENTS PROFESSIONALLY

The types of responses given to the request for an image begin to build an understanding of how the researchers manifested their professional selves after the catastrophe. A pattern seemed to emerge of a greater balance among their narrower subdisciplinary science specialty roles, their roles as general social scientists, their roles as engaged change agents and their roles as catastrophe survivors.

FIGURE 15.2

The Wall of Vehicles
They had used cranes to stack the vehicles to completely block the four-lane road between the Lower 9th Ward and St. Bernard Parish. The wall was generated by bigotry—to protect the whites of St. Bernard Parish from the Blacks of the Lower 9th, even though they were all dead or gone

Note, above (in the Findings section,) that the images that respondents described were mostly not technical representations of narrow social-science specialties that many of them practiced before the catastrophe. Three respondents linked the image/event to their research specifically; the rest did not, although requested to do so by me. This observation suggests that their intellectual/emotional immersion in the event was greater than merely doing research on it.

Calling to activism and identity integration

While many of the respondents had been advocates for social justice change pre-Katrina, all were university professors or about to become professors (ie., the two doctoral candidates). However, after Katrina their self-images were modified. Respondents embraced the activist role, but as one of their identities that was now being integrated. One respondent clearly noted his/her commitment to this self-image: "We are sociologists. If we can't come up with something useful, helpful, what are we doing? The question was constantly in my mind as I did my research." Another respondent firmly stated their activist involvement by describing their methods: "I applied feminist methodology—agitation and compassion. I was being called to be an advocate." Even though the respondent no longer lives in the *ground zero* area, another expressed a continuing commitment to improve the area: "Your psychology will always be tied to what happened there[...] to make it a better experience there." Finally one respondent described the way in which their different identities became integrated: "Soon after [becoming immersed in the event] there was no difference between my personal, political response—what we can do—and my research agenda." Using the same theme of connecting identities, another concluded, "Now, doing my research has become a way of connecting identities—my teaching identity, service work identity for the university and being a researcher."

Commitment to professionalism

Self-assessments post-Katrina, just as with the evaluation by the respondents of their own research, were very positive: "Katrina made me a better person and helped me to follow my 'Ten Commandments' for academic life."; "It has made me more discerning. My tolerance for frivolousness is zero."; "It has changed me in the sense that I feel much more passionate than I have for years."; "I have grown professionally and become more invested in the research I am interested in doing."; "I'm feeling as if I am doing something (research) that no one else is doing."; "I have been invigorated with new ideas, so that I am not bored with the area of research [...] I am very focused on new ideas, new questions." Another respondent reinforced the same idea of improvement of their research:

> Researchers who do research this way [at ground zero] gain confidence and thus are more willing to push themselves into situations and trust themselves to figure it out. It gives us confidence in our capacity.

IMPLICATIONS FOR RESEARCHERS AND DISCIPLINE

Social-science disaster research conducted by survivors adds more "data" to the case currently being made more frequently; that the social sciences have been wrong in their perceptions of how to be respected sciences. At least in this case of dramatic, life-altering, community-endangering challenges, such as those linked to atmospheric hazards including climate change, research conducted by scientists who have had personal relevant experiences can contribute to the quality of research, the ability to have the research "be heard", and the ability to advocate for useful social change including through policy. Rather than adhering to narrow, distanced research methods, usually advocated as producing the best science, Katrina *ground zero* social scientists followed different, broader paths to successful scientific studies and responsible humane engagement that perhaps even they had not expected to pursue and/or to be successful.

The survivor researcher respondents in this study were responsive to the specific catastrophe conditions they experienced. They were not resistant to refocusing their research, and were energized to select important and interesting (their assessment) research topics and to do high quality research (their assessment), despite the challenges they faced both personally and professionally. Elliott (2015), in the forward of a 2015 book on Katrina research done by survivor researchers,

Rethinking Disaster Recovery, concurs with these findings: "Such conditions simultaneously invite and complicate empirical research for all investigators, but they are especially acute for those living and working in disaster zones" (p. ix).

The respondents in this study wore the hat of their discipline more consciously and wanted their research to be relevant, of a high caliber, and evaluated both inside and outside of the academy as such. They had a broader view of social science—pre-Katrina specialties and subspecialties were not so tightly adhered to; they (re)gained confidence in their basic research ability and were energized in their research, something that some reported had waned. They believed in the contribution that their disciplines could make to address the challenges that emerged from the catastrophe and wanted to make such contributions. Similar conclusions were drawn by Marks (2008), a *ground zero* survivor/researcher of his discipline, social work, as did a multidisciplinary group (anthropologists, geographers, political scientists, psychologists, and sociologists; 5 of the 18 were *ground zero* survivor/researchers) which convened at Mississippi State University at the edge of the impact area (Gill et al., 2007).

The *ground zero* researchers studied in this project also were at peace about "honoring" different dimensions of their professional selves—research, teaching, applied efforts, activism—and were pleased that the event helped them to practice and integrate these different elements. (They added or emphasized the last two elements more than before the catastrophe. Respondents in this study appeared very comfortable in their multiple identity "skins", as one respondent called it (O'Neil, 1972).

This research raises several further questions. For example: Is the study biased toward those social scientists that have been or are inclined to multiple identities? Are social scientists that select environmental and disaster topics, either before or during a catastrophe, self-selected as advocates and activists and comfortable with research done "up close and personal?" This would be similar perhaps to the research experiences of gender specialists. My answer to the latter question is yes: I believe that researchers in disaster and environmental specialties who are subject to such a horrific catastrophe were able to dismiss the "tabooed" boundaries between quality research done with objective distance and quality research done "up close and personal." But those not in these subspecialties also responded similarly. They also queried the historic requirement of separation of research from the advocacy/activism/humanity roles that emerged as so important for the *ground zero* researchers. It is more likely that their experiences impacted them more than their original subspecialties. Sadly, it appears that being "up close and personal" with suffering, discrimination, injustices, some of which respondents have personally experienced, sharpens our humanity and commitment to quality research, outcomes that may be more difficult to achieve without such experiences.

Three additional questions not addressed in this research emerge for me from this research, and from the response to it that I have received including during the Southern Sociological Society Lectures. First, is it possible that the changes in the researchers observed in this study were even greater than have been reported here? Does being a *ground zero* social scientist affect survivor researchers' perceptions of the very nature of scientific understanding and the research/interpretation enterprise? And could this happen because the social scientists have lived through experiences which cause them to interpret disasters differently, achieve critical insights, and come to value new, different research/interpretation methods? My colleague Richard Krajeski believes this to be the case. He suggests that we might go even further to speculate that the respondents have significantly changed their worldview (personal communication, n.d.). These are suggestions for even more powerful impacts than my interview themes elicited.

The second question is about possible major differences between the reception of the disaster survivor researchers who remained in the area and those who left. The work of Haney and Barber (2012) describe a very different experience by those who moved. They found that the researchers who moved are self-conscious of their Katrina experiences and the degree to which they were close to their subjects and their experiences, thus, possibly hinting that their research was not appropriately objective. Does it take remaining at *ground zero* to maintain respect for the *ground zero* researchers and their research within their disciplines, whether within the universities in which they teach or when they go to conferences and submit research manuscripts to journals and for book publications? Not a single respondent in my study mentioned such professional suspicion or disrespect of them or their work by their discipline colleagues. The different findings between the Haney and Barber work and this research warrant more research because of the core scientific values upon which the theme touches.

Third, did some of the researchers who came in from outside the geographic area have similar transformative experiences that affected their research and if so, what was it about their experiences or themselves that caused it to happen, even though they were not survivors of the catastrophe? Are there ways to prepare, sensitize out-of-area disaster researchers before they go into the catastrophe field so that they can approximate the survivor-researcher capacity described by my respondents? Some of the respondents expressed concern about lack of sensitivity, and therefore quality, of the research conducted by them. Given the limitations of space for this article, I did not report those findings.

How to achieve the highest quality professional work on a catastrophe is a sensitive topic—as it should be. Catastrophes warrant the best research and policy recommendations that social scientists can achieve. We are sociologists. If we can't come up with something useful and helpful, then what are we doing?

REFERENCES

Belkhir, Jean Ait (2015). New Orlean's Katrina recovery for whom and what?: A race, gender and class approach. In Jeannie Haubert (Ed.), *Rethinking disaster recovery: A Hurricane Katrina retrospective* (pp. 121–138). Lanham, MD: Lexington Books.

Elliott, J. R. (2015). Forward: Ten years later. In Jeannie Haubert (Ed.), *Rethinking disaster recovery: A Hurricane Katrina retrospective* (pp. vii–x). Lanham, MD: Lexington Books.

Gill, D. A., Clarke, L., Cohen, M. J., Ritchie, L. A., Ladd, A. E., Meinhold, Stephen, & Marshall, Brent K. (2007). Post-Katrina guiding principles of disaster social science research. *Sociological Spectrum, 27,* 789−792.

Haney, T. J., & Barber, K. (2012). Reconciling academic objectivity and subjective trauma: The double consciousness of sociologists who experienced Hurricane Katrina. *Critical Sociology, 39*(1), 105−122.

Haney, T. J., & Elliott, J. R. (2013). The sociological determination: A reflexive look at conducting local disaster research after Hurricane Katrina. *Sociology Mind, 3*(1), 7−15.

Lifton, R. (1997). *Death in life: Survivors of Hiroshima.* New York: Basic Books.

Laska, S, Peterson, K, Rodrigue, C, Cosse', T, Philippe, R, Burchett, O, & Krajeski, R (2015). Layering of natural and human-caused disasters in the context of sea level rise: Coastal Louisiana communities at the edge. In Michele Companion (Ed.), *Disaster's impact on livelihood and cultural survival: Losses, opportunities and mitigation* (pp. 225−238). New York: Taylor and Francis Press.

Marks, R. E. (2008). Canoeing home: A personal and professional journey through Murky Waters. *Traumatology [special issue about Hurricane Katrina], 14*(4), 14−20.

O'Neill, John (1972). *Sociology as skin trade: Essays towards reflective sociology.* NY: Harpers.

CHAPTER 16

How barefoot scholars were deployed: The good, the bad, the ugly

Nancy B. Mock
Tulane University, New Orleans, LA, United States

CHAPTER OUTLINE

Introduction	345
The Lead-up to Katrina	346
Katrina and Early Recovery Efforts	347
Developing the Barefoot Scholar Initiative	348
The Return to New Orleans and the Evolution of RALLY	352
Lessons Learned	353
Post Script: The Crowd and the Cloud	354

INTRODUCTION

This chapter discusses the deployment of barefoot scholars from the perspective of my own experience. We use the term "barefoot scholars" to refer to a largely volunteer workforce of academics, local citizens and students who conducted applied field research in support of the recovery efforts. We use the term "barefoot" to reflect the largely voluntary and field oriented nature of the work. Like the barefoot doctors of China (discussed in Chapter 1), many of the barefoot scholars had minimal training in research methods (eg., graduate students and community members). Others, such as academics, had considerable research training in specific sectors, but not much cross-sectoral research experience. Most of us were participating in experiential learning about the recovery process. In this chapter, we discuss the evolution of a nonprofit organization, Recovery Action Learning Laboratory (RALLY), which provided financial and other support for the barefoot scholars. RALLY was engaged in providing its services for a variety of other nonprofits active in the recovery of New Orleans. In the final part of the chapter, lessons learned about the deployment and organization of barefoot scholars are presented.

When Katrina struck, those affected were initially in a state of chaos. No one really knew what to do. But New Orleans is an education hub in the region, with 5 four-year degree granting institutions and several community colleges. For many of us who were part of major research institutions, the need to bring our skills and local knowledge to bear on the city's recovery was an imperative. Thus, many faculty members from the local and regional area fielded a number of initiatives to bring knowledge in support of recovery efforts. In this chapter, I describe my own efforts as a case study utilizing my email log and field notes as data sources. I discuss the moments running up to the disaster, initial and early recovery efforts to provide evidence in support of recovery efforts, and end with lessons learned for the deployment of barefoot scholars in the future. This chapter will focus on how those efforts unfolded, the barriers and errors made, the successes and finally lessons learned for future disasters.

THE LEAD-UP TO KATRINA

Beginning in the mid-1980s, I developed an intellectual and emotional attraction to the study of disasters and humanitarian work. Much of my early academic work concerned the nature of disasters in low-income countries. It wasn't a stretch for me to see many parallels between low-income countries and low-income cities in the United States, as many of the problems are similar, such as socioecological fragility and chronic high levels of poverty. Perhaps the major difference was that the "more advanced" levels of our governance systems should have been able to provide a robust response to an event such as Katrina. Prior to the storm, I had been doing work with a colleague who had transitioned from international development work at the United States Agency for International Development to working on global health security in the Department of Homeland Security (DHS). We worked together on a project to promote better cross learning from disaster experiences around the world. So when, on August 26, it seemed highly likely that New Orleans might be affected by Hurricane Katrina, I was mentally prepared to engage. I knew from my international experiences that information for decision support would be critical to a successful response and recovery. I also understood New Orleans' vulnerabilities and the potential to address its vulnerabilities during the recovery process. These were all conscious thoughts even before the storm grazed New Orleans on August 29.

My husband, four young children, and I evacuated early to Hammond, Louisiana, a municipality across Lake Ponchartrain. I had been through the evacuations previously and understood the challenges of finding lodging, getting caught in traffic jams and running out of fuel. Having four young children, we decided to do a two-stage evacuation. We left the city on Friday night, overnighted in Hammond, and then reevaluated on Saturday morning.

On Saturday morning, it was becoming clear that we needed to move farther from New Orleans. The forecasts were not looking good as the storm was growing in size and strength. We decided to drive to the town where my husband's uncle lived, well out of range of any major storm effects. We arrived in Nacogdoches, Texas on Saturday evening. It was then that I started planning. I sent an email to students who had taken my most recent disaster class to let them know that I planned to reenter New Orleans and assist in the recovery efforts as soon as possible, but at that time, I had no idea what the outcome of the hurricane would be. I made calls to some of my doctoral students who also were concurrently in the armed services, informing them of my efforts and interest in involving them in the recovery efforts.

KATRINA AND EARLY RECOVERY EFFORTS

On August 29, as the unthinkable unfolded, we sat watching television in our Texas motel room and saw that our city had flooded extensively. Tulane University's internet and email server were completely brought down by the storm and its sequelae. I had already registered on the community forum (internet) page set up by the *Times Picayune*, our local newspaper. I knew about community forums because we owned a house in Pensacola when New Orleans had been affected by Hurricane Ivan, just 1 year prior to Katrina. On-line community forums were the way that we were able to piece together some semblance of a common operating picture of damage to our property and communities. Therefore, I knew to post on-line immediately after the storm struck, providing my coordinates and soliciting volunteers to work with me on recovery efforts. Through the on-line forum, two Tulane faculty members contacted me. One asked what I knew about the situation, noting, "It is incredibly difficult to get any information about Tulane University and its state (8/30/2005)."

One faculty colleague offered his contacts and interest in helping in any efforts that I initiated. He also shared my contacts with other faculty members in our Department of International Health and Development (IHD), including the chair of the department, who also put students in touch with me for recovery work and practicum opportunities. The department chair encouraged students to consider recovery work under my guidance in fulfillment of their capstone experience. By September 2, Tulane had canceled courses for the semester and encouraged students to take courses elsewhere (many universities offered to take Tulane students for the Fall semester of 2005).

This decision (while very sound), along with the federal work-study program administration's decision to link work-study allocations to the campus where the students were taking courses that semester, made it more challenging to engage students in the early recovery efforts. Our senior departmental administrator noted "Right now the Feds are holding firm on the regulation that if our students are attending class at another institution we must transfer their records to that school and then they can work there. I am doing this non-stop (9/13)."

Although I didn't have an outside email service before the storm, I set up a Hotmail account on August 30. Also, I purchased a Verizon subscription and mobile internet connection using our second home in Florida's 850-area code exchange so that I could both send and receive calls and have mobile internet access. I knew that I would be mobile for the foreseeable future. Our 504 area code exchange was not functioning, and in fact was not fully functional until nearly 1 month after the Katrina event. Tulane's internet and email services were also down during the month of September.

Through the on-line community forum, we learned that our home had not been flooded. The community forum provided a more balanced perspective than did main stream media, which focused on sensationalizing the events in the city. The community forum, at least during the first 2 weeks, was a way for citizens to share information about what was actually happening on the ground. The media coverage of the sequelae of the storm was horrifying. There was much speculation by the media that New Orleans would never be rebuilt. We were thankful to know about the community forum. The forum made clear that the "old" city (older neighborhoods) were in pretty good shape: Katrina was not a major wind event in New Orleans and the old neighborhoods were built above sea level.

DEVELOPING THE BAREFOOT SCHOLAR INITIATIVE

In the first weeks post-disaster, I kept focus on recruiting volunteers and maintaining situational awareness for potential partnerships and funding streams. I knew that I could function as a barefoot scholar with very little funding as soon as I could return to my home in New Orleans, although I was also aware that funding would be needed to support students and possibly other faculty members whose homes had been destroyed. I realized that I had to support my students, and it worried me at night, as I lay awake, contemplating how I would accomplish this. I also contacted a few of my disaster management scholar friends (both those who were preeminent scholars and those who were particularly entrepreneurial). They also didn't have a clear path to funding as stated by one of my close colleagues: "Nance for the moment I don't know if I can access any money anywhere. Hold on until I figure out what's possible (9/8)," and "Nancy, there is no capacity to initiate cash flow and won't be for some time (9/9)."

I knew what was needed was what we barefoot scholars could provide: assessment and analysis support to organizations trying to respond. I knew, like many of my colleagues and neighbors, that we would somehow make it work. It was the right thing ... somehow funding would fall in place. I thought what we were doing should have some type of identity, because early on it was obviously a project. On September 3, we came up with the name "Recovery Action Learning Laboratory". It had the acronym RALLY, and it really stuck. The use of the name RALLY started to appear in communications beginning September 6. Little did I know that later, because we got more hits than a similarly named Atlanta-based fund raising nonprofit organization, we would end up being threatened with a law suit because we unknowingly used the same acronym.

The first few weeks after Katrina struck were very frustrating for us because we couldn't return to New Orleans, as the City was shut down. We still had poor communications, our staging area, Baton Rouge, was suffering a deluge of responders and "disaster-relief" tourists who occupied all housing. "Housing in BR is a real problem. I've heard that all available apartments/houses are taken, and realtors are getting 100 calls a day from people looking to buy. Do you know of anyone who is helping students to find housing (from LSU for example)? (9/5)."

Despite the housing crunch, I had decided to organize my RALLY team in Baton Rouge because I had disaster research colleagues at Louisiana State University (LSU) whom I contacted on September 2. They invited me to participate in recovery projects with them, even offering me office space. My problem was finding lodging for my students and myself. I found various temporary options: a flea bag motel for a few nights (had rats and roaches) outside of Baton Rouge or the home of a faculty colleague's friend for a couple nights. My husband had a good law school friend who had bought a small house for his children attending LSU, and he offered me a room in the house and couch space for my students. We were able to find limited lodging, but we still asked students to cover their transportation and didn't yet have any funding identified to support their basic costs of living. Many of them qualified for Federal Emergency Management Agency (FEMA) support, but FEMA was yet not clear about what support would be provided and how long it would continue.

A core group of volunteers began to materialize as early as August 31. The first student volunteer, a new student from New Mexico, began helping me on August 31, by setting up a chat forum for faculty. By September 5, I had a core team of three volunteers working virtually to organize the RALLY team. They were working from different physical locations around the country and in Italy. On September 5, I sent out the first group communication to students who had requested to be considered part of our team:

Dear Folks,

I want you to know each other as I begin to move more rapidly now on RALLY (Recovery Action Learning Laboratory). M is a new student, currently in New Mexico, who jump started communications and has helped a great deal. K is a veteran CE/DM student. B has provided my primary information technology support at the Payson Center.

My first need from the team is to organize a more permanent solution to team building and connectivity among the RALLY community. B, I am hoping that you can work with M on this. M and K, I would like to ask you to begin to organize a data base of student volunteers. I want all of their contact information as well as the list of skills. I also want to know if each student is able to relocate.

Any progress you can make on this would be greatly appreciated. Nancy (9/5).

My initial efforts were focused on three levels:

- Institutionally based efforts through Tulane University, largely pursuing traditional funding sources such as the DHS and the National Science Foundation (NSF).

- A project that was already on the burner to develop a joint disaster resilience PhD program between Tulane and the Department of Defense.
- Small-scale research assistance to organizations working in the recovery efforts locally.

This was the Katrina agenda, though I still had a plate full of international work that I was doing through Tulane. International projects did not stop even though the university was not operating at full capacity during the fall semester. I had four international projects and was also a member of two international expert groups on disasters and humanitarian assistance that were active that semester. Thus, the early recovery period was hectic. During the month of September, we primarily worked with LSU to develop proposals for DHS and the NSF. We were given much encouragement from funding agencies, though the proposal processes were quite long and not adapted to the rapid needs of early recovery. Even the funding agencies were struggling to deal with the novelty of the Katrina situation.

A number of problems surfaced during our pursuit of traditional national research funding. First, agencies gave no specific guidance for the inclusion of local research institutions on research proposals. Tulane University was a tier one research institution. In my experience working internationally, our funding agencies and our teams gave priority to local research teams wherever possible, both because it resulted in better research and because it helped build the local economy. This was not the approach in the domestic research arena.

Related to this, and because American universities are highly competitive, many of us found that our research colleagues in other institutions were somewhat exploitative of our compromised position to compete (no infrastructure, library access, etc.). Many of us were surprised that various prestigious institutions had planted their own flags on our soil without considering the merit of including a local tier one research institution. When combined with the poaching of students and faculty by various institutions, this greatly handicapped local researchers.

However, the most successful and useful Katrina recovery efforts we undertook were those focused on small-scale research assistance to organizations working on the ground. This is what I considered to be the real success of barefoot scholarship. Although it took a while to establish relationships with organizations, RALLY turned out to be a dynamic small nonprofit organization, well known in the city of New Orleans, by public and private agencies, for providing useful decision support, largely by the students that made up our team.

By September 10, we had 35 students who wanted to volunteer and return to the region. They were scattered over the country, so most needed transportation and lodging costs covered in order to return to New Orleans. At that point, they fell into two groups: one decided to pursue course work where they were located (usually their hometowns, with other family members or hometowns of their friends). A smaller group decided to come and had faith that somehow we would find the resources to do our work. Box 16.1 shows the enthusiasm, concerns and geographic spread of student volunteers as reflected in emails they sent to me during the first few days after the Katrina event. All of the students were highly enthusiastic about helping out in the relief effort, and one of these students felt a great attachment to New Orleans. Students were located across the United States, and some were returning from overseas internships. Several students were trying to decide if they would remain at Tulane University, and others were concerned that they would need financial assistance in order to volunteer. One student was most concerned about applying what had been learned at Tulane to a real-life crisis as reflected below.

BOX 16.1 A SAMPLE OF EMAILS FROM STUDENT VOLUNTEERS

9/2 "I am an IHD student. I spent this past summer in Sri Lanka doing my capstone on post-tsunami work. I am very interested in offering assistance for the Hurricane Katrina relief operations. Please let me know if there is any way that I could get involved. Thank you."

9/2 "I am a student of INHL department, now I am in Atlanta, GA as evacuee of Tulane, I want to participate in assisting the disaster response efforts of New Orleans with all my physical and intellectual capacity."

9/3 "I am a student in nutrition/food security, [...] will do ANYTHING. Please let me on board, though I haven't even read what you sent out I want to help!!!"

9/3 "I am a IHD student at Tulane. Dr. ... told us to contact you if we were interested in possible opportunities to help with relief efforts. I would love the opportunities to put to use things we have learned and help the people and the city I have grown to love over the last year. Please keep me abreast of anything that comes up."

9/3 "Dr. Mock, I hope you are well and dry, and that you and your family have weathered the storm as well as possible. Like most folks who have left N.O. I am weighing my options for the future. The most appealing of these is of course assisting in the relief effort. If you have any information concerning IHD students returning to help the people of New Orleans, please drop me a line as soon as you can."

9/3 "I am looking for any work down in New Orleans, I just returned from Sri Lanka, and of course a bit overwhelmed right now, but need to do something. I only have 3 classes left to graduate and I refuse to relocate to another school. I want to finish in New Orleans! Hopefully this spring, but for now I am looking for work down there. I obviously can't afford to volunteer; I was depending on financial aid for school and living, so if you know of anything or if I can work for you again I would do it in a moment's notice. Please help, I am sure you are overwhelmed right now and could use the help. Please be in contact. I am staying with my parents in Oklahoma and my number is ...."

9/9 "knew you were out there somewhere working with the relief efforts, and I'm happy to have finally tracked you down. I'm in Houston right now and I've been working with Red Cross shelters as the first aid person and also with the Harris County Public Health and Environmental Services to conduct rapid health assessment of the evacuees to monitor trends in diseases. I'd like to get involved with your team in the relief efforts, and I'm trying to make my decision as to what I'm going to do this semester. I've got the option to take classes here at UTSPH and continue doing what I'm doing volunteer wise, go to Columbia University (all I need to do is register for classes), or work with you and your team. Please get back to me ASAP. Please review the attached resume, it's been condensed to 1 page."

By late September, I had decided that the RALLY approach should be neighborhood-oriented initially. This was because New Orleans was organized into neighborhoods with identities, and it was clear that neighborhoods were manageable units to understand. This meant that my student teams also would be able to concentrate their work in small areas.

THE RETURN TO NEW ORLEANS AND THE EVOLUTION OF RALLY

On September 27, I was allowed back into New Orleans. Residents were allowed to reenter New Orleans according to the level of damage in their neighborhoods. Mine was among the first zip codes to be allowed back into the city. I remember driving into the city on River Road from Baton Rouge. The early evening was hazy, and as I drove into the edge of the city, I drove past Cooter Browns, a local bar and eatery. It was full of people like me that had just returned. It also had internet, so it was a draw to local residents. However, there were still lots of fallen trees and debris along the sidewalks. I arrived at my house. Our block was abandoned except for two other houses. I walked into my house, and, to my disbelief, all utility services worked. We were good to go in New Orleans, and from that day, I invited any student volunteers who needed housing to stay in my house. I set up my dining room as our office, and we began to work from there beginning on September 28. We started to do rapid assessments of neighborhoods, looking to identify needs and potential partners.

Meanwhile, one of my colleagues based in Italy was able to set up a small campus for students working on our joint international projects. These students remotely contributed to RALLY. They were paid and supported by our joint international project, but they used their free time to support RALLY.

By mid-October, we zeroed in on the Treme as our first neighborhood project. Through a contact with a friend of one of my close colleagues, I was able to identify a neighborhood partner in St. Augustine's Catholic Church. St. Augustine's had a dynamic pastor, Jerome LeDoux, who was enthusiastic about using the church as a base for community recovery. At the same time, a former student of mine working for Second Harvest contacted me to find venues for establishing food pantries. It was an excellent match with St. Augustine Church, so RALLY collaborated with Second Harvest to set up a food pantry in St. Augustine's Church. It was very energizing for the students and it gave us presence in the Treme. We set up a client feedback system at the entry of the pantry and, based on that feedback, we developed various recovery services for clients. The pantry operated successfully until March of 2006, when the Catholic Church decided to decommission St. Augustine's Church as a parish (a contentious decision that the Catholic Church would later reverse). We began to monitor the recovery of Treme through door to door neighborhood surveys.

As an outcome of our work in Treme, with our focus on collaboration with the City of New Orleans to estimate early population recovery and our participation in recovery coordination meetings, word of mouth about RALLY began to produce many clients for our applied research services. Beginning in early 2006, we secured small contracts from local and international organizations who were doing or planning recovery work in New Orleans. Because of my international work, organizations such as International Medical Corps and World Vision reached out to me to discuss collaborative work. Because of the growing profile of our work, Save the Children and MercyCorp and other global organizations also contacted us to undertake survey and evaluation work to inform their efforts.

A large number of local organizations also engaged our services in many sectors including housing, criminal justice, neighborhood safety, education, community surveys, and others. The activity level accelerated when we contacted Baptist Community Ministries, the largest granting organization in New Orleans, who became an excellent client of and donor to RALLY work. RALLY generated small grants/contracts between 2006 and 2010 that enabled students and recent graduates to participate in New Orleans' recovery. A core group of 20 students participated at various times over the 4-year period. We decided that RALLY would never be a large organization and to phase it out after the early recovery phase of Katrina.

The RALLY project supported recovery in many ways, including:

- counting New Orleans's repopulation,
- assessing community needs in numerous neighborhoods,
- following the recovery of two neighborhoods,
- assisting neighborhood organizations to collect and utilize household data,
- assessment of psychosocial interventions in trailer parks,
- helping test the applicability of psychosocial interventions applied outside the United States to our local context,
- developing collaboration metrics for network building initiatives,
- evaluating pilot initiatives, such as disaster case management and interventions aimed to improve the efficacy of the criminal justice system, and
- assessment of public housing client needs.

This barefoot scholar work was among the most rewarding of my career. Our volunteers, who ultimately received subsistence pay, were able to live in and contribute analytical skills to the recovery of New Orleans. Interestingly, none of the larger initiatives that we pursued with other partners materialized into substantive efforts.

LESSONS LEARNED

Barefoot scholarship is an important element in the recovery process. Local scholars in affected communities provide continuity through the recovery process.

They are low cost, but, for various reasons, it is challenging to implement these types of initiatives. These reasons include:

- They may be hindered by poor communications in post disaster events.
- Outside institutional policies and practices lead to poaching local talent and competition with local universities. This could be remedied to some degree if granting agencies provided incentives for including local research institutions in recovery research.
- There is no central place where funding opportunities for applied research can be found during the recovery process. This might be addressed by identifying applied research as a priority for the recovery process and ensuring that there is a central clearing house for applied research funding.

There was little preparedness for early recovery applied research among local academics. This meant that we were designing and implementing in an ad hoc fashion, which slowed the process of securing resources for work after the disaster event. Preparedness activities that would have greatly facilitated our work include

- maintaining a data base of students who have taken disaster related classes, including contact information and their interests in virtual and field oriented work;
- maintaining a roster of faculty members and community leaders who would serve as leaders in disaster recovery research; and
- making contacts with major local donor agencies prior to disaster events to make sure they are aware of resources and contact persons for disaster recovery applied research.

Early engagement is key for establishing presence and profile after the disaster event. While it is important not to interfere with early recovery interventions, it is also important to establish evidence-based management as best practice. Barefoot scholars can do much with few financial resources. Early engagement on the ground is more important than writing grant proposals. Our team probably lost 6 weeks of field work by trying to write proposals that ultimately were not productive. Because of the delay in early field work, we lost many potential volunteers.

Disaster recovery applied research provides a phenomenally rich venue for active student and faculty learning, as well as tremendous opportunities for academia-community engagement. Because of its Katrina related experience, Tulane University developed one of the most progressive service learning programs in the country. Hopefully in the future, programs like that can be deployed for local disaster recovery.

POST SCRIPT: THE CROWD AND THE CLOUD

Katrina was the last major disaster event before the application of crowd sourcing and cloud computing to disaster management/recovery. Subsequent to that time, digital disaster volunteers have emerged as a new form of barefoot scholar assistants. These volunteers are instantaneously mobilized when disaster events occur. In 2008, crisis-mapping applications were developed to map post-election violence in Kenya. By 2010, crisis mapping combined with global microtasked translation and natural language processing resulted in the development of near real-time crisis mapping of the Haiti earthquake and its sequelae. The incorporation of digital volunteers as barefoot scholar assistants is important, but not without challenges. Digital volunteers need trusted eyes and ears on the ground to triangulate findings. Incorporating digital volunteers into barefoot scholarship is one of the frontier issues for evidence-based disaster management.

CHAPTER 17

Lessons learned from New Orleans on vulnerability, resilience, and their integration

Michael J. Zakour

West Virginia University, Morgantown, WV, United States

CHAPTER OUTLINE

Economic Inequality Causal Chain	359
Migration of the Poor	359
Housing and Health Crises	360
Lack of Economic Resources	360
Social Stratification Causal Chain	361
Lack of Formal Education	361
Lack of Human Capital	362
Lack of Economic Growth	362
Structures of Domination Causal Chain	362
Trusted Media	363
Media Controlled by Elites	363
Lack of Political Partnerships	364
Racial Ideology Causal Chain	365
Marginalization	365
Segregation	366
Separate and Unequal	366
Geographic Distance Causal Chain	367
Smaller Geographic Service Ranges	367
Low Volunteer Capacity	368
Environmental Ideology Causal Chain	368
Population Growth	369
Rapid Urbanization	369
Community Empowerment and Social Development	370
Community Empowerment Causal Chain	370
Evacuation Experience	372
Client-Centered Services	372
Place Attachment	373

| Human and Social Capital 374
| Flexible Disaster Plans 374
| **Social Development Causal Chain** 375
| Public Funding 375
| Population Programs 376
| Health Programs 377
| Building Codes 377
| Economic Growth 378
| **Assumptions of V+ Theory** 378
| Unsafe Conditions Assumption 379
| Root Causes Assumption 379
| Disaster Causal Chain Assumption 380
| Assumption of Capabilities, Liabilities, and Susceptibility Relationships 381
| **Conclusion** 381
| **References** 382

Vulnerability, resilience, and their relationships are at the forefront of new developments in disaster theory and research. The complex relationships between vulnerability and resilience is an area of interest to disaster theorists, researchers, and practitioners (Etkin, 2016). In this chapter, we review the lessons learned about vulnerability and resiliency, and their relationships, in the context of Vulnerability-plus (V+) theory (Zakour & Gillespie, 2013). First, we examine evidence that supports V+ theory. We examine the empirical literature on vulnerability and resiliency in the Katrina disaster, with a special emphasis on evidence from the chapters in the present volume. Support for the assumptions of V+ theory are identified and summarized.

We explore the empirical evidence supporting V+ theory by discussing each of the root societal causes of disaster vulnerability: (1) economic inequality, (2) social stratification, (3) structures of domination, (4) racial ideologies, (5) geographic distance, (6) environmental ideologies, (7) community empowerment, and (8) social development. The first six root causes are liabilities, while the last two are capabilities (Zakour & Gillespie, 2010; Zakour, 2010). For each of these eight root causes in the V+ model, we then examine the causal chain of the progression to vulnerability originating in that root cause. We organize this discussion of causal chains by the structural constraints that initiate each causal chain. Causality is generally from left to right, except that disasters and resilience resources have symmetrical causality, and hazards interact with safety of conditions to produce disasters. Hazards and disasters are also able to damage resilience resources.

The level of vulnerability interacts with the hazards associated with Hurricane Katrina to produce the disaster. The hazards in this volume are the natural hazard of Hurricane Katrina, the technological failures of levees and floodwalls, and the human-caused hazard of the breakdown of the emergency management system.

For each of the eight causal chains, losses in disaster are demonstrated to be the result of the level of vulnerability represented by unsafe conditions. Finally, the resilience resources and capabilities in the V+ model that moderate the effects of the Katrina disaster in each causal chain are identified and described.

ECONOMIC INEQUALITY CAUSAL CHAIN

The first root societal cause of vulnerability we examine is economic inequality (Fig. 17.1). Economic inequality is an important root cause of disaster vulnerability in New Orleans and Southeast Louisiana, and is guided and justified by a capitalist ideology. Both inequitable distribution of income and the ideology of inequality have promoted a highly skewed wealth distribution in New Orleans, making Southeast Louisiana more disaster vulnerable, especially for low-income populations (Bassett, 2009; Zakour & Swager, This Volume, Chap. 3).

MIGRATION OF THE POOR

Over generations and even centuries, the inequitable distribution of wealth in New Orleans has pushed poor and African-American residents off the Natural Levee of the Mississippi and onto marginal, low-lying land (Campanella, 2006). The structural constraint of migration corresponds with the movement of the population, especially the poor, onto low-lying land either through necessity (i.e., cheaper land and rents) or force. Unsafe conditions included substandard housing located on flood-prone land contaminated by toxic substances.

Because of the high level of vulnerability, the hazards of Hurricane Katrina, levee failures, and breakdown of the response system led to the deaths of nearly 1600 people in New Orleans and Southeast Louisiana and flooding of 80% of the

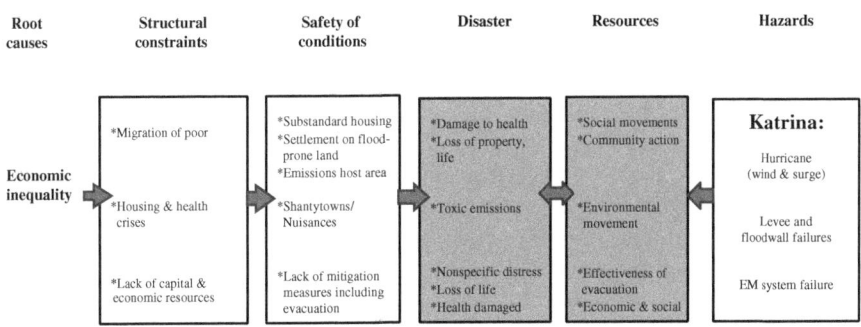

FIGURE 17.1

Economic Inequality Causal Chain

Source: *Adapted from Zakour, M. J., & Gillespie, D. F. (2013)*. Community disaster vulnerability: Theory, research, & practice. *New York: Springer Science.*

city (van Heerden, This Volume, Chap. 6). A resilient recovery was fostered among some populations by the resources of social movements and community action (Laska, Howell, & Jerollerman, This Volume, Chap. 5). Some social aid organizations expanded into social movements (Adams, 2013; Solnit, 2009; Zakour & Swager, This Volume, Chap. 3).

HOUSING AND HEALTH CRISES

Along with income inequality, the structural constraint of housing and health crises (Honoré, 2009) has promoted the growth of unsafe conditions, including shantytowns or slums in New Orleans. The environment in these slums has been noxious, including heavy metal contamination resulting in health crises (Campanella, 2006; Zakour & Grogg, This Volume, Chap. 7). Environmental movements with a social justice component provided a needed resource for recovery.

LACK OF ECONOMIC RESOURCES

A lack of economic resources is a third structural constraint caused by income inequality. Some studies show that income inequality has had an important effect on the unique culture of New Orleans (Bullard & Wright, 2009). The high levels of poverty and income instability acted as a structural constraint driving vulnerability (Zakour & Swager, This Volume, Chap. 3). This caused unsafe conditions, especially a lack of adequate disaster mitigation through public systems or household evacuation (Haney, Elliott, & Fussell, 2007; Jones-Deweever, 2011). African Americans, Hispanics, and Asians were more likely to experience distress after disasters, largely because these populations were disproportionately poor and lacked resources for evacuation (Joseph, Matthews, & Myers, 2014).

Those stranded in the flooding were exposed to the worst of Katrina and were less likely to experience a resilient recovery (Colton, 2006; Haney et al., 2007). Lacking a private automobile or other private means of evacuation meant that African Americans and the poor were unable to return home to rebuild, because busses and airlifts took survivors to distant locations and separated families.

In general, the loss of life, property, and labor during the Katrina disaster was not moderated by a system of resource distribution. Economic aid produced little growth after Katrina (Downey, 2016). Poor African Americans lacked the capital and other capabilities needed for a resilient recovery (Cutter et al., 2006). Because of the failures of the recovery system, many middle-income and poor households went further into debt (Adams, 2013; Zakour & Grogg, This Volume, Chap. 7). Finch, Emrich, and Cutter (2010) found that both vulnerability and flood exposure were negatively associated with recovery.

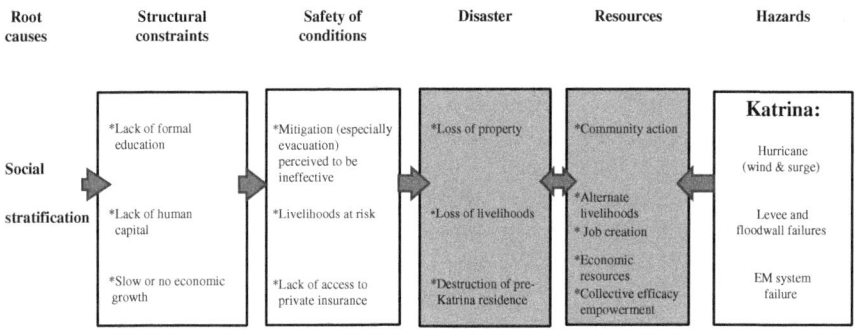

FIGURE 17.2

Social Stratification Causal Model

Source: *Adapted from Zakour, M. J., & Gillespie, D. F. (2013).* Community disaster vulnerability: Theory, research, & practice. *New York: Springer Science.*

SOCIAL STRATIFICATION CAUSAL CHAIN

Since the city's founding, social stratification has been high in New Orleans. Both populations and neighborhoods have been highly stratified. Slavery and Jim Crow laws were among the worst manifestations of stratification (Fig. 17.2). More affluent whites have resided mostly on the Natural Levee of the Mississippi, the higher ground near Lake Pontchartrain, and in suburbs in Jefferson Parish adjacent to and west of Orleans Parish (Colton, 2006; Zakour & Grogg, This Volume, Chap. 7). In 2000, New Orleans was highly stratified, and ranked second among all US cities for concentrated poverty (Downey, 2016).

LACK OF FORMAL EDUCATION

Lack of formal education in some neighborhoods was a structural constraint interacting with social stratification in New Orleans (Fig. 17.2). People in neighborhoods with low levels of formal education mistrusted the governmental response system, and accurately perceived public mitigation measures, especially evacuation, to be ineffective. These neighborhoods were less likely to be able to evacuate during the voluntary and mandatory evacuation period in New Orleans.

After Katrina struck, when 80% of the area in Orleans Parish was flooded, sometimes for up to 6 weeks (van Heerden, This Volume, Chap. 6), property and other losses were severe. Fortunately, Social Aide and Pleasure Clubs and community organizations emerged to provide recovery resources (Adams, 2013; Solnit, 2009). Church-based volunteers and social movements provided vulnerable people with labor and resources necessary for collective efficacy empowerment (Adams, 2013; Solnit, 2009).

LACK OF HUMAN CAPITAL

Another structural constraint caused by social stratification was the low number of skilled workers in some neighborhoods and other geographical areas of New Orleans (Zakour & Harrell, 2003). Livelihoods were fragile because most of the metropolitan area's economy was dependent on service jobs (Downey, 2016), leading to unsafe conditions. The livelihoods of lower income people, including racial and ethnic populations, were at high risk from economic downturns and disasters (Zakour & Swager, This Volume, Chap. 3).

During the Katrina disaster, many people employed in the tourism and service industries lost their livelihoods (Downey, 2016). Because of unemployment, fewer people were able to return to New Orleans. For those returning to New Orleans, alternate livelihoods were limited, given Southeast Louisiana relied on industries disrupted by Hurricane Katrina, such as resource extraction and tourism industries. The percentage of people without a high school education in a neighborhood was negatively and significantly ($P < .01$) related to new job creation (Downey, 2016).

LACK OF ECONOMIC GROWTH

The high level of social stratification among neighborhoods (census tracts) in the New Orleans Metropolitan area was due in part to the structural constraint of stagnant local economies, and the related high levels of unemployment (Downey, 2016). As a result, many in the poorest neighborhoods in New Orleans could not afford private homeowners or renters insurance, a situation that led to fragile livelihoods and unsafe conditions. Fussell and Harris (2014) found that after Katrina, low-income African-American homeowners and renters, especially if they had private insurance, were more likely to be able to return to their pre-Katrina residences, especially compared to subsidized housing residents (Section 8 and public housing). In contrast, those with destroyed residences were less likely to return (Fussell, Sastry, & VanLandingham, 2010; Kim & Oh, 2014; Zakour & Grogg, This Volume, Chap. 7).

Sadly, for most neighborhoods and populations, resilience resources were not accessible. Low-status neighborhoods received inadequate economic resources during the recovery. This lack of needed resources was partly because corporations benefited by keeping a large portion of governmental aid while subcontracting-out work (Adams, 2013; Downey, 2016; Freudenburg, Gramling, Laska, & Erikson, 2009; Zakour & Grogg, This Volume, Chap. 7).

STRUCTURES OF DOMINATION CAUSAL CHAIN

The root cause of structures of domination in New Orleans and Southeast Louisiana led to a high level of disaster vulnerability (Fig. 17.3). Structures of

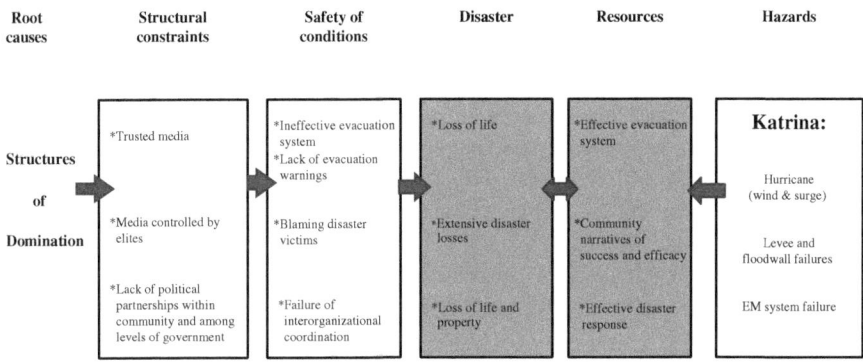

FIGURE 17.3

Structures of Domination Causal Model

Source: *Adapted from Zakour, M. J., & Gillespie, D. F. (2013).* Community disaster vulnerability: Theory, research, & practice. *New York: Springer Science.*

domination refer to unequal relationships among categories of people in the household (e.g., men and women), the community (e.g., employers and employees), and in society (e.g., Whites and Nonwhites). It also refers to the nature of governance, the relationships among government and the people they govern, and relationships among levels of government (Wisner, Blaikie, Cannon, & Davis, 2004).

The root cause of the Katrina disaster was based, in part, on the failures of government at all levels, especially the federal government. These failures had been intensifying since 9/11 and the Bush Administration (Parker, Stern, Paglia, & Brown, 2009). Failure was severe for the poor, the elderly, and persons with mobility limitations and other disabilities (Colton, 2006). Government wished to avoid the obligation to aid people forced to evacuate in a random manner after Katrina. This is part of the neoliberal tide in cities like New Orleans (Kadetz, This Volume, Chap. 13).

TRUSTED MEDIA

Distrust of the media, a structural constraint, made evacuation warnings less credible. The lack of a public or private household means of transportation left the poor very vulnerable when Katrina approached New Orleans, and when flooding and elite panic began (Colton, 2006). The failure of evacuation, including postdisaster evacuation, exposed populations to all of the horrific conditions of Katrina, and led to a high death toll.

MEDIA CONTROLLED BY ELITES

Related to distrust of the media is the structural constraint of an elite-controlled media. The news media had favored elites and helped reduce elite liability during a crisis. In contrast, the media portrayed AfricanAmericans and the poor negatively (Parker et al., 2009; Zakour & Grogg, This Volume, Chap. 7). If African-American, poor, or elderly persons did not evacuate, and experienced more severe losses from disasters, these vulnerable populations were blamed for inaction, a culture of fatalism, or stubbornness. A narrative that blamed the victim developed (Laska et al., This Volume, Chap. 5). When they were evacuated, vulnerable populations were given the negative label of "evacuees" (Kadetz, This Volume, Chap. 13). The evacuee label absolved government from responsibility in helping people to return, and blamed them for failure to return and rebuild.

After Hurricane Katrina hit New Orleans and the levees failed, the Katrina disaster caused massive loss of life and of property. Emergency managers were not prepared for a major disaster (Parker et al., 2009). Community narratives of success in the face of the Katrina disaster were dampened by blaming the victims, and by elite panic. The media and economic elites supported the image of poor African Americans as savage, murderous, looters who destroyed large amounts of property for their own benefit. Stories of altruism among survivors were largely ignored (Solnit, 2009; Zakour & Grogg, This Volume, Chap. 7). Community narratives after Katrina were shaped by the media, and by the structural inequalities embedded in bureaucratic relief organizations. In an example of positive community narratives, shared narratives of survival and frames that viewed relationships as hierarchical allowed the Vietnamese Community to recover resiliently (VanLandingham, This Volume, Chap. 11). Spirituality and faith-based organizations also fostered community narratives of success and efficacy (Airriess, Li, Leong, Chen, & Keith, 2008; Kreutziger, Blue, & Zakour, This Volume, Chap. 12).

LACK OF POLITICAL PARTNERSHIPS

The structural constraints increasing vulnerability to Hurricane Katrina were primarily the lack of political partnerships among government at all levels, especially the local level (Parker et al., 2009). A lack of coordination among disaster organizations was partly because of the failure of the New Orleans network to establish a partnership with Emergency Management Assistance Compacts (EMACs) with extensive past experience and capacity, such as the Florida EMAC (Kapucu, Augustin, & Garayev, 2009).

The ability of forcibly evacuated survivors to meet their basic needs and return to their pre-Katrina homes was met with structural violence by government and well-connected elites (Kadetz, This Volume, Chap. 13, Laska et al., This Volume, Chap. 5). This was part of the emergency management system's collapse during Katrina. Because of standard operating procedures of emergency management organizations during the George W. Bush administration, there was a complete failure of coordination among all levels of government and local actors. In addition, there was a process of denial and distraction regarding major disasters at the federal level, lack of coordination between federal and local levels and among all governmental actors, and a shift at the federal level to homeland security, wars in Iraq and Afghanistan, and the fight against terrorism (Parker et al., 2009).

RACIAL IDEOLOGY CAUSAL CHAIN

New Orleans has been highly vulnerable since 1960, partly because of the high level of poverty and percentage of poor African-American residents. It ranked among the top 3% of all parishes in vulnerability in 1960. Even after Katrina, Orleans, St. Bernard, and Plaquemines Parishes still are more vulnerable than 85% of all counties in the United States (Cutter et al., 2006).

MARGINALIZATION

Racial ideologies about the inferiority of African Americans have guided and justified the sociopolitical marginalization of African Americans in New Orleans (Fig. 17.4). Before the Civil War, and during the Jim Crow Era, African Americans were disenfranchised socially and politically, and they have experienced this structural constraint since 2000 (Fussell et al., 2010). Organizations and bureaucratic procedures enforced the idea that African Americans were not deserving of public help, and this exercised structural violence against African Americans (Laska et al., This Volume, Chap. 5). African-American neighborhoods in Orleans Parish (county) suffered from a lack of mitigation policies and projects, especially compared to predominately White Jefferson Parish (Zakour & Harrell, 2003; Zakour & Grogg, This Volume, Chap. 7).

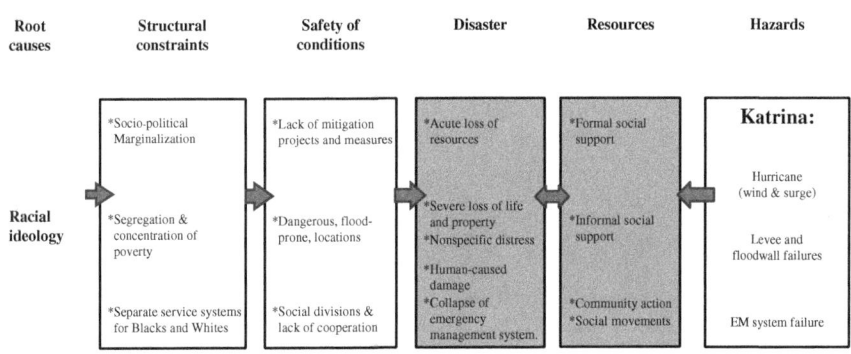

FIGURE 17.4

Racial Ideology Causal Model

Source: *Adapted from Zakour, M. J., & Gillespie, D. F. (2013).* Community disaster vulnerability: Theory, research, & practice. *New York: Springer Science.*

High levels of vulnerability for low-income African Americans and other poor populations lead to severe losses from Katrina, including the acute loss of resources. However, there were limits to the quantity and quality of social support (see Kaniasty & Norris, 2009). Women were more likely than men to receive all types of emotional support, and they were more likely to receive housing from individuals rather than community sources (Roberto, Kamo, & Henderson, 2009).

SEGREGATION

Sociopolitical marginality has been reinforced by segregation and concentration of African Americans in eastern areas of New Orleans (Campanella, 2006; Campanella, This Volume, Chap. 2). In New Orleans, segregation was prevalent into the 21st century (Campanella, 2006; Zakour & Grogg, This Volume, Chap. 7). Segregation resulted in location of African Americans on marginal land exposed to long-term flooding from Katrina (Fussell et al., 2010).

With many African Americans concentrated in eastern New Orleans, their tax base has been relatively poor. Their homes were unlikely to be engineered or retrofitted for wind and flood resistance (Fussell et al., 2010). The eastern areas of New Orleans in which African Americans predominate have been more exposed to hazards such as flooding than the rest of the metropolitan area. The major canals in eastern New Orleans are all connected to the Gulf of Mexico. This area is next to the highly degraded and eroded wetlands that allow surges to pass with little friction (Campanella, This Volume, Chap. 2).

African-American, low-income, female-headed households with young children experienced the greatest shortage of recovery resources (Joseph et al., 2014). During Katrina, African Americans, the poor, elders, and persons with a disability experienced a high loss of life, especially in the Lower Ninth Ward when the floodwalls were overtopped and collapsed. Given severe damage to homes and neighborhoods, levels of distress were very high (Curtis, Mills, & Leitner, 2009; Fussell et al., 2010; Haney et al., 2007). In a study of older African Americans who relocated after Katrina, age (range = 60–97) was negatively associated with help from community sources. Despite a strong level of informal social support, additional help was needed from community organizations and programs (Roberto et al., 2009).

SEPARATE AND UNEQUAL

A final aspect of racism and racist ideologies has been the institutionalization of separate service systems for African Americans and Whites. Separate systems in the half-century before Katrina included separate Boards of Regents for higher education systems, and a similar separation of schools and healthcare into public and private. This separation of service systems for the White and African-American populations in New Orleans made the coordinated evacuation of the population difficult, and caused evacuation to fail for around 100,000 poor

African Americans. Social divisions produced a lack of cooperation among organizations and systems serving only African Americans or only Whites. Because of racial ideologies and the political marginalization of African Americans in New Orleans, the emergency management system was very fragile before Katrina (Cutter & Emrich, 2006).

After Katrina, there was a massive increase in community action and social movements. Social action was aimed at a "just" recovery, and inequalities were revisited to assure social justice for the poor and African Americans. Community action and social movements were among the most helpful resilience resources (Adams, 2013; Kreutziger, Blue, & Zakour; Solnit, 2009; Zakour & Grogg, This Volume, Chap. 7).

GEOGRAPHIC DISTANCE CAUSAL CHAIN

Geographic distance in New Orleans represents a formidable barrier for resource redistribution in disasters (Fig. 17.5). Marginal geographic areas with vulnerable populations also mirror the sociopolitical marginality of these populations. Because most disaster organizations lack metro-wide geographic service ranges, high personnel capacity, and programs of trained volunteers, geographic distance from central New Orleans limits service delivery of all types to peripheral geographic areas, such as New Orleans East and the Lower Ninth Ward (Zakour & Harrell, 2003).

SMALLER GEOGRAPHIC SERVICE RANGES

An important structural constraint is that fewer organizations serve the high-risk areas in which vulnerable populations reside. This is both because organizations have a smaller service range (catchment area), and because of the lower disaster

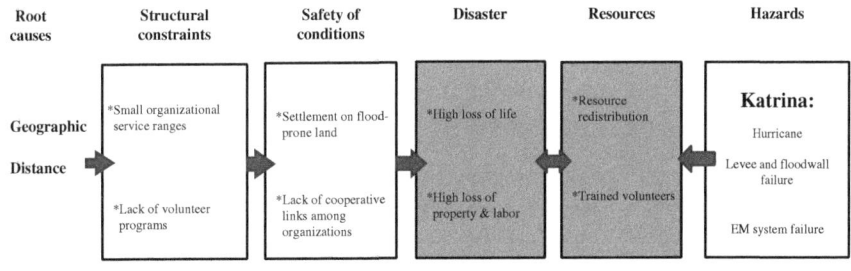

FIGURE 17.5

Geographic Distance Causal Model

Source: *Adapted from Zakour, M. J., & Gillespie, D. F. (2013).* Community disaster vulnerability: Theory, research, & practice. *New York: Springer Science.*

services capacities of organizations. This lack of disaster organizations is a particular problem for areas like the Ninth Ward of New Orleans, because it is located next to canals, Lake Pontchartrain, and areas that channel water into New Orleans. The areas in which vulnerable populations reside are dangerous locations and have a higher biophysical risk of disasters such as flooding and hurricanes (Campanella, 2006; This Volume, Chap. 2; van Heerden, This Volume, Chap. 6). The land is of lower elevation, having sunk from draining swamps for housing developments. Settlement in areas such as the Lower Ninth Ward, separated from the Industrial Canal by only a floodwall, led to a high loss of life when water overtopped floodwalls and they collapsed.

LOW VOLUNTEER CAPACITY

Most disaster services organizations serving vulnerable populations lack adequate service capacities including well-managed volunteer programs, and this represents an important structural constraint. The poverty and the poor tax base of the Ninth Ward helps explain the lower disaster service capacity and the lack of volunteer personnel capacity (Zakour & Harrell, 2003). In addition, the level of interorganizational cooperative links among disaster organizations is low for poor populations. These organizations also are more likely to be isolates (no cooperative links) or peripheral (one link only).

During the Katrina disaster, marginal geographical areas experienced a high loss of life, and response resources were not rapidly available for these areas. Because of a lack of coordination of disaster organizations, and a lack of trained volunteers, distribution of response resources was slow and ineffective (Kapucu et al., 2009). Redistribution of resources, especially trained volunteers from Emergency Management Assistance Compacts (EMACs), was lacking in the Katrina response in New Orleans. Florida's disaster organizations had a substantial number of well-trained volunteers. Cooperative links would have facilitated the flow of resources, including personnel, more effectively in Southeast Louisiana, making the response to Katrina much more coordinated (Kapucu et al., 2009).

ENVIRONMENTAL IDEOLOGY CAUSAL CHAIN

In New Orleans, as in much of the developed world, the cultural construction of nature–society relations has been the separation and opposition of humans and nature, and the ideology that human beings have dominion over nature. In this ideological system, humans are not concerned with their effects on nature, and believe that rational methods will bring order to nature (Oliver-Smith, 2004). Related to these beliefs is the idea that nature can be conquered by technology, and the view that barriers preventing residence on swampy areas could be neutralized (Campanella, This Volume, Chap. 2). This ideology has both guided and justified urban land use and the degradation of the wetlands in Southeast

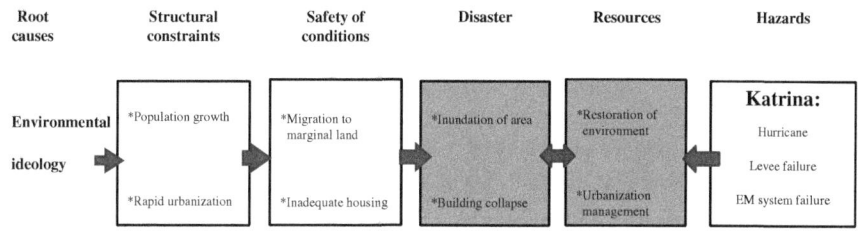

FIGURE 17.6

Environmental Ideology Causal Model.

Source: Adapted from Zakour, M. J., & Gillespie, D. F. (2013). *Community disaster vulnerability: Theory, research, & practice.* New York: Springer Science.

Louisiana (Fig. 17.6). For example, oil wells have been drilled since the discovery of oil in the early part of the 20th century (Marks, 2010b, 2010b; Zakour & Grogg, This Volume, Chap. 7). These practices were part of the urbanization of New Orleans that extended the city into former wetlands between the Mississippi River and Lake Pontchartrain (Campanella, 2006).

POPULATION GROWTH

The structural constraint of rapid population growth began as immigration of freed slaves from Louisiana and other areas of the South into New Orleans after the Civil War. Growth continued in the late 19th century with waves of immigrants arriving in this port city (Freudenburg et al., 2009). Beginning in the early part of the 20th century, the urban population and especially African Americans of New Orleans have migrated away from higher land to low-lying land in dangerous locations (Campanella, This Volume, Chap. 2). By 2000, only 38% of people lived above sea level.

As a result of population growth and movement to low-lying land, inundation from Hurricane Katrina's surge after failure of the levees affected a high percentage of the population, particularly among AfricanAmericans (Campanella, This Volume, Chap. 2). Sixty-four percent of homes in predominately African-American neighborhoods were inundated for a week or more. Despite the damage due to environmental degradation and inappropriate land use, during the recovery period little was done to restore the environment. The rebuilding of the levees to the 100-year storm level is the only action reducing vulnerability in New Orleans and Southeast Louisiana. Subsidence, sea level rise, and the condition of the wetlands have changed little (Campanella, This Volume, Chap. 2).

RAPID URBANIZATION

Urbanization sped up during industrialization in New Orleans and the United States. The areas in which poor and African-American populations concentrated had a higher biophysical risk of disasters such as flooding and hurricanes

(Campanella, 2006; van Heerden, This Volume, Chap. 6). Housing before Katrina was nonengineered and inadequate, because homes were not built to resist high winds, nor were they elevated (Freudenburg et al., 2009). When flooding occurred after Hurricane Katrina, the floodwater created tremendous pressure on buildings, causing homes and infrastructure to be badly damaged (Campanella, 2006).

Urbanization management and shrinking the footprint of New Orleans never happened, given the lack of a supporting policy and funding to fairly compensate homeowners and landlords. Under these conditions, preventing residents from returning to their homes and neighborhoods by converting their neighborhoods into green zones would have been a violation of human rights (Oliver-Smith, 2009; van Heerden, This Volume, Chap. 6). The Road Home program provided funding to encourage residents to rebuild on their pre-Katrina locations. However, one type of successful urbanization management after Katrina was elevating homes that had flooded, rebuilding homes with elevation such as piers, and establishing new building codes (Campanella, This Volume, Chap. 2).

COMMUNITY EMPOWERMENT AND SOCIAL DEVELOPMENT

Two causal chains consist of resilience capability, or resource, variables. Predisaster resilience is called general or inherent resilience. Resilience resources reduce the level of vulnerability if they are of high quality and quantity. As with the other six causal chains above, postdisaster resources moderate the effects of disasters. These two causal chains also provide continuity between the predisaster progression to vulnerability, and the networking of postdisaster resiliency capacities.

Sadly, in the years before Hurricane Katrina, community empowerment and sustainable social development were not adequate. The structural pressures of capabilities, when they were present, did not interact to increase the safety of conditions. This absence of community empowerment and social development resulted in much greater losses in disaster than would have otherwise occurred.

COMMUNITY EMPOWERMENT CAUSAL CHAIN

Community empowerment is a process through which people lacking an equal share of resources gain greater access to and control over resources (Norris, Stephens, Pfefferbaum, Wyche, & Pfefferbaum, 2008). Community empowerment capability is unevenly distributed, and many communities in New Orleans have suffered from a lack of empowerment (Fig. 17.7). Historically, the low status of poor African Americans, beginning in slavery, disempowered this population. Their undeserving status deprived them of public social insurance in times of crisis (Laska et al., This Volume, Chap. 5). High levels of poverty in many African-

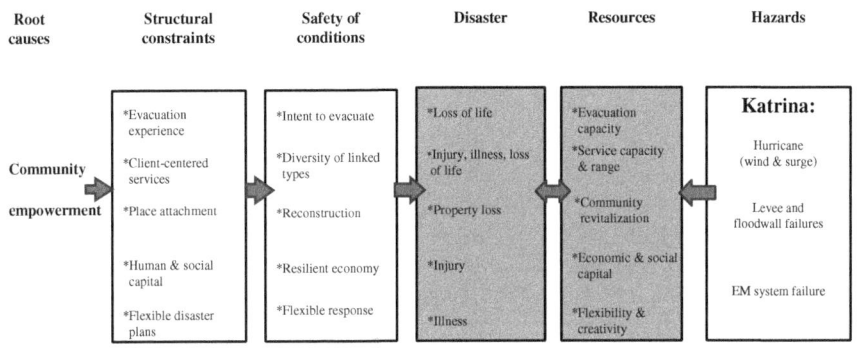

FIGURE 17.7

Community Empowerment Causal Model

Source: *Adapted from Zakour, M. J., & Gillespie, D. F. (2013).* Community disaster vulnerability: Theory, research, & practice. *New York: Springer Science.*

American neighborhoods in Orleans Parish have deprived them of needed resilience resources (Campanella, 2006; Zakour & Swager, This Volume, Chap. 3).

More specific examples of lack of community empowerment are the Central City / Hoffman Triangle and the Tremé neighborhoods. Historically, neither the Central City / Hoffman Triangle nor the Tremé neighborhoods had experienced collective efficacy empowerment. This was related to the high percentages of African Americans in Tremé and of working-class people in Hoffman Triangle, and their high poverty rates (Mock, Kadetz, Papendieck, & Coates, This Volume, Chap. 14).

Two communities, one a geographical community and the other a community of interest, have experienced some level of empowerment. The geographical community is the Vietnamese community in Village d'Est (Kadetz, This Volume, Chap. 13). A cultural source of empowerment has been the hierarchical frame of the Vietnamese community, and it facilitated rapid rebuilding after Katrina. Two noncultural sources of empowerment are the barrio and the stereotype effects. The barrio effect, from being surrounded by ethnic colleagues, and the stereotype effect, being viewed by the larger society as worthy, helped the Vietnamese community to recover resiliently (VanLandingham, This Volume, Chap 11). The New Orleans musicians' community also experienced some degree of community empowerment. This root cause contributes to both general resilience, and postdisaster resilience through social support from friends, community, and involvement in spirituality (Morris & Kadetz, This Volume, Chap. 10).

Structural constraints driving the level of risk include (1) evacuation experience, (2) client-centered services, (3) place attachment, (4) human and social capital, and (5) flexible disaster plans. Each of these constraints interacts with the community empowerment root cause to determine the safety of conditions for New Orleans and its populations. In the remainder of this section, we examine causal chains beginning with the community empowerment root cause, and each of the related structural constraints.

EVACUATION EXPERIENCE

The first structural constraint is the community and population level of evacuation experience. Low-income African Americans have had little evacuation experience, because of a lack of resources such as funds and an automobile, inadequate public evacuation services, and past voluntary evacuations that did not end in worst case scenarios. This population in New Orleans has tended not to evacuate, and along with the poor (Ferreira & Figley, This Volume, Chap. 8) has less evacuation experience than other populations.

Positively associated with actual evacuation, intent to evacuate is a variable indicating safety of conditions (Arlikatti, Lindell, Prater, & Zhang, 2006). Evacuating together and maintaining social support ties as a household was a predictor of low distress after Katrina (Haney et al., 2007). People who were unable to evacuate and remained in New Orleans during Katrina were exposed to severe disaster conditions. They were more likely to be killed during Hurricane Katrina and the flooding that followed. They also died from exposure during the evacuation waiting period they were forced to endure (Honoré, 2009; Solnit, 2009).

The poorly planned and inadequate public evacuation program after the flooding of Katrina relocated people in a random fashion and made it difficult for them to return from their distant destinations (Honoré, 2009). Though an efficient evacuation system was not available before or in the immediate aftermath of the Katrina disaster, a new system has been created in the past few years, called Evacuteer (Laska et al., This Volume, Chap. 5).

CLIENT-CENTERED SERVICES

The second structural constraint caused by community empowerment is the availability of client-centered services. Zakour (2008), in a study of organizations providing disaster services in New Orleans, found that those organizations offering client-centered services had higher evacuation capacities. However, New Orleans East, and the Lower Ninth Ward, as well as St. Bernard Parish, lacked social services organizations serving them (Zakour & Harrell, 2003). Both pre- and post-Katrina, the Central City / Hoffman Triangle and the Tremé neighborhoods lacked social services, especially client-centered services (Mock et al., This Volume, Chap. 14).

The second safety variable is an organization's diversity of linked types. Diversity of linked types refers to the number of different types of organizations with which a focal organization has cooperative links (Harrell & Zakour, 2000). Organizations with a greater diversity of linked types have higher disaster services capacities, including a higher level of evacuation capacity (Zakour & Harrell, 2003).

With a lack of an adequate plan and response at the local level, the City of New Orleans was not able to use school buses to move survivors out of the city. The city had not adequately planned to recruit drivers for this evacuation, and the buses sat in their parking lot until they were flooded and useless. During the waiting period for evacuation of survivors in the Superdome and Convention Center, many people suffered from exposure, lack of food and water, and relentless heat (Honoré, 2009). Reduced access to healthcare because of the disaster caused a deterioration of health for many after Katrina, including musicians (Morris & Kadetz, This Volume, Chap. 10).

Organizations with greater service capacities, and a larger geographic range of service delivery, were better able to prepare clients to evacuate, and to help them evacuate before a hurricane. This level of readiness potentially saved lives during Katrina and the flooding (Zakour & Harrell, 2003). Disaster services organizations active in the Katrina disaster were few in number and had a smaller service capacity and range, particularly for eastern New Orleans and St. Bernard Parish. Their smaller services range is exemplified by the Tremé food bank (Mock et al., This Volume, Chap. 14).

PLACE ATTACHMENT

The third structural constraint associated with community empowerment is place attachment. Because of the unique culture of New Orleans and Southeast Louisiana, communities located there have a strong level of place attachment. There is also a strong sense of place among most communities, neighborhoods, and populations in New Orleans. A majority of people living in New Orleans had families located there for generations (Campanella, 2006).

Reconstruction skills and abilities are associated with sense of place and safety of conditions. Communities in Southeast Louisiana had a pattern of reconstruction after disasters that occurred periodically in New Orleans and surrounding areas. However, during Katrina low- and middle-income neighborhoods were not provided with adequate resources for reconstruction (Downey, 2016; Finch et al., 2010). For example, reconstruction to make educational, health, and social services available in both Tremé and the Hoffman Triangle neighborhoods was slow to occur (Mock et al., This Volume, Chap. 14). Property loss was widespread, and badly damaged homes throughout New Orleans took years to rebuild, or were demolished.

Because of the strong place attachment of so many communities, there has been substantial but uneven revitalization of neighborhoods and communities since Katrina. Affluent neighborhoods have been revitalized, as have been communities with strong levels of social capital. However, the media focused so much on homeowner reconstruction that African Americans, the poor, and renters became almost invisible. Revitalizing their neighborhoods and their neighborhood cultures was much more difficult (Laska et al., This Volume, Chap. 5). An example of this is that community revitalization was not strong in either Tremé or the Hoffman Triangle area, because recovery resources were difficult to access (Mock et al., This Volume, Chap. 14). On a positive note, some qualities of New Orleans musician subculture helped in revitalization: spirituality, perseverance, flexibility, creativity, and positivity (Morris & Kadetz, This Volume, Chap. 10).

HUMAN AND SOCIAL CAPITAL

The fourth structural constraint in this causal chain of community empowerment is human and social capital. For the African American and poor population in New Orleans, particularly after Katrina, education was less available for children for an extended period of time, and healthcare was not accessible (Mock et al., This Volume, Chap. 14). Unlike most African Americans and the poor in New Orleans, the Vietnamese community provides an example of a high level of human and social capital. Cultural frames and culture confounders, including the selection and barrio effects, fostered human and social capital (Kadetz, This Volume, Chap. 13; VanLandingham, This Volume, Chap. 11). Social capital for the New Orleans musician community was in the form of social support networks providing emotional, psychological, spiritual, and general community support (Morris & Kadetz, This Volume, Chap. 10).

A resilient economy is a safety variable caused by community empowerment and human and social capital. Except for affluent neighborhoods, local economies in New Orleans have not been resilient either (Finch et al., 2010). New job creation has been weak for most parts of the city (Downey, 2016). Though the New Orleans economy was not diversified, New Orleans musicians have been able to participate in a resilient economy and achieve renewed income stability (Morris & Kadetz, This Volume, Chap. 10).

Vulnerability because of a lack of human and social capital, and a resilient economy, caused a variety of losses in Katrina, including injury and illness. If people were to return to their pre-Katrina homes, they faced great difficulty in trying to recover and rebuild. The Road Home program and other recovery programs did not provide economic and social capital for the typical survivor. Instead, these programs focused on helping those who had substantial property before Katrina. Renters, people in subsidized housing, and others who didn't own their homes were ignored. This contributed to a difficult and less resilient recovery (Laska et al., This Volume, Chap. 5). Economic capital was reduced in Tremé and the Hoffman Triangle neighborhoods, because twice as many households post-Katrina lost income compared to the number that gained income (Mock et al., This Volume, Chap. 14). As a more positive example, members of the musician community gained economic and social capital from community support and resources for financial management. Musical performance fostered sharing and trust, and ultimately enhanced social cohesion (Morris & Kadetz, This Volume, Chap. 10).

FLEXIBLE DISASTER PLANS

The fifth structural constraint and risk driver associated with community empowerment is flexible disaster plans. Flexible disaster plans were absent from the Katrina disaster because the emergency management system collapsed early in the disaster. Though the Hurricane Pam simulation was held a few months before

Katrina, the lessons from this simulation were not incorporated into planning for future hurricanes such as Katrina. A robust communications system was not available for use in the disaster response plan (Honoré, 2009). The absence of a flexible disaster response created unsafe conditions. Because of deficiencies in disaster planning and a lack of joint exercises, the emergency response system in New Orleans was not able to execute a flexible response when Hurricane Katrina made landfall, and instead the system became completely ineffective (Parker et al., 2009).

The lack of community empowerment and a flexible plan and response caused a pattern of loss including loss of life, injury, and illness during Katrina. The larger emergency management system collapsed and could not exercise flexibility or creativity. Only a few organizations, such as the Coast Guard and the National Guard, exercised the flexibility to rescue stranded survivors before Federal Emergency Management Agency personnel arrived in force. The Coast Guard had a tradition of taking the initiative on the ground before receiving exact instructions from the command hierarchy (Laska et al., This Volume, Chap. 5).

SOCIAL DEVELOPMENT CAUSAL CHAIN

Social development is a capability that can slow the progression to vulnerability, and through creation of resilience resources lessen losses in disasters (Fig. 17.8). New Orleans has experienced little social development aimed at improving quality of life for the last 50 years (Honoré, 2009). Instead, money has been dedicated to improving tourism, and navigation on the Mississippi River (Passavant, 2011). Elites with political connections have focused on economic development that has benefited the "growth machine" to a modest degree and distributed residual risk to the general public and especially the poor. The growth machine is a process built and set in motion by persons focusing on profit, without sensors to observe the environmental damage growth causes (Freudenburg et al., 2009).

PUBLIC FUNDING

A lack of public funding for social development projects is a structural constraint influencing the rest of this causal chain. Because of the neoliberal abandonment of cities including New Orleans, public funding for social development, including mitigation projects, has been very limited (Johnson, 2008). The levee system was disjointed and suffered from lack of maintenance. Floodwalls were poorly built, including those protecting the Lower Ninth Ward from the Industrial Canal.

Partly because of lack of public funding for disaster preparedness, New Orleans was not prepared for a disaster of the magnitude of Katrina. The levees had neither been joined into a system, nor maintained. Because of the severity of Hurricane Katrina, the failure of the levees and floodwalls, and the collapse of

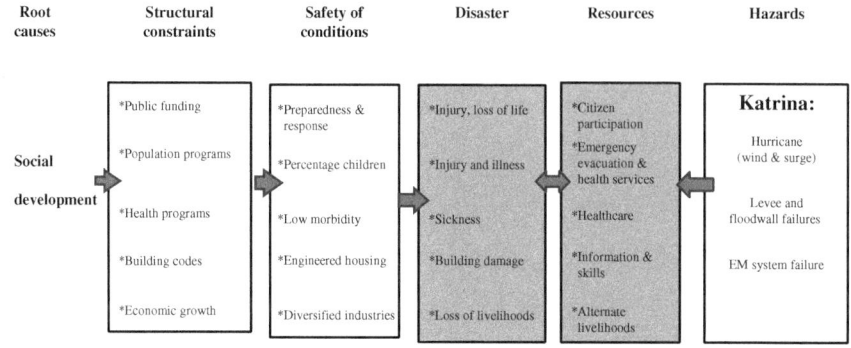

FIGURE 17.8

Social Development Causal Model

Source: Adapted from Zakour, M. J., & Gillespie, D. F. (2013). Community disaster vulnerability: Theory, research, & practice. New York: Springer Science.

the emergency management system (Parker et al., 2009; Tierney, 2014), loss of life and injury were severe (Honoré, 2009; van Heerden, This Volume, Chap. 6).

Citizen participation was limited because of structural violence, much of it at the local level. Citizen participation was quickly forced on residents of areas that the city determined to be green areas. Though these citizens, who were mostly poor African Americans, were exhausted from their effort to survive and rebuild, they were required by the Mayor to defend rebuilding their neighborhoods. The poor, African Americans, and especially female-headed households with young children, saw their participation in the rebuilding process dampened by structural violence and attacks on their personal agency (Laska et al., This Volume, Chap. 5).

POPULATION PROGRAMS

There have been few population programs to improve the health of mothers, infants, or children in New Orleans (Honoré, 2009). Rates of infant mortality have been high, and high rates of child poverty resulted in the poor health of many African-American children in Southeast Louisiana. A significant percentage of households in New Orleans has been single-parent, female-headed households with young children under 5 years old, so that the percentage of children in New Orleans was substantial before Katrina (Zakour & Harrell, 2003).

Young female-headed, single-parent, households with young children are highly vulnerable as a population. They were less likely to have evacuated than the general population, and they were less likely to have the resources, such as an automobile, to evacuate (Laska et al., This Volume, Chap. 5; Zakour & Harrell, 2003). Because they remained in flooded New Orleans, and with the lower resistance to illness of children, members of these households experienced injury and illness in the Katrina disaster. During evacuation, children were separated from families and neither parents nor children knew what their destination was to be. There were no health services for at least a week after landfall of Hurricane Katrina for parents and children stranded in New Orleans.

HEALTH PROGRAMS

Charity Hospital had long been one of the few public sources of healthcare for the poor, though nonprofit organizations and programs also provided this healthcare. For example, New Orleans musicians were able to access significant and diverse health and mental health resources and services through programs such as Sweet Home New Orleans (Morris & Kadetz, This Volume, Chap. 10). Partly because of a scarcity of healthcare, life expectancy in New Orleans is 68, and is 11 years lower than the national average in the United States. For people in some sections of New Orleans, such as the Lower Ninth Ward, life expectancy is only 54, comparable to sub-Saharan Africa (Somosot, 2012). Twelve percent of infants suffer from low-birth weights (compared to 8.2% for the United States), a risk factor for many health problems and for infant mortality (N.O. Health Department, 2013).

Though the Katrina disaster brought additional illness to people, it also badly damaged the public charity healthcare system, especially Charity Hospital. Healthcare for the poor and African-American populations of New Orleans was lacking after Charity Hospital was closed, at a time when healthcare needs were much higher (Laska et al., This Volume, Chap. 5). Healthcare was in limited supply in the recovery period, and young mothers frequently complained of a lack of healthcare services for themselves and their children (Joseph et al., 2014).

BUILDING CODES

Before Katrina there were no building codes that required elevating homes built on flood-prone ground (Campanella, This Volume, Chap. 2), a structural constraint promoting vulnerability. During the reconstruction phase of Katrina city officials permitted homeowners to rebuild without elevating their homes. Officials deprived homeowners of information needed to make a sound decision about reducing their risk to future disaster (Laska et al., This Volume, Chap. 5). Housing in poor and African-American neighborhoods was not engineered to be resistant to flooding or high winds. This situation produced unsafe conditions. Neither buildings nor land were elevated, even though neighborhoods were developed on low-lying land near bodies of water or canals (Freudenburg et al., 2009; Zakour & Harrell, 2003).

Because of high winds, flooding, and lack of engineered buildings, over 100 thousand residences were badly damaged or destroyed (van Heerden, This Volume, Chap. 6), especially in persistently flooded areas. Unfortunately, it was difficult for some survivors to obtain information and skills needed for a resilient recovery, and this was a liability making recovery more problematic. For example, a lack of scientific knowledge regarding the Murphy Oil Spill in St. Bernard Parish, including the nature of spill, prevented community members, emergency managers, and other to respond appropriately. Public anxiety was also heightened (Button, 2010). A more positive example is that after Katrina, information and skills were eventually disseminated to the New Orleans community regarding flood-risk and required elevation of homes according to risk. This information has been codified into new building codes, so that homes have been rebuilt or retrofitted as engineered housing with piers or other elevation (Campanella, This Volume, Chap. 2). Also, migration selection in the Vietnamese community in New Orleans meant that on average people had a high level of skills needed in recovery (VanLandingham, This Volume, Chap. 11).

ECONOMIC GROWTH

Because of its reliance on shipping and mineral extraction industries, economic growth in New Orleans was historically slow, a structural constraint caused by a lack of social development. The few industries in New Orleans, such as oil and tourism, periodically declined during oil busts and national economic downturns (Freudenburg et al., 2009). The economy of New Orleans has never been well diversified (Marks, 2010a, 2010b).

The severe disruption in the tourism, seafood, and oil industry after Katrina was a significant threat to livelihoods in New Orleans. For example, the disaster lead to disruption of tourism, and an exacerbation of income stability for both performers and teachers of music. People in New Orleans, however, had some access to alternate livelihoods after Katrina (Downey, 2016). Mentoring for musicians helped in the formation of new opportunities in music, through traditional brass bands, performances at restaurants and clubs, and performances during tours (Morris & Kadetz, This Volume, Chap. 10).

ASSUMPTIONS OF V+ THEORY

In this section, empirical support for 4 of the 12 assumptions of V+ theory is examined. These four assumptions are identified and discussed because they currently have the least supporting evidence of all of the assumptions of vulnerability theory (Zakour & Gillespie, 2013). Each assumption below is labeled with the number assigned in Zakour and Gillespie (2013). The first three assumptions discussed are concerned with unsafe conditions, root causes, and the causal chain of the progression to vulnerability, as well as hazards and the resulting disaster. The final assumption concerns the complex relationships among liabilities, capabilities, and disaster susceptibility. Disaster vulnerability is defined as the product of the level of susceptibility (liabilities) of a community in relation to its level of resilience (capabilities).

UNSAFE CONDITIONS ASSUMPTION

Unsafe conditions in which people live and work are the most proximate and immediate societal causes of disaster (Assumption 7, Zakour & Gillespie, 2013).

A number of studies indicate that, when a hazard occurs, unsafe conditions are the immediate causes of disaster. In order to make oil production and import possible, the coastal environment in Louisiana has been degraded through wetland drainage, canal dredging, levee construction, and settlement of worker populations in low-lying areas. This is one immediate cause of vulnerability and disasters (Marks, 2010a, 2010b; van Heerden, This Volume, Chap. 6). A lack of mitigation projects, location in dangerous areas near tidal lakes and the Gulf of Mexico, and use of substandard housing (e.g., slabs instead of piers), are other immediate and proximate causes of the Katrina disaster (Zakour & Swager, This Volume, Chap. 3).

Fussell et al. (2010) found that many homes in poor African-American neighborhoods were substandard and succumbed to high winds and flooding. Fragile livelihoods, including low levels of formal education, and high levels of morbidity and infant mortality in New Orleans, were found to represent unsafe conditions and high levels of vulnerability (Downey, 2016; Honoré, 2009; N.O. Health Department, 2013; Somosot, 2012).

Several studies point to a lack of evacuation capabilities as a critical part of unsafe conditions (Bassett, 2009; Colton, 2006; Ferreira & Figley, This Volume, Chap. 8; Haney et al., 2007; Kadetz, This Volume, Chap. 13). A number of other studies describe the environmental conditions of dangerous locations, and several of these studies point to inadequate preparedness as an unsafe condition leading to disaster (Bullard & Wright; 2009; Cutter & Emrich; 2006; Kapucu et al., 2009). Unsafe conditions continue after Katrina, raising the level of vulnerability to future disasters. Resources and capital to rebuild after disaster were not redistributed or available for educational, health, and social services for low-income populations and neighborhoods (Laska et al., This Volume, Chap. 5; Mock et al., This Volume, Chap. 14).

ROOT CAUSES ASSUMPTION

Root causes, the socio-cultural characteristics of a community or society, historically and in the present, are the ultimate causes of disasters (Assumption 8, Zakour & Gillespie, 2013).

Numerous studies describe the root causes that are the ultimate causes of disaster. The structure of vulnerability has been slowly and implacably established on Southeast Louisiana's coastline, economy, and ecology for the past century. The economy of Louisiana, and the funds that support the development and maintenance of its built infrastructure, depend on sources from oil, shipping, and related industries such chemical manufacture (Marks, 2010a, 2010b; van Heerden,

This Volume, Chap. 6). Related to Western ideologies about human—nature relationships is the idea that nature can be conquered by technology (Campanella, This Volume, Chap. 2; Freudenburg et al., 2009).

Research on structural inequalities related to race and poverty show this to be a root cause of disaster (Bassett, 2009; Cutter et al., 2006; Fussell et al., 2010; Zakour & Swager, This Volume, Chap. 3). Other research, on communities and neighborhoods, has identified the liabilities that act as root societal causes of disaster, such as social stratification (Downey, 2016), a lack of collective efficacy empowerment (Kadetz, This Volume, Chap. 13; Mock et al., This Volume, Chap. 14), and geographic distance from the city center (Zakour & Harrell, 2003).

DISASTER CAUSAL CHAIN ASSUMPTION

Disasters occur because of a chain of causality in which root causes interact with structural pressures to produce unsafe conditions. Hazards then interact with unsafe conditions to trigger a disaster (Assumption 9, Zakour & Gillespie, 2013).

A few studies document the causal chain leading from root causes to disaster and the resilience resources moderating losses. Downey (2016) shows that social stratification (a root cause) interacts with a lack of human capital and formal education (structural constraints) to produce unsafe conditions, especially fragile livelihoods. With the three hazards of Katrina, (the hurricane winds and surge, levee failures/flooding, and relief organization failure), a disaster occurred in which livelihoods were lost. Alternate livelihoods and adequate and effective economic resources are needed to restore lost livelihoods, especially for lower socioeconomic status populations. Parker et al. (2009) analyzed the nature of government emergency management from federal to local levels before Katrina. This interdisciplinary policy study provides evidence of a chain of causality from root causes, structural constraints, and unsafe conditions during the progression to vulnerability. This study provides further evidence for a hazard interacting with unsafe conditions to produce a disaster, and a lack of adequate evacuation and response resources intensifying Katrina into a catastrophe. The end result was an uneven return to wellness after Katrina. A lack of interorganizational coordination is a key theme in this chain of causality.

Through an historical perspective, we see the progression of vulnerability beginning with economic inequality resulting in high loss of life and emotional distress. The root cause of economic inequality interacted with the migration of the poor and a crisis in housing to produce unsafe conditions. Unsafe conditions included substandard housing and slums, and a lack of adequate mitigation through evacuation. The hazard of Hurricane Katrina and the levee failures caused a high loss of life, and accompanying emotional distress, during the Katrina disaster (Haney et al., 2007; Zakour & Grogg, This Volume, Chap. 7). In the Katrina disaster, the root cause of social stratification interacted with the structural pressures and risk drivers of lack of formal education, human capital, and local economic growth for the poor, African Americans, and other populations. These populations lived in unsafe conditions, accurately perceiving that mitigation through evacuation was ineffective, and they largely worked in the service sector, making their livelihoods risky. The unsafe conditions interacted with natural, technological, and organizational hazards, causing the Katrina disaster (Zakour & Swager, This Volume, Chap. 3). Racial ideologies began a chain of causality, leading to sociopolitical marginality of African Americans, resulting in a relative lack of mitigation projects. Resource redistribution was ineffective as a moderating variable, failing to prevent loss of life, property, and labor (Zakour & Grogg, This Volume, Chap. 7).

ASSUMPTION OF CAPABILITIES, LIABILITIES, AND SUSCEPTIBILITY RELATIONSHIPS

Environmental capabilities and liabilities, and disaster susceptibility, are related in complex ways to produce the level of community vulnerability (Assumption 11, Zakour & Gillespie, 2013).

In his discussion of the relationships among vulnerability and resilience, Etkin (2016) proposes a complex set of relationships among the two phenomena. Several studies show the complex relationships among capabilities, liabilities, and disaster susceptibility. Finch et al. (2010) found a curvilinear relationship between the Social Vulnerability Index (SoVI) and rate of recovery in Orleans Parish. The most vulnerable and least vulnerable neighborhoods recovered the fastest, while moderate vulnerability neighborhoods had the slowest rate of recovery. Jones-Deweever (2011) found that African-American women had a high level of disaster susceptibility, but that they were likely to receive high levels of social support, including help in finding housing after Katrina (Roberto et al., 2009).

Solnit (2009) found that elite panic after the breakdown of the emergency management system in Katrina led to transformative resilience, so that social movements emerged to deal with issues of social justice and reform. The manifestation of liabilities and disaster susceptibility paradoxically caused latent social movements (capabilities) to arise. Morris & Kadetz (This Volume, Chap. 10) found that long-term liabilities for musicians, such as income instability, both increased disaster susceptibility, and gave rise to coping mechanisms (capabilities) such as mentoring and social support, leading to a resilient recovery.

CONCLUSION

This chapter has presented research that is consistent with the causal chains in V+ theory. Though few of the studies cited in this chapter were longitudinal and able to prove causality, their research results add to the impressive body of empirical support for V+ theory (Zakour & Gillespie, 2013). Chapter 3, by Zakour & Swager (this volume), summarizes V+ theory; Chapter 7, by Zakour & Grogg (this volume), applies V+ theory to historical data from New Orleans; and the current chapter provides additional empirical support for this theory. These chapters, taken together, help identify and explicate the relationships between vulnerability and resilience theory and research, and advance and integrate recent developments in disaster research on vulnerability and resilience (Etkin, 2016).

REFERENCES

Adams, V. (2013). *Markets of sorrow, labors of faith: New Orleans in the wake of Katrina.* Durham, NC: Duke University Press.

Airriess, C. A., Li, W., Leong, K. J., Chen, A. C. C., & Keith, V. M. (2008). Church-based social capital, networks and geographical scale: Katrina evacuation, relocation, and recovery in a New Orleans Vietnamese American Community. *Geoforum, 39*(3), 1333–1346.

Arlikatti, S., Lindell, M. K., Prater, C. S., & Zhang, Y. (2006). Risk area accuracy and hurricane evacuation expectations of coastal residents. *Environment and Behavior, 38*(2), 226–247.

Bassett, D. L. (2009). The overlooked significance of place in law and policy: Lessons from Hurricane Katrina. In R. D. Bullard, & B. Wright (Eds.), *Race, place, and environmental justice after Hurricane Katrina: Struggles to reclaim, rebuild, and revitalize New Orleans and the Gulf Coast* (pp. 49–62). Boulder, CO: Westview Press.

Bullard, R. D., & Wright, B. (2009). Race, place, and the environment in post-Katrina New Orleans. In R. D. Bullard, & B. Wright (Eds.), (2009). Race, place, and environmental justice after Hurricane Katrina: Struggles to reclaim, rebuild, and revitalize New Orleans and the Gulf Coast (pp. 19–47). Boulder, CO: Westview Press.

Button, G. (2010). A gulf of uncertainty: "New oil in old barrels". In G. Button (Ed.), *Disaster culture: Knowledge and uncertainty in the wake of human and environmental catastrophe (Chap. 11).* Walnut Creek, CA: Left Coast Press.

Campanella, R. (2006). *Geographies of New Orleans. Urban fabrics before the storm.* Lafayette, LA: Center for Louisiana Studies, University of Louisiana at Lafayette.

Colton, C. E. (2006). Vulnerability and place: Flat land and uneven risk in New Orleans. *American Anthropologist, 108*(4), 731–734.

Curtis, A., Mills, J. W., & Leitner, M. (2009). Katrina and vulnerability: The geography of stress. In V. M. Brennan (Ed.), *Natural disasters and public health: Hurricanes Katrina, Rita, and Wilma* (pp. 101–116). Baltimore, MD: Johns Hopkins University Press.

Cutter, S. L., & Emrich, C. T. (2006). Moral hazard, social catastrophe: The changing face of vulnerability along the hurricane coasts. *Annals of the American Academy of Political and Social Science, 604,* 102–112.

Cutter, S. L., Emrich, C. T., Mitchell, J. T., Boruff, B. J., Gall, M., Schmidtlein, M. C., Burton, C. G., & Melton, G. (2006). The long road home: Race, class, and recovery from Hurricane Katrina. *Environment, 48*(2), 9–20.

Downey, D. C. (2016). Disaster recovery in black and white: A comparison of New Orleans and Gulfport. *American Journal of Public Administration, 46*(1), 51–74.

Etkin, D. (2016). *Disaster theory: An interdisciplinary approach to concepts and causes.* Boston, MA: Butterworth-Heinemann, Elsevier.

Freudenburg, W. R., Gramling, R. B., Laska, S. B., & Erikson, K. T. (2009). *Catastrophe in the making: The engineering of Katrina and the disasters of tomorrow.* Washington, DC: Island Press.

Finch, C., Emrich, C. T., & Cutter, S. L. (2010). Disaster disparities and differential recovery in New Orleans. *Population & Environment, 31*, 179–202.

Fussell, E., & Harris, E. (2014). Homeownership and housing displacement after Hurricane Katrina among low-income African-American mothers in New Orleans. *Social Science Quarterly, 94*(4), 1086–1100.

Fussell, E., Sastry, N., & VanLandingham, M. (2010). Race, socioeconomic status, and return migration to New Orleans after Hurricane Katrina. *Population & Environment, 31*(1–3), 20–42.

Haney, T. J., Elliott, J. R., & Fussell, E. (2007). Families and hurricane response: Evacuation, separation, and the emotional toll of Hurricane Katrina. In D. L. Brunsma, D. Overfelt, & J. S. Picou (Eds.), *The sociology of Katrina: Perspectives on a modern catastrophe* (pp. 71–90). New York: Rowman & Littlefield Publishers.

Harrell, E. B., & Zakour, M. J. (2000). Including informal organizations in disaster planning: Development of a range-of-type measure. In M. J. Zakour (Ed.), *Disaster and traumatic stress research and intervention. Tulane studies in social welfare* (Vols. 21–22, pp. 61–83). New Orleans, LA: Tulane University, School of Social Work.

Honoré, R. (2009). *Survival: How a culture of preparedness can save you and your family from disasters.* New York: Atria Books.

Johnson, G. S. (2008). Environmental justice and Katrina: A senseless environmental disaster. *The Western Journal of Black Studies, 32*(1), 42–52.

Jones-Deweever, A. (2011). The forgotten ones: Black women in the wake of Katrina. In C. Johnson (Ed.), *The neoliberal deluge: Hurricane Katrina, late capitalism, and the remaking of New Orleans* (pp. 300–326). Minneapolis, MN: University of Minnesota Press.

Joseph, N. T., Matthews, K. A., & Myers, H. F. (2014). Conceptualizing health consequences of Hurricane Katrina from the perspective of socioeconomic status decline. *Health Psychology, 33*(2), 139–146.

Kaniasty, K., & Norris, F. H. (2009). Distinctions that matter: Received social support, perceived social support, and social embeddedness after disasters. In Y. Neria, S. Galea, & F. H. Norris (Eds.), *Mental health and disasters* (pp. 175–200). Cambridge: Cambridge University.

Kapucu, N., Augustin, M. ,, & Garayev, V. (2009). Interstate partnerships in emergency management: Emergency Management Assistance Compact in response to catastrophic disasters. *Public Administration Review, 69*(2), 297–313.

Kim, J., & Oh, S. S. (2014). The virtuous circle in disaster recovery: Who returns and stays in town after disaster evacuation? *Journal of Risk Research, 17*(5), 665–682.

Marks, B. (2010a). Louisianas, oil & petro-addiction. *Against the Current, 147*, 4–6.

Marks, B. (2010b). The Gulf disaster: No end in sight. *The Independent, 153*, 6–7, June 23.

New Orleans Health Department (2013). *Child and family health in New Orleans: A life course perspective of child and family health at a neighborhood level.* Retrieved from:

https://www.nola.gov/getattachment/Health/Data-and-Publications/Child-and-Family-Health-in-New-Orleans-December-2013.pdf/

Norris, F. H., Stevens, S. P., Pfefferbaum, B., Wyche, K. F., & Pfefferbaum, R. L. (2008). Community resilience as a metaphor, theory, set of capacities, and strategy for disaster readiness. *American Journal of Community Psychology*, *41*(1/2), 127−150.

Oliver-Smith, A. (2004). Theorizing vulnerability in a globalized world: A political ecological perspective. In G. Bankoff, G. Frerks, & D. Hilhorst (Eds.), *Mapping vulnerability: Disasters, development & people* (pp. 10−24). London: Earthscan.

Oliver-Smith, A. (2009). Anthropology and the political economy of disasters. In E. C. Jones, & A. D. Murphy (Eds.), *The political economy of hazards and disasters* (pp. 11−28). Lanham, MD: Altimira Press.

Parker, C. F., Stern, E. K., Paglia, E., & Brown, C. (2009). Preventable catastrophe? The Hurricane Katrina disaster revisited. *Journal of Contingencies and Crisis Management*, *17*(4), 206−220.

Passavant, P. A. (2011). Mega-events, the Superdome, and the return of the repressed in New Orleans. In C. Johnson (Ed.), *The neoliberal deluge: Hurricane Katrina, late capitalism, and the remaking of New Orleans* (pp. 87−129). Minneapolis, MN: University of Minnesota Press.

Roberto, K. A., Kamo, Y., & Henderson, T. (2009). Encounters with Katrina: Dynamics of older adults' social support networks. In K. E. Cherry (Ed.), *Lifespan perspectives on natural disasters: Coping with Katrina, Rita, and other storms* (pp. 133−152). New York: Springer.

Solnit, R. (2009). *A paradise built in hell. The extraordinary communities that arise in disaster.* New York: Penguin Books.

Somosot, M. (2012). Life expectance is low in some parts of New Orleans. *The Times-Picayune.* June 21. Retrieved from http://www.nola.com/health/index.ssf/2012/06/life_expectancy_is_low_in_some.html.

Tierney, K. J. (2014). The social roots of risk: Producing disaster, promoting resilience *(High reliability and crisis management series)*. Stanford, CA: Stanford University Press.

Wisner, B., Blaikie, P., Cannon, T., & Davis, I. (2004). *At risk. Natural hazards. people's vulnerability and disasters* (2nd ed.). New York: Routledge.

Zakour, M. J. (2008). Social capital and increased organizational capacity for evacuation in natural disasters. In D. F. Gillespie (Ed.). *Disasters and development [Special Issue]. Social Development Issues 30*(1), 13−28.

Zakour, M. J. (2010). Vulnerability and risk assessment: Building community resilience. In D. F. Gillespie, & K. Danso (Eds.), *Disaster concepts and issues: A guide for social work education and practice* (pp. 15−60). Alexandria, VA: Council on Social Work Education.

Zakour, M. J., & Gillespie, D. F. (2010). Recent trends in disaster vulnerability and resiliency research: Theory, design, and methodology. In D. F. Gillespie, & K. Danso (Eds.), *Disaster concepts and issues: A guide for social work education and practice* (pp. 35−60). Alexandria, VA: CSWE Press.

Zakour, M. J., & Gillespie, D. F. (2013). *Community disaster vulnerability: Theory, research, and practice.* New York: Springer.

Zakour, M. J., & Harrell, E. B. (2003). Access to disaster services: Social work interventions for vulnerable populations. *Journal of Social Service Research*, *30*(2), 27−54.

Epilogue: Back to the future?

Paul Kadetz[1], Nancy B. Mock[2] and Michael J. Zakour[3]
[1]Drew University, Madison, NJ, United States [2]Tulane University, New Orleans, LA, United States [3]West Virginia University, Morgantown, WV, United States

INTRODUCTION

On the early morning of August 26, 2017 a category 4 hurricane, Hurricane Harvey, made landfall in coastal Texas. Breaking all previous records for rainfall in the United States, Harvey left more than 50 in. of water in its wake, resulting in the National Weather Service to label this event "unprecedented." Although the hurricane affected much of Southwestern Texas, Houston, the fourth largest city in the United States, was particularly affected. The damage from Harvey is estimated to be the most expensive in the history of all disasters on American soil (Rice, 2017). A week after Hurricane Harvey, Hurricane Irma, the strongest Atlantic hurricane on record, with winds topping 185 mph, devastated much of the Caribbean before impacting 90% of the state of Florida. Less than 2 weeks after Irma, Hurricane Maria, the strongest hurricane to hit Puerto Rico in a century, left the island without communications, water, power, and with little infrastructure. Local officials described the conditions in Puerto Rico as "apocalyptic." Coinciding with the 12th anniversary of Hurricane Katrina, this series of recent hurricanes reinforces how important it is to consider the lessons that have been learned in the period between these disasters.

LESSONS NOT LEARNED: PREPAREDNESS

There are many problematic similarities between the factors impacting the devastation left in Harvey, Maria, and Katrina's wake. First, as in the case of New Orleans, those in Texas and Puerto Rico with the fewest resources, who were the most socially marginalized and the most vulnerable, were the least likely to have been evacuated before the hurricane (Misra, 2017). In the case of Harvey, this was magnified by the refusal of local government to issue a mandatory evacuation (Misra, 2017). Furthermore, these same residents were most likely to live in the lowest lying areas that were most vulnerable to flooding. Similar to the precipitating factors in pre-Katrina New Orleans (see Chapter 7: Three centuries in the making: Hurricane Katrina from an historical perspective), subsidence and uncontrolled urban development on what formerly were absorbent prairie and wetlands have contributed to rising sea levels, of up to 6 in., and markedly decreased drainage capacity in already storm and flood-prone regions of Texas (Fernandez & Fausset, 2017; Harden, 2016). Given these irrefutable vulnerabilities and given that Houston was impacted by Hurricane Katrina, with the influx and permanent residence of evacuees from New Orleans, many who were also affected by Hurricane Harvey (Misra 2017), it is especially egregious that the management of Harvey (and Maria) seems ill-prepared. This was witnessed by the massive evacuation effort *after* the event, and the numerous evacuation shelters, that themselves needed to be evacuated due to flooding, along with areas that were either left with no water at all or water that was not safe for drinking.

What is especially worrisome to consider for the appropriate preparation for disaster events are the ensuing health hazards that, in the instance of Harvey, are specific to the many oil refineries and chemical plants concentrated in the Houston area. Two Exxon Mobil refineries released hazardous pollutants as a result of storm damage (Mufson, 2017). Burlington Resources Oil and Gas spilled over 30,000 gallons of crude in DeWitt County, Texas with an additional 8500 gallons of wastewater spilled (Miami Herald, 2017). And there were numerous explosions and subsequent long burning fires from the Arkema Chemical plant in Crosby, Texas, that released undisclosed, but toxic, chemicals into the surrounding environment (New York Times, 2017). There have also been other, smaller, petrochemical spills and fires reported throughout the area during the hurricane. A lack of accountability and a relaxing of legal regulations for the petrochemical industries of Texas have contributed to these incidents that will undoubtedly be deleterious in their impact to local health over time.

For Houston, Miami, Puerto Rico, and other "climate vulnerable" areas, it is clear that by examining and heeding the factors that have impacted the disastrous consequences of Hurricane Katrina and its aftermath over time, this volume may serve as a guide to avoid going "back to the future" in the preparation, prevention, and management of subsequent environmental hazards. Many of the contributors to this volume have emphasized the need for changes in disaster policy, research, and interventions. The so-called "softer" areas of social science need to be prioritized in disasters, particularly to understand specific sociocultural circumstances of a given context. According to Ferreira and Figley (Chapter 8: The resilience in the shadows of catastrophe: Addressing the existence and implications of vulnerability in New Orleans and Southeastern Louisiana), the vulnerable "are often marginalized or ignored by policy makers and postdisaster service planners because the lack of a comprehensive understanding and estimation of disaster vulnerability. Government leaders responsible for disaster mitigation must [better address social vulnerability both in terms of mitigation and response for] the special needs of vulnerable communities." Post-Katrina New Orleans raised an alarm of the need for political accountability and the fostering of resilient communities through the development of social infrastructure.

FOSTERING SOCIAL INFRASTRUCTURE

The contributors to this volume maintain that the development of community resilience that is able to withstand social and environmental shocks can be achieved by shifting the sole focus of rebuilding technical infrastructure to include the fostering of social infrastructure. Social infrastructure can be developed, first, through the identification of resilience capacities of communities and from understanding the intricate interplay between vulnerability and resilience at the local, metropolitan, state, and higher levels (Chapter 17: Lessons learned from New Orleans on vulnerability, resilience, and their integration). Vulnerability can be identified and best understood as an outcome of structural violence by tracing the social inequalities and inequities that result in the exclusion of certain groups and their intersectionalities from resources needed for survival and agency; particularly in a situation of social shock. Almost all chapters in this volume discuss various mechanisms that contributed to pre- and postdisaster vulnerability.

However, in order to build social infrastructure, existing strengths and community assets need to be identified, encouraged, and leveraged. There are several community assets to be fostered for strengthening social infrastructure including social cohesion, social capital, trust, agency (and as Laska, Chapter 5: "Built-in" structural violence and vulnerability: A common threat to resilient disaster recovery, specifies; "the honoring of agency"), and collective efficacy; all of which can be grouped under an umbrella of "culture". For example, both Van Landingham (Chapter 11: Resilience among vulnerable populations: The neglected role of culture) and Kadetz (Chapter 13: Collective efficacy, social capital and resilience: An inquiry into the relationship between social infrastructure and resilience after Hurricane Katrina) present the Vietnamese community of East New Orleans as an unusual case study of community resilience in post-Katrina New Orleans. However, Kadetz identifies resilience in this community as an outcome of the abovementioned community assets, while Van Landingham frames this community's resilience as an outcome of culture (and cultural confounders). Furthermore, culture and cultural activities can produce the community assets identified, as Morris and Kadetz (Chapter 10: Culture and resilience: How music has fostered resilience in post-Katrina New Orlean) illustrate in the case of the culture of New Orleans musicians and the cultural activity of music's ability to strengthen social infrastructure.

It is also time to acknowledge that in the 21st century social infrastructure can also be fostered virtually. Mock (Chapter 16: How barefoot scholars were deployed: The good, the bad, the ugly) identifies that the application of crowd sourcing and cloud computing along with digital volunteering may offer marked future directions for disaster management and recovery.

LOOKING FORWARD TO THE FUTURE: LESSONS FOR RECOVERY

As Mock et al. (Chapter 4: A systems approach to vulnerability and resilience in post-Katrina New Orleans) note in their analysis of the New Orleans Index, monitoring recovery results proved useful to stakeholders working on the recovery of New Orleans. The indicators employed clearly demonstrated ways in which New Orleans recovery was more or less robust. A few lessons from this New Orleans indicator project, however, should be considered by Houston, Puerto Rico, and others in similar circumstances:

- Monitor financial flow from the beginning of the recovery process. It was more than a year before the Katrina index began to include any financial flow information. Accountability to all stakeholders is key and the lack of transparency reflected by the lack of financial flows was problematic.
- Collect more granular information about recovery and the recovery process. One of the limitations of the New Orleans Index was that most measures reflected aggregate data for New Orleans as a whole. Neighborhood and subneighborhood level indicators need to be considered to better reflect differential progress in recovery and its determinants.

Given that much of New Orleans is ecologically vulnerable and that many areas of the city have not yet recovered 12 years later, Houston, Miami, and San Juan should heed the choices concerning where to rebuild. One of the difficulties faced during Katrina reconstruction and the "smaller foot print" debate was that many home owners in areas that were being considered to be retired to green spaces faced poor prospects for being able to rebuild or purchase a house on higher ground. An argument can be made for a "new deal" for those in vulnerable areas that provide positive incentives for individuals and communities to move. Though New Orleans lost an opportunity to reconfigure itself into an ecologically and socially more resilient and robust community, Houston and other postdisaster communities need not repeat this pattern of recovery. Similarly, other than the experiment in K-12 education, very little was done during the early Katrina recovery process to address one of the greatest vulnerabilities of New Orleans, deep and widespread poverty and social inequity. Many missed opportunities include not establishing programs to integrate the poor into facets of the economy. While there was some attempt to develop economic recovery zones within the city, integrating these strategies with focused approaches to transition households out of poverty was a missed opportunity.

In summary, New Orleans fought many of the battles with our national policy constraints to enable disaster affected communities to reinvent themselves and to rebuild themselves into better and more resilient communities. Houston and others should retain this spirit and resolve to build more structurally, ecologically, and socially improved communities.

LOOKING FORWARD TO THE FUTURE: ADOPTING NEW PERSPECTIVES

Throughout this volume, we have explored a number of theoretical and value-centered views of disaster resilience. But there are practical lessons to heed. First are the inclusion of local perspectives and understandings and the hybridization of this emic knowledge in conjunction with scientific knowledge. As many of the chapters in this volume have offered, the inclusion of barefoot scholars—who can embrace both emic and etic perspectives simultaneously and provide imperative insight into the particular of local communities—is essential for engendering disaster management and interventions that are appropriate to a given context.

Similar to the importance of community perspectives is the appropriate decentralization of financial, material, and knowledge resources to local levels. Both response efforts in Texas and Puerto Rico were predominantly led by local governments and community volunteers, who led proactively, rather than passively awaiting for federal government intervention, as was often the case after Hurricane Katrina (though Katrina had many local volunteers involved in the early search and rescue phases). A lesson to be learned from these events is the value and needed inclusion of well-prepared decentralized disaster management. This is especially important when the inadequacies of the federal response are replicated from disaster to disaster. Both Hurricanes Harvey in Texas and Maria in Puerto Rico demonstrated that many lessons from Katrina had not be learned and/or heeded by the federal government. For example, once again there were communication and energy blackouts that could have been prevented. The federal system should have guaranteed emergency telecommunication and energy, as we learned to prevent these in Katrina's wake. Though in general, emergency response may have improved since Katrina, the vulnerable in Hurricanes Harvey, Irma, and Maria were no less vulnerable than those in Hurricane Katrina. For example, the same issues of people dying in long-term care were replicated post-Irma in Florida and nonfunctioning hospitals were witnessed throughout Puerto Rico, post-Maria.

In general, there is a need for a deeper understanding of vulnerability beyond universalized categories of, for example, gender, age, and race. Vulnerability transcends categories, including the myriad combinations of exclusion and marginalization that are specific to a given sociocultural context and that are socially enforced through the structural violence directed toward the members of these groups. In America, elderly and poor, low-income and African-American, low-income women and African-Americans, and Hispanics were groups whose resilience development during the Katrina response was neglected. Yet younger men of all races had a chance of being launched into resilience trajectories by the recovery economy. It is critical that we analyze and understand vulnerability and the myriad ways in which disaster risk reduction and recovery efforts can launch people, households, and communities from vulnerability into resilience trajectories. This volume emphasizes that the human management of disasters is a complex, open-system, nonlinear and multi-directional, natural, built, and social environment systems problem. Yet, it is society that must build community resilience and reduce vulnerabilities to future hazards.

So, are we, in fact, going back to the future with the recovery from Hurricanes Harvey, Irma, and Maria? As the disaster that ensued from Hurricane Katrina was an outcome of a series of poor human decisions over time that thwarted resilience, this volume maintains that these outcomes can ultimately be prevented by heeding the myriad lessons learned.

REFERENCES

Fernandez, M., & Fausset, R. (2017). A storm forces houston, the limitless city, to consider its limits. *The New York Times*.

Harden, J. (2016). For years, the Houston area has been losing ground. *Houston Chronicle*.

Harvey live updates: In Crosby, Texas, blasts at a chemical plant and more are feared. August 31, 2017. *The New York Times*.

Miami Herald. (2017). The latest: Death toll 31 as 6 more fatalities confirmed. *Miami Herald*.

Misra, T. (2017). Harvey has hit the poorest and most vulnerable texans the hardest. Actually, disasters do discriminate. *Mother Jones*.

Mufson, S. (2017). ExxonMobil refineries are damaged in Hurricane Harvey, releasing hazardous pollutants. *The Washington Post*.

Rice, D. (2017). Harvey to be costliest natural disaster in U.S. history, with an estimated cost of $160 billion. *USA Today*.

Index

Note: Page numbers followed by "*f*" and "*t*" refer to figures and tables, respectively.

A

Absorptive capacity, 11
Access to resources model of vulnerability, 50–51
   capabilities, 50
   dynamic nature of, 50–51
   influence of root causes on, 51
Action research, 12–13, 20, 188, 308–317
Adams, Vincanne, 17
Adaptive capacities, 11, 58
Adaptive resilience, 11
Agency, 100–102
   community, 102
   disrespect as structural violence, 103–104
   exacerbation of trauma and, 109–110
   individual, 102
   levels of, 102*f*
   recovery, 114–118
Altruistic community, 181, 268–269
Army Corps of Engineers, 32–33, 87–88, 137–138, 297–298
Assets, 49–50, 220–222, 247, 290, 309, 323, 387
   cultural, 9–10
Assets-based approaches, 286–290
Atchafalaya River delta distributary system, 140
Attractors, 85

B

Bandura, A., 284, 293
Barefoot doctors, 14–15
Barefoot scholars, research experience of, 14–15, 267–268, 274, 324, 345, 353–354
   developing an initiative, 348–352
   Katrina events, 346–347
      early recovery efforts, 347–348
   lessons learned, 353–354
   on RALLY approach, 352–353
   return to New Orleans, 352–353
Barrio effects, 263
Base Flood Elevation (BFE), 33–35, 101
Black population of New Orleans, 165–169, 176
BP Horizon oil spill, 152, 268, 332–333
BP oil catastrophe, 332–333
Build It Back program, 106–108
Building codes, 370, 377–378

C

Canizaro, Joseph, 108
Capabilities, 50, 381
   collective action, 55–56
   defined, 11
   information and communication adaptive, 54–55
Capacity, 10–11, 47–48
Carondelet Canal, 166
Case management, 116
Catastrophe
   Katrina's damage as, 160, 175–178, 180–181, 184, 188
   New Orleans and Southeastern Louisiana, 198–207
Causal chain, 48–50
   community empowerment, 370–371
   domination, 362–365
   disaster, 380–381
   economic, 359–360
   environmental ideology, 368–370
   geographic, 367–368
   racial ideology, 365–367
   social development, 375–378
   social stratification, 361–362
   in V+ model, 68–72
Central City, 305, 310–312, 311*f*, 312*f*, 313*f*
   demographic data, 315–316
   distribution of Hispanic people, 319*f*
   dynamics in the Hoffman Triangle area, 316–317
   estimated occupancy rate, 318*f*
   immigrants of, 311–312
   neighborhood dynamics and change in, 312–317
Chain of causality, 68–72. *See also* Vulnerability-plus (V+) theory
   characteristics of disaster, 71
   hazard types, 70–71
   leverage points in PAR model, 49–50
   Pressure and Release (PAR) model, 48
   resilience resources, 71–72
   root causes of vulnerability, 48, 68–69
   structural constraints, 48–49, 69
   sustainable livelihoods, 49
   unsafe conditions, 49, 69–70

City-assisted evacuation, 119, 119f
Climate change, 4–5, 11, 47–48, 70–71, 100, 125, 154, 340
Coastal wetlands, management of
  control of the Mississippi, 139–141
    consequences, 139–140
    political and economic, 139
  exploration and mining of oil and gas, 141
Coastal wetlands in the United States, 154
Collective action capabilities, 55–56, 178, 182–183
Collective efficacy, 284
empowerment, 55
Communication, 178, 180–181
Community agency, 102–103
Community competence, 55, 258–259
Community disaster resilience, 57
Community empowerment causal chain, 370–371, 371f
  client-centered services, 372–373
  community of interest, 371
  evacuation experience, 372
  flexible disaster plans, 374–375
  of geographical community, 371
  human and social capital, 374
  place attachment, 373
  structural constraints, 371
Community narratives, 55
Community resilience, 197, 284. *See also* Vietnamese community of New Orleans
Complex Adaptive Social System (CASS), 79–80, 85–86
Complex Adaptive Systems (CAS), 80
  information processing in, 86–87
  properties, 80
Complex systems, 4, 11, 15–17, 79–85, 88, 92, 197, 216–217, 219, 285–286, 300–301
  The New Orleans Index, 82t
  in post-Katrina recovery, 81–85
    cross-level and cross-scale interactions, 84
    feedback and signals, 84
    level and scale of system components, 81–83
    nested hierarchies, 84
    nonlinearity of system change, 85
    self-organization, 85
    social networks, 84–85
    threshold and tipping points, 84
    vertical networking, 84–85
Conservation of Resources (COR) theory, 236, 251
Culture, 8–10, 68, 93, 125, 219, 234–235, 237–238, 248–253, 261–264, 294, 308, 360, 373–374
  postdisaster recovery, 261–262, 262f

Culture Confounders, 261–263
Cursillo, or Cursillos de Cristiandad, 271
Cynafin, 80–81

D

Decentralized execution, 124
Decentralized interventions, 389
Delay, 105–106, 114, 121
Development
  economic, 53, 60–61, 258–261
  housing, 172
  industrial, 169
  postdevelopment theorists, 287–288
  post-Katrina "redevelopment", 117
  social, 47–49, 370, 375
Disaster, 17, 46–48, 55–56, 70, 134, 199–200, 272
  capitalism, 6
  causal chain assumption, 380–381
  characteristics, 71
  classifying, 4–5
  community disaster resilience, 51, 57
  flexible disaster plans, 374–375
  hybridizing knowledge for research, 13–14
  management, 46
  nature of the Katrina disaster and catastrophe, 175–176
  pattern of disaster damage and loss, 176–177
  perspectives and paradigms, 12–13
    etic perspectives, 12–13
  postdisaster gentrification, 123
  postdisaster housing, 113–114
  recovery, 5, 15–17, 19–21, 85, 259–260, 306
  requiring multiple perspectives and paradigms, 12–13
  risk management, 11
  vulnerability and resilience, 3–4, 18–19, 162, 187
Disaster capitalism, theory of, 293–294
Domination causal chain, 362–365, 363f
  distrust in media, 363
  political partnerships, lack of, 364–365
  structural constraint of an elite-controlled media, 363–364
  Dynamic pressures, 204–205

E

Economic development, 53, 121, 258–259
Economic inequality, 359–360, 359f
  housing and health crises, 360
  lack of economic resources, 360
  migration of poor, 359–360

Economic resources, 178–180, 187, 360
Economic stability, 259
Elevation, 31–33, 39–42. *See also* Base Flood Elevation (BFE)
  Advisory Base Flood Elevation maps, 26
  high-resolution, 27
  Lee Circle area, 27
  LIDAR, 32–33
  topographic, 27–29, 32
Emergency manager, 103
Environmental degradation, 144–146, 161, 169–171
Environmental ideology causal chain, 368–370, 369*f*
  population growth, 369
  rapid urbanization, 369–370
Environmental justice, 6–7, 164
Environmental migrant, 292
Evacuation, 360–361, 363, 366–367, 372, 376–377, 379
Evacuees, 13, 15, 120–121, 285, 291–292, 308, 363–364
Evacuspots, 118–119
Evacuteers, 118–119, 372
Experiential learning, 345
Exposure, 25–26, 35–36, 41
Exxon oil issues, 169

F

Faith-based organizations in fostering resilience, 268
  Cursillo, or Cursillos de Cristiandad, 271
  effects on individual Katrina survivors, 277–280
    recovery imbued with spirituality, 277
  Hagar's House, 274–275
  Katrina-related outreach and ministries, 271, 274
  Luke's House, 275–276
  United Methodist Church (UMC), 267–277
    decision process for mergers and closures, 273–274
    Early Response Teams (ERT), 269–270
    Mt. Zion UMC, 276
    nonhierarchical and feminist approach, 273–274
    provision of resources and capabilities, 270–271
    transformative resilience, 271–272
  United Methodist Committee on Relief (UMCOR), 268, 270
  United Methodist Men (UMM), 277
  United Methodist Women (UMW), 276–277

Farmer, Paul, 17–18, 292–293
Federal Emergency Management Agency (FEMA), 11, 101, 106, 109–110, 114
  Advisory Base Flood Elevation maps, 26
  Operation Blue Roof, 179–180
  requirements for floodplain development, 190
  settlement in flood zones, 32–36, 34*f*
    catagorizing zones, 33
    communication to public, 32
    computation of BFE, 33–35
    demographic composition, 35, 36*f*
    determining, 32–33
    flood-zone residency, 35
    metrics of social vulnerability, 35
Female vulnerability, 218–220
  in New Orleans, 222–228
    education, 224–225, 225*f*
    financial, 223, 223*f*
    healthcare, 224–225, 225*f*
    housing, 223–224, 224*f*
    intersectionality, 226–228
    political economy and neoliberal vulnerabilities, 225–226
    transportation, 224–225, 226*f*
Financial vulnerability, 223
Flexible disaster plans, 374–375
Flood Control Act, 1965, 139–140, 148–149
Flood Insurance Rate Maps (FIRMs), 32–33
  digital (D-FIRMs), 32–33
Flooding, 26–27, 31–35, 37, 39–40

G

Galtung, Johan, 292–293
Gender, 215–220, 222–228
Gendered vulnerability, 216–217
Geographic distance causal chain, 367–368, 367*f*
  low volunteer capacity, 368
  smaller service range, 367–368
Geographic resilience, 8
Geography, 32
Grant-making process, 107
Great Footprint Debate, 41
Greenspacing, 26, 93–94, 109, 285, 291, 295
Ground zero impact, study of, 331–332
  calling to activism and identity integration, 339–340
  findings, 335–340
  implications of, 340–342
  methodology, 333–334
  reasons, 332–333
  research
    assessment of, 336–337
    bias assessment, 337

Ground zero impact, study of (*Continued*)
    challenges and benefits, 337
    quality assessment, 336
    topic selection, 335–336
  responses, 338–340
  social scientists as witness, 337–338
Growth Machine, 164–165
Gulf Intracoastal Waterway (GIWW), 144, 170–171
Gulf Oil Spill disaster, 46

H

Habitat sustainability, 154
  recommendations for, 151–154
    restoring of coast, 152
    risk-based engineered levee design process, 153
    risk-based management planning, 153–154
Hagar's House, 274–275
Haiti earthquake, 46
Hazards, 32, 42
Heuristic approach to disaster resilience and future vulnerability, 189–190
Historical analysis, 160–161
Housing stability, 259
Housing vulnerability, 223–224
Houston, 5, 292, 385–389
Human capital, 362
Human-caused hazards, 4
Human/community-centered resilience, 8
Human development, 46–47, 57, 69–70
Hurricane & Storm Damage Risk Reduction System (HSDRRS), 25–26, 32–33, 41
Hurricane Betsy, 148–149, 167–169
Hurricane Irma, 385, 389–390
Hurricane Isaac, 332
Hurricane Harvey, 5, 385–386, 389
Hurricane Katrina, 3–5, 46, 104
  as case of elite panic, 177–178
  as catastrophe, 177–178
  causal process, 201–207
    dynamic pressures, 204–205
    failure of Levee system during, 205
    root causes, 203–204
    unsafe conditions, 205–207
  characteristics of, 186
  evacuation of people, 178
  flooding and floodwater depths, 174*f*, 175
  historical and longitudinal analysis, 160–161
    environmental degradation and damage, 161
  historical events
    migration and population displacement since 1960, 171–172
    neglect of New Orleans area, 172–173
    oil, canals, and environmental degradation, 169–171
    political and economic marginalization of people of color, 165–169
    progression to vulnerability, 165
  immediate impact of, 6
  post-Katrina recovery. *See* Post-Katrina recovery
  secondary effects of, 201–202
  soil contamination, 177
  structural constraints, 160–161
  substantive implications of, 188–189
  vulnerability and resilience during, 162
    access to resources, 163
    causal chain in vulnerability-plus theory, 162
    community resilience, 162
    natural, technological, and organizational failure, 173–175
    nature of damages, 175–176
    pattern of damages and loss, 176–177
    progression of vulnerability, 203*f*
    root causes, 163
    structural constraints, 163
    unsafe conditions, 163
  waves of, 160
Hurricane Maria, 87, 385, 389
Hybrid knowledge, 13–14
Hyogo Framework for Action 2006–2015, 10

I

Indian Ocean Tsunami, 46
Individual agency, 102
Individual resilience, 196–197
Industrial Canal, 169–170, 188–189
Information, 178, 180–181
Inner Harbor Navigation Canal (IHNC), 144
Institutional racism, 103
Internally displaced persons, 13, 222, 291–292
Intersectionality, 17–18, 84–85, 217–218, 220, 222–228

J

Jim Crow laws, 167

K

The Katrina Index, 87, 89*t*, 91–93, 218
Klein, Naomi, 6, 104, 225–226, 287, 292–294

L

Lafayette Cemetery Research Project, 199–200
Lake Borgne, 144

Lake Pontchartrain & Vicinity Hurricane Protection Project (HPS), 148
Laska, Dr. Shirley, 42
Levee failures, 149–154
 accounting for subsidence and sea-level rise, 150–151
 polder, 149–150
 recommendations for sustainability, 151–154
Level of analysis, 20, 88, 308
 aggregated, 217–218, 228
 disaggregated, 228
 granularity, 94–95
Liabilities, 381
Lifetime costs, 101
Local interventions, 389
Long-term evacuee housing, 118–121
Louisiana Road Home program, 106–108
Louisiana's coastal wetlands
 distributaries of the Mississippi River, 134–135
 Holocene delta development, 136f
 "natural" cycle of wetland development and maintenance, 134–137
 origin and function of, 134–139
 relative sea-level rise (RSLR), 137–138
  rapid rate of, 138
  sediments compact and dewater, 138
  USACE data set, 138
 storm-surge and hurricane winds, reduction in, 138–139
 suspended sediment distribution, 137f
Lower-income survivors
 evacuation response to, 111–112
 health care for, 112–113
 housing, 113–114
Low-income New Orleans neighborhoods
 Central City, 305, 310–312, 311f, 312f, 313f
  demographic data, 315–316
  distribution of Hispanic people, 319f
  dynamics in the Hoffman Triangle area, 316–317
  estimated occupancy rate, 318f
  immigrants of, 311–312
  neighborhood dynamics and change in, 312–317
 female vulnerability, 319
 loss of community and connectivity, 318–319
 neighborhood residents' perceptions and social infrastructure, 317–321
 performance of school system, post-Katrina, 319–320
 Recovery Action Learning Laboratory (RALLY) project, 308–317
  findings of, 321–323
  recovery efforts, 306–307
 assessment and monitoring of recovery process, 307–308
 challenges, 307
 in terms of reconstructing, 306
 Tremé, 305, 310–312, 311f
  demographic data, 316t
  Katrina flooding in, 314f
  location, architecture, and cultural appeal, 315
  neighborhood dynamics and change in, 312–317
  storm damage in, 315f
Luke's House, 275–276

M

Marginalization, 165–169, 365–366
Mississippi River Gulf Outlet (MRGO) project, 141–144, 142f, 143f, 175
 direct impacts of, 142–144
 economic development, 141–142
 effects of Reach 2 on waves, 146–148, 147f
 environmental degradation, 170–171
 funnel, 144–145
 saltwater intrusion associated with construction, 142
Multidimensionality of resilience, analysis
 assets-based approaches, 286–290
 context, 285–286
 impacts of neoliberalism and creation of need, 287
 needs-based approach, 286–287
 normative community development, 287–289
 post-Katrina New Orleans, 291–299
  African-American and Vietnamese evacuees, comparison of, 295–296
  African-American community, resilience of, 293, 295–298
  assets-based positive deviance assessment, 290, 295–297
  disaster capitalism and forced disaster migration, 292–295
  impact of inequality, 298–299
  middle-income housing, 293–294
  New Orleans Public School system, 294
  relationships of social capital and collective efficacy, 299
  representations of structural violence, 291–292
  social cohesion and social capital, 298–299
Murphy Oil spill incident, 4, 71, 176, 186, 377–378
Music, 241–251. *See also* New Orleans musicians, study of Katrina effects on

N

National Flood Insurance Program (NFIP), 32
  zones and rates, 32–33
National Guard and Army, 106, 112, 375
National Response Plan, 11
Needs-based approaches, 286–287
Neighborhood action research, 308–317
Neoliberalism, 8, 287–289, 292–295, 298–299
New Orleans area, 6–10, 198
  at-risk residential areas, 93–94
  Black neighborhoods, 166–167, 169–170
  canals, levees, and floodwalls, 167–169, 168f, 171
  evacuation system, 8–9
  livelihoods of residents, 177
  migration from, 27
  music and cultural community in, 248
  neglect of
    health status of residents, 173
    poverty rates, 172–173
    rate of infant mortality, 173
  political ecology of, 163–165
  population, 1960, 27
  post-Hurricane Katrina, settlement patterns of, 28–40
    by FEMA, 32–36
    within flooded zone, 31
    in horizontal space, 36–40, 38f, 39f
    in low-lying areas, 31, 31f
    spatial extent of below-sea-level areas, 31
    in vertical space, 28–32, 29f, 30f
  pre-Katrina, 198–199
  pre-Katrina New Orleans, 215–216
  principle of environmental justice and development projects, 164
  resilience paradigm of, 200–201
  root causes of vulnerability
    economic crisis, 172
    French settlements, 165
    human modification of the landscape, 165
    inequitable access to political power and economic wealth, 165–169
    migration and population displacement since 1960, 171–172
    oil, canals, and environmental degradation, 169–171
  settlement patterns of, 26–28
  social and urban geography, 165–169
  structural constraints and risk drivers, 169–171
  topographic changes, 27–28
  vulnerability of, 6–7, 161, 175, 200–201, 220–222
    addressing, 208–209
    gendered, 222–228
    immediate impact of Hurricane Katrina, 6
    political economic trends, post-Katrina, 6–7
    provision of health and human services, 7
    and resilience theory, 7–9
    of urban areas, 220–222, 221f
New Orleans Disaster Recovery Partnership (NODRP), 92–93
The New Orleans Index, 82t, 87–91, 94–95
  application of, 88
  changes in, 89t
  monitoring of core measures, 91
  value of, 92–93
New Orleans musicians, study of Katrina effects on, 237
  bond between place and identity, 238
  data analysis, 240
  data collection, 240
  ethics, 241
  findings
    conservation of resources theory, 251
    music community relationships and mentors, 248–249
    music performance, impact of, 249–251
    performative culture, conception of, 250
    role in community healing, 250
    social support, impact of, 247–248
    themes based on informant experiences, 251
  identification of factors, 238–239
  intergenerational transmission of culture, 237–238, 248–249
  locating resilience in performance, 241–251
    mental health, 241–247
    New Orleans Resilient Musicians Scale (Norms), 242t
    protective factors and assets, 247
    risk factors and stress, 241–247
  sample, 239–240
  study design, 239
  VGA protocol, 240
New Orleans Musicians Foundation (NOMAF), 239–240
Non-governmental organizations (NGOs), 9, 182–183, 234, 268–269
  New Orleans Musicians Clinic (NOMC), 234, 248
  The New Orleans Musicians Relief Fund, 248
  Sweet Home New Orleans (SHNO), 234–236, 248
  Sweet Relief, 248
  Tipitina's Office Co-op, 248

O

Oil industry of New Orleans, 169–171
Oil spills, 201–207
    BP oil spill, 152, 268, 332–333
    Exxon oil issues during Harvey, 386
    Murphy oil Spill, 4, 176, 186, 377–378
Orleans Parish, 201
Outreach ministry, 274
Owner-occupied housing recovery, 114–117

P

Polder levee failures, 149–150
Political ecology of Katrina and Southeast Louisiana, 183–184
Population centroid, 37, 38*f*, 39*f*
    Asian, 40
    white and black, 39–40
Population migration, 169, 171–172
Positive deviance, 290, 295–297
Postdevelopment, 287–289
Postdevelopment theorists, 287–288
Postdisaster gentrification, 123
Postdisaster housing, 113–114, 223–224, 224*f*, 293–294
Post-Katrina recovery, 3–4
    complex systems, 81–85
        cross-level and cross-scale interactions, 84
        feedback and signals, 84
        level and scale of system components, 81–83
        nested hierarchies, 84
        nonlinearity of system change, 85
        self-organization, 85
        social networks, 84–85
        threshold and tipping points, 84
        vertical networking, 84–85
    diaspora of musicians and tradition bearers, 237
    information blackout, 87
    levels of poverty in, 215–216
    misguided survivor agency, example of, 101
    "New Deal" type program, 93–94
    policy implications, 189–190
    recommendations for sustainability, 151–154
    resilience, 93
        collective action capabilities, 182–183, 188
        economic resources, 178–180, 187
        information and communication adaptive capabilities, 180–181, 187
        market-driven recovery, 179
        nonprofit grass-roots organizations, role of, 183
        political–economic relationships in socio-ecological systems, 178–179
        resources, 178–183
        Road Home Program, 179–180, 182, 188
        Small Business Administration (SBA) funding, 179
        social capital, 181–182, 187
    signals and information in, 86–94
Pressure and Release (PAR) model, 48, 194, 202
    dynamic pressures, 204–205
    leverage points in, 49–50
    for New Orleans and Southeastern Louisiana, 206*f*
    progression of vulnerability, 203*f*
    unsafe conditions, 205–207
Professionalism, 340
Protective factors, 247

R

Racial ideology causal chain, 365–367, 365*f*
    marginalization, 365–366
    segregation, 366
    separation of service systems, 366–367
RALLY. *See* Recovery Action Learning Laboratory (RALLY) project
Rapidity of a resource, 53
Recovery, 3–11, 13, 15–16, 18–19, 51–52, 54–55, 58–60, 72, 80, 92, 108, 113–118, 178, 188–189, 235, 258–260, 268, 275–276, 306, 321–323
    as Complex Adaptive Social System (CASS) issue, 85–86
    defined, 306
    efforts, 347
    imbued with spirituality, 277
    links between resilience, 259, 259*f*
    monitoring, 87–91
    owner-occupied housing recovery, 114–117
    planning, 117–118
    post-Katrina. *See* Post-Katrina recovery
    resilience and, 259, 259*f*
    resilient recovery, 163, 182, 271–272, 278–280, 307, 320, 360, 381
Recovery Action Learning Laboratory (RALLY) project, 20, 92–93, 308–317, 320–323, 345, 348–350, 352–353. *See also* Central City; Tremé
Recovery planning, 92
Redundancy, 52
Refugees, 262–263, 291–292, 331–332. *See also* Vietnamese community of New Orleans
Religious inspiration and revitalization movements, 269

Resilience/resiliency, 46, 51–56, 92–93, 102–103, 258–259
  access to resources model, 50–51
  adaptive resilience, 11
  classifying resilience, 10–11
    social infrastructure and capacity, 10–11
  community, 197
  contrasting with vulnerability, 56–61
    complementarity, 57–59
    continuity, 59–61
  defined, 196–198
  general or inherent, 370
  individual, 196–197
  links between recovery, 259, 259f
  on a macrolevel, 197–198
  non-governmental organizations (NGOs), 234
    New Orleans Musicians Clinic (NOMC), 234, 248
    The New Orleans Musicians Relief Fund, 248
    Sweet Home New Orleans (SHNO), 234–236, 248
    Sweet Relief, 248
    Tipitina's Office Co-op, 248
  research, 236–239
    context, 236–238
    impact on musicians, 236–237. *See also* New Orleans musicians, study of Katrina effects on
  resilience capabilities, 52
  resilient recovery, 51
  resources for resilience, 52–53
    collective action capabilities, 55–56, 182–183
    economic, 53, 178–180
    information and communication adaptive capabilities, 54–55, 180–181
    quality of needed resources, 53–56
    rapidity, 53
    redundancy, 52
    robustness, 52
    social capital, 53–54, 181–182
  systems approach to, 79–85
  theory, 7–9
  wellness, 51–52
Risk factors, 241–247
Rita, Hurricane, 268
Robustness of resources, 52
Rockefeller Foundation, 115–116
Root causes, 48, 51, 68–69, 163, 165–172, 203–204, 379–380
Rorschach Test, 41
Rule of Hand, 85

S

Safety of conditions, 49–50, 56, 59–60, 64–70, 66f, 73
Sandy, Hurricane, 104, 108
Sea, Lake and Overland Surges from Hurricanes (SLOSH) model, 149–150
Sea-level rise, 4, 137–138, 150–151
Segregation, 366
"Shrinking the urban footprint", 26
Social capital, 53–54, 181–182, 187, 258–259, 284–285
  attachment, 54
  citizen participation, 54
  received and perceived support, 54
Social determinants of vulnerability, 217–218
Social development causal chain, 375–378, 376f
  absence of building codes, 377–378
  health problems, 377
  population programs, lack of, 376–377
  public funding, 375–376
  slow economic growth, 378
Social infrastructure, 6, 8, 10–11, 20, 181, 216–217, 251–252
  and capacity, 8, 10–11
  neighborhood residents' perceptions and, 317–321
  resilience and, 10, 20, 297–298
Social marginalization, 58–59
Social role adaptation, 259
Social stratification causal model, 361–362, 361f
  economic growth, lack of, 362
  formal education, lack of, 361
  human capital, lack of, 362
Social support, 247–248
Social systems, 79–80
  recovery as a complex adaptive, 85–86
Social vulnerability, 195, 217
  factors influencing, 195–196
  in Southeast Louisiana, 207
Southeast Louisiana, 198–200
  addressing vulnerability in, 208–209
  disasters and impact on human psyche, 199–200
  earliest reported disasters in, 199, 200t
    oil spills, 201–202
  political ecology of, 183–184
  population and racial composition of, 198t
  predictors of social vulnerability in, 207
  Urban Flood Control Project, 33
Spatial targeting of vulnerability, 227–228
St. Bernard Parishes, 172, 181, 183–184, 186
St. Bernard polder, 144
Standard Project Hurricane (SPH), 149–150

Structural constraints causing disaster, 48–49, 69
Structural violence, 17–18, 84–85, 103–105, 180–182, 237–238, 364–365, 376
 agency disrespect as, 103–104
 capacity of, 8, 10–11
 delay as, 105–106
 examples of, 106–110
  evacuation plan, issues with, 111–112
  health care scarcity, 112–113
  housing lower-income survivors, 113–114
  mistreatment of Sandy survivors, 108
  planning approaches, 108–110
 Farmer's concept of, 17–18, 103, 292–293
 Galtung's concept of, 292–293
 on lower income disaster survivors, 111–114
 neoliberalism and, 292–295
 representations of, 291–292
 resilience and, 10, 20, 297–298
 stopping
  evacuation and long-term evacuee housing, 118–121
  health care immediately after the disaster, 121
  owner-occupied housing recovery, 114–117
  planning for the community's recovery, 117–118
  postdisaster gentrification, 123
  restoring agency to disaster survivors, 121–123
  survivor aid, 121–123
 trauma, 104–105
Structures of domination, 362–365
Subsidence, 150–151
Surge reduction, 145
Survivor researchers, 337, 341
Susceptibility, 381
Sustainability, 3–4, 11, 16–17, 25–26, 91, 95, 102–103, 151–154. *See also* Habitat sustainability
Sustainable livelihoods, hazards and, 49
Swamp Land Act, 139–140
Systems thinking, 80. *See also* Complex systems approach to vulnerability and resilience, 79–85

T

Temporal domination, 105–106
Theory Integration, 61–68. *See also* Vulnerability-plus (V+) theory
Theory of disaster capitalism, 293–294
Toxic gumbo, 201–202
Transformative capacity, 11
Trauma, 104–105
 long-term, 105

Tremé, 305, 310–312, 311*f*
 demographic data, 316*t*
 Katrina flooding in, 314*f*
 location, architecture, and cultural appeal, 315
 neighborhood dynamics and change in, 312–317
 storm damage in, 315*f*
Trust, 20, 116–117, 181–182, 249–250, 253, 284–286, 296–301, 322
Trusted media, 363

U

United Methodist Church (UMC), 267–277
 decision process for mergers and closures, 273–274
 Early Response Teams (ERT), 269–270
 Mt. Zion UMC, 276
 nonhierarchical and feminist approach, 273–274
 provision of resources and capabilities, 270–271
 transformative resilience, 271–272
United Methodist Committee on Relief (UMCOR), 268, 270
United Methodist Men (UMM), 277
United Methodist Women (UMW), 276–277
United Nations Guiding Principles on Internal Displacement, 216
United States Army Corps of Engineers (USACE), 32, 87–88, 137–144, 146, 148–151, 153–154
Unsafe conditions, 49–50, 69–70, 163, 186, 205–207, 379
Urban Land Institute, 117
Urban risk, 25
US disaster research, 216–217
USACE. *See* United States Army Corps of Engineers (USACE)

V

V+ theory. *See* Vulnerability-plus (V+) theory
Vietnamese Americans, 260–261, 263
Vietnamese community of New Orleans, 84–85, 258, 295–296, 298
 Culture Confounders, 262–263
 postdisaster recovery, 260
  cultural influences on, 263
  economic stability, 260
  housing stability, 260
  physical and mental health, 260
  social role adaptation, 260
 rapid and robust recovery of, 262
 resilience of, 260–261
  community competence, 260–261

Vietnamese community of New Orleans (*Continued*)
  economic development, 260–261
  information and communication flow, 260–261
  social capital, 260–261
Voluntary or volunteer organizations, 59
Volunteers, 361, 367–368
Vulnerability, 3–7, 11, 35, 40–41, 46, 154, 162, 165, 200–201, 203–204, 218–222, 248, 285, 358–359, 374, 378. *See also* Female vulnerability; Gendered vulnerability; Social vulnerability
  adaptation to the natural environment, 57
  in the context of New Orleans, 220–222
  culture of, 10
  defined, 194–196
    operational definition of, 196
  financial vulnerability, 223
  gendered, 226–228
  heuristic approach to, 189–190
  housing vulnerability, 223–224
  nature of, 194–195
  of New Orleans area, 6–7
  in New Orleans and Southeastern Louisiana, 208–209
  political economy and neoliberal, 225–226
  predictors of social vulnerability in Louisiana, 207
  root causes of, 48, 165–166
  social, 35, 195
  social processes influencing effects of hazards and disasters, 47
  in a social system, 8
  of socio-ecological systems, 47
  spatial targeting of, 227–228
  as a system problem, 7–9, 16–17
  urban vulnerability, 220–222
Vulnerability and resilience theories, 57–59, 162, 173–178
  comparing and contrasting, 56–61
  continuity between, 59–61
  New Orleans vulnerability and resilience paradigm, 200–201
  systems approach to, 79–85

Vulnerability-plus (V+) theory, 46–47, 58, 61–68, 358
  access to resources model, 50–51
  assumptions of, 49–50, 62–63, 378–381
    of capabilities, liabilities, and susceptibility relationships, 381
    disaster causal chain assumption, 380–381
    root causes of vulnerability, 379–380
    unsafe conditions, 379
  causal relationships among variables, 67
  chain of causality, 48–50, 68–72, 162
    characteristics of disaster, 71
    collective action resources, 188
    economic resources, 187
    hazard types, 70–71
    information and communication resources, 187
    leverage points in PAR model, 49–50
    Pressure and Release (PAR) model, 48
    resilience resources, 71–72
    root causes of vulnerability, 48, 68–69, 184–185, 379–380
    social capital resources, 187
    structural constraints, 48–49, 69, 185–186
    sustainable livelihoods, 49
    unsafe conditions, 49, 69–70, 186, 379
  empirical support for, 64–68, 65t
  implications for, 184–188
  as integration of theories, 61–68
  progression to vulnerability, 62f
  social development perspective, 47–51
  theoretical model, 66f
  variables and causal relationships, 61, 67

W

Wetland storm reduction value, 145–148
  computer modeling, 145–146
    "Neutral MRGO" condition, 146
    "no MRGO" condition, 146
  role in surge reduction, 145
  sheltering effect of forested wetlands, 148